Semi-Markov Processes: Applications in System Reliability and Maintenance

Semi-Markov Processes: Applications in System Reliability and Maintenance

Franciszek Grabski
Polish Naval University
Gdynia, Poland

AMSTERDAM • BOSTON • HEIDELBERG • LONDON • NEW YORK • OXFORD
PARIS • SAN DIEGO • SAN FRANCISCO • SYDNEY • TOKYO

Elsevier
Radarweg 29, PO Box 211, 1000 AE Amsterdam, Netherlands
The Boulevard, Langford Lane, Kidlington, Oxford OX5 1GB, UK
225 Wyman Street, Waltham, MA 02451, USA

Notice
No responsibility is assumed by the publisher for any injury and/or damage to persons or
property as a matter of products liability, negligence or otherwise, or from any use or operation
of any methods, products, instructions or ideas contained in the material herein.

Library of Congress Cataloging-in-Publication Data
A catalog record for this book is available from the Library of Congress

British Library Cataloguing in Publication Data
A catalogue record for this book is available from the British Library

ISBN: 978-0-12-800518-7

For information on all Elsevier publications
visit our web site at store.elsevier.com

This book has been manufactured using Print on Demand technology. Each copy is produced
to order and is limited to black ink. The online version of this book will show color figures
where appropriate.

ELSEVIER   **Book Aid**
             International

Working together
to grow libraries in
developing countries

www.elsevier.com • www.bookaid.org

Dedication

To my wife Marcelina

Contents

Preface

The semi-Markov processes were introduced independently and almost simultaneously by Levy [70], Smith [92], and Takacs [94] in 1954-1955. The essential developments of semi-Markov processes theory were proposed by Pyke [85, 86], Cinlar [15], Koroluk and Turbin [60–62], Limnios [72], Takacs [95]. Here we present only semi-Markov processes with a discrete state space. A semi-Markov process is constructed by the Markov renewal process, which is defined by the renewal kernel and the initial distribution or by other characteristics that are equivalent to the renewal kernel.

Semi-Markov Processes: Applications in System Reliability and Maintenance consists of a preface, 15 relatively short chapters, a summary, and a bibliography.

Chapter 1 is devoted to the discrete state space Markov processes, especially continuous-time Markov processes and homogeneous Markov chains. The Markov processes are an important class of the stochastic processes. This chapter covers some basic concepts, properties, and theorems on homogeneous Markov chains and continuous-time homogeneous Markov processes with a discrete set of states.

Chapter 2 provides the definitions and basic properties related to a discrete state space semi-Markov process. The semi-Markov process is constructed by the so-called Markov renewal process. The Markov renewal process is defined by the transition probabilities matrix, called the renewal kernel, and by an initial distribution or by other characteristics that are equivalent to the renewal kernel. The concepts presented are illustrated by some examples. Elements of the semi-Markov process statistical estimation are also presented in the chapter. Here, the estimation of the renewal kernel elements is considered by observing one or many sample paths in the time interval, or given number of the state changes. Basic concepts of the nonhomogeneous semi-Markov processes theory are also introduced in the chapter.

Chapter 3 is devoted to some characteristics and parameters of the semi-Markov process. A renewal kernel and an initial distribution contain full information about the process and they allow us to find many characteristics and parameters of the process, which we can translate on the reliability characteristics in the semi-Markov reliability model. The cumulative distribution functions of the first passage time from the given states to a subset of states, and expected values and second moments corresponding to them, are considered in this chapter. The equations for these quantities are presented here. Moreover, the chapter discusses a concept of interval transition probabilities and the Feller equations are also derived. Karolyuk and Turbin theorems of the limiting probabilities are also presented here. Furthermore, the reliability and maintainability characteristics and parameters in semi-Markov models are considered in the chapter.

Chapter 4 is concerned with the application of the perturbed semi-Markov processes in reliability problems. The results coming from the theory of semi-Markov processes perturbations allow us to find the approximate reliability function. The perturbed semi-Markov processes are defined in different ways by different authors. This theory has a rich literature. In this chapter we present only a few of the simplest types of perturbed SM processes. All concepts of the perturbed SM processes are explained in the same simple example. The last section is devoted to the state space aggregation method.

In Chapter 5 the random processes determined by the characteristics of the semi-Markov process are considered. First is a renewal process generated by return times of a given state. The systems of equations for the distribution and expectation of them have been derived. The limit theorem for the process is formulated by the adoption of a theorem of the renewal theory. The limiting properties of the alternating process and integral functionals of the semi-Markov process are also presented in this chapter. The chapter contains illustrative examples.

The semi-Markov reliability model of two different units of a renewable cold standby system and the SM model of a hospital electrical power system are discussed in Chapter 6.

In Chapter 7, the model of multistage operation without repair and the model with repair are constructed. Application of results of semi-Markov process theory allowed to calculate the reliability parameters and characteristics of the multistage operation. The models are applied for modeling the multistage transport operation processes.

In Chapter 8, the semi-Markov model of the load rate process is discussed. The speed of a car and the load rate of a ship engine are examples of the random load rate process. The construction of discrete state model of the random load rate process with continuous trajectories leads to the semi-Markov random walk. Estimating the model parameters and calculating the semi-Markov process characteristics and parameters give us the possibility to analyze the semi-Markov load rate.

Chapter 9 contains the semi-Markov model of the multitask operation process.

Chapter 10 is devoted to the semi-Markov failure rate process. In this chapter, the failure rate is assumed to be a stochastic process with nonnegative and right-continuous trajectories. The reliability function is defined as an expectation of a function of that random process. Particularly, the failure rate can be defined by the discrete state space semi-Markov process. The theorem concerning the renewal equations for the conditional reliability function with a semi-Markov process as a failure rate is presented. The reliability function with a random walk as a failure rate is investigated. For Poisson failure rate process and Furry-Yule failure rate process the reliability functions are presented.

In Chapter 11, time to a preventive service optimization problem is formulated. The semi-Markov model of the operation process allowed us to formulate the optimization problem. A theorem containing the sufficient conditions of the existing solution is formulated and proved. An example explains and illustrates the presented problem.

In Chapter 12, a semi-Markov model of system component damage is discussed. The models presented here deal with unrepairable systems. The multistate reliability functions and corresponding expectations, second moments, and standard deviations are evaluated for the presented cases of the component damage. A special case of the model is a multistate model with two kinds of failures. A theorem dealing with the inverse problem for a simple damage exponential model is formulated and proved.

In Chapter 13, some results of investigation of the multistate monotone system with components modeled by the independent semi-Markov processes are presented. We assume that the states of system components are modeled by the independent semi-Markov processes. Some characteristics of a semi-Markov process are used as reliability characteristics of the system components. In the chapter, the binary representation of the multistate monotone systems is discussed. The presented concepts and models are illustrated by some numerical examples.

The semi-Markov models of functioning maintenance systems, which are called maintenance nets, are presented in Chapter 14. Elementary maintenance operations form the states of a SM model. Some concepts and results of Semi-Markov process theory provide the possibility of computing important characteristics and parameters of the maintenance process. Two semi-Markov models of maintenance nets are discussed in the chapter.

In Chapter 15, basic concepts and results of the theory of semi-Markov decision processes are presented. The algorithm of optimizing a SM decision process with a finite number of state changes is discussed here. The algorithm is based on a dynamic programming method. To clarify it, the SM decision model for the maintenance operation is shown. The optimization problem for the infinite duration SM process and the Howard algorithm, which enables us to find the optimal stationary strategy are also discussed here. To explain this algorithm, a decision problem for a renewable series system is presented.

The book is primarily intended for researchers and scientists dealing with mathematical reliability theory (mathematicians) and practitioners (engineers) dealing with reliability analysis. The book is a very helpful tool for scientists, Ph.D. students, and M.Sc. students in technical universities and research centers.

Acknowledgments

I would like to thank Professor Krzysztof Kołowrocki for mobilization to write this book, the council for its efforts to release it and valuable comments on its content. I am grateful to Dr Agata-Załęska Fornal for her friendly favor in insightful linguistic revision of my book and valuable comments. I gratefully acknowledge Dr Erin Hill-Parks, Associate Acquisitions Editor, for kind cooperation and assistance in meeting the requirements of Elsevier to sign a publishing contract. I am grateful to Cari Owen, Editorial Project Manager, for her care and kind assistance in dealing with the editors. Finally, I thank my wife Marcelina for her help in computer problems in the process of writing a book as well as her patience and great support.

Franciszek Grabski

Discrete state space Markov processes

1

Abstract

The Markov processes are an important class of the stochastic processes. The Markov property means that evolution of the Markov process in the future depends only on the present state and does not depend on past history. The Markov process does not remember the past if the present state is given. Hence, the Markov process is called the process with *memoryless property*. This chapter covers some basic concepts, properties, and theorems on homogeneous Markov chains and continuous-time homogeneous Markov processes with a discrete set of states. The theory of those kinds of processes allows us to create models of real random processes, particularly in issues of reliability and maintenance.

Keywords: Markov process, Homogeneous Markov chain, Poisson process, Furry-Yule process, Birth and death process

1.1 Basic definitions and properties

Definition 1.1. A stochastic process $\{X(t) : t \in \mathbb{T}\}$ with a discrete (finite or countable) state space S is said to be a Markov process, if for all $i, j, i_0, i_1, \ldots, i_{n-1} \in S$ and $t_0, t_1, \ldots, t_n, t_{n+1} \in \mathbb{R}_+$ such that $0 \leqslant t_0 < t_1 < \cdots t_n < t_{n+1}$,

$$
\begin{aligned}
&P(X(t_{n+1}) = j \mid X(t_n) = i, X(t_{n-1}) = i_{n-1}, \ldots, X(t_0) = i_0) \\
&= P(X(t_{n+1}) = j \mid X(t_n) = i).
\end{aligned}
\tag{1.1}
$$

If $t_0, t_1, \ldots, t_{n-1}$ are interpreted as the moments from the past, t_n as the present instant, and t_{n+1} as the moment in the future, then the above-mentioned equation says that the probability of the future state is independent of the past states, if a present state is given. So, we can say that evolution of the Markov process in the future depends only on the present state. The Markov process does not remember the past if the present state is given. Hence, the Markov process is called the stochastic process with **memoryless** property.

From the definition of the Markov process it follows that *any process with independent increments is the Markov process.*

If $\mathbb{T} = N_0 = \{0, 1, 2, \ldots\}$, the Markov process is said to be a *Markov chain*, if $\mathbb{T} = \mathbb{R}_+ = [0, \infty)$, it is called the *continuous-time Markov process*. Let $t_n = u$, $t_{n+1} = \tau$. The conditional probabilities

$$
p_{ij}(u, s) = P(X(\tau) = j \mid X(u) = i), \quad i, j \in S
\tag{1.2}
$$

are said to be *the transition probabilities* from the state i at the moment u, to the state j at the moment s.

Semi-Markov Processes: Applications in System Reliability and Maintenance. http://dx.doi.org/10.1016/B978-0-12-800518-7.00001-6

Definition 1.2. The Markov process $\{X(t) : t \in \mathbb{T}\}$ is called homogeneous, if for all $i, j \in S$ and $u, s \in \mathbb{T}$, such that $0 \leqslant u < s$,

$$p_{ij}(u, s) = p_{ij}(s - u). \tag{1.3}$$

It means that the transition probabilities are the functions of a difference of the moments s and u. Substituting $t = s - u$ we get

$$p_{ij}(t) = P(X(s - u) = j | X(u - u) = i) = P(X(t) = j | X(0) = i), \quad i, j \in S, \ t \geqslant 0.$$

The number $p_{ij}(t)$ is called *a transition probability from the state i to the state j during the time t.*

If $\{X(t) : t \in \mathbb{R}_+\}$ is a process with the stationary independent increments, taking values on a discrete state space S, then

$$\begin{aligned}
p_{ij}(t) &= P(X(t + h) = j | X(h) = i) = \frac{P(X(t + h) = j, \ X(h) = i)}{P(X(h) = i)} \\
&= \frac{P(X(t + h) - X(h) = j - i, \ X(h) = i)}{P(X(h) = i)} \\
&= \frac{P(X(t + h) - X(h) = j - i) \, P(X(h) = i)}{P(X(h) = i)} \\
&= P(X(t + h) - X(h) = j - i).
\end{aligned}$$

Therefore, any process with the stationary independent increments is a homogeneous Markov process with transition probabilities

$$p_{ij}(t) = P(X(t + h) - X(h) = j - i). \tag{1.4}$$

For the homogeneous Markov process with a discrete state space, the transition probabilities satisfy the following conditions:

(a) $p_{ij}(t) \geqslant 0, \quad t \in \mathbb{T}$,
(b) $\sum\limits_{i \in S} p_{ij}(t) = 1$,
(c) $p_{ij}(t + s) = \sum\limits_{k \in S} p_{ik}(t) p_{kj}(s), \quad t \in \mathbb{T}, \ s \geqslant 0$. $\qquad (1.5)$

The last formula is known as the Chapman-Kolmogorov equation.

1.2 Homogeneous Markov chains

As we have mentioned, a Markov chain is a special case of a Markov process. We will introduce the basic properties of the Markov chains with the discrete state space. Proofs of presented theorems omitted here may be found in Refs. [3, 9, 22, 47, 88, 90].

1.2.1 Basic definitions and properties

Now let us consider a discrete time homogeneous Markov process $\{X_n : n \in \mathbb{N}_0\}$, having a finite or countable state space S that is called a homogeneous Markov chain (HMC). Recall that for each moment $n \in \mathbb{N}$ and all states $i, j, i_0, \ldots, i_{n-1} \in S$ there is

$$P(X_{n+1} = j \mid X_n = i, X_{n-1} = i_{n-1}, \ldots, X_1 = i_1, X_0 = i_0)$$
$$= P(X_{n+1} = j \mid X_n = i) \tag{1.6}$$

whenever

$$P(X_n = i, X_{n-1} = i_{n-1}, \ldots, X_1 = i_1, X_0 = i_0) > 0.$$

Transition probabilities of the HMC

$$p_{ij}(n, n+1) = P(X(n+1) = j \mid X(n) = i), \quad i, j \in S, \ n \in \mathbb{N}_0 \tag{1.7}$$

are independent of $n \in \mathbb{N}_0$:

$$p_{ij}(n, n+1) = p_{ij}, \quad i, j \in S, \ n \in \mathbb{N}_0. \tag{1.8}$$

The square number matrix

$$\mathbf{P} = \begin{bmatrix} p_{ij} : i, j \in S \end{bmatrix} \tag{1.9}$$

is said to be a *matrix of transition probabilities or transition matrix* of the HMC $\{X(n) : n \in \mathbb{N}_0\}$. It is easy to notice that

$$\forall_{i,j \in S} \ p_{ij} \geqslant 0 \quad \text{and} \quad \forall_{j \in S} \ \sum_{j \in S} p_{ij} = 1. \tag{1.10}$$

The matrix $\mathbf{P} = \begin{bmatrix} p_{ij} : i, j \in S \end{bmatrix}$ having the above-mentioned properties is called a *stochastic matrix*. There exists the natural question: Do the stochastic matrix $\mathbf{P}$ and discrete probability distribution $\boldsymbol{p}(0) = [p_i = i : i \in S]$ define completely the HMC? The following theorem answers this question.

Theorem 1.1. Let $\mathbf{P} = \begin{bmatrix} p_{ij} : i, j \in S \end{bmatrix}$ be a stochastic matrix and $\mathbf{p} = [p_i : i \in S]$ be the one-row matrix with nonnegative elements, such that $\sum\limits_{i \in S} p_i = 1$. There exists a probability space $(\Omega, \mathcal{F}, \mathcal{P})$ and defined on this space HMC $\{X(n) : n \in \mathbb{N}_0\}$ with the initial distribution $\mathbf{p} = [p_i : i \in S]$ and the transition matrix $\mathbf{P} = \begin{bmatrix} p_{ij} : i, j \in S \end{bmatrix}$.

Proof: [9, 47].

From (1.6) we obtain

$$P(X_0 = i_0, X_1 = i_1, \ldots, X_n = i_n) = p_{i_0} p_{i_0 i_1} p_{i_1 i_2} \ldots p_{i_{n-1} i_n}. \tag{1.11}$$

A number

$$p_{ij}(n) = P(X(n) = j \mid X(0) = i) = P(X(n+w) = j \mid X(w) = i) \tag{1.12}$$

denotes a transition probability from state i to state j throughout the period $[0, n]$, (in n steps). Now, the Chapman-Kolmogorov equation is given by a formula

$$p_{ij}(m+r) = \sum_{k \in S} p_{ik}(m) p_{kj}(r), \quad i, j \in S \tag{1.13}$$

or in matrix form

$$\mathbf{P}(m+r) = \mathbf{P}(m)\mathbf{P}(r),$$

where $\mathbf{P}(n) = \begin{bmatrix} p_{ij}(n) : i, j \in S \end{bmatrix}$.

From the above equation we get

$$P(n) = P(1 + 1 + \cdots + 1) = P(1) \cdot P(1) \cdot \cdots \cdot P(1) = P \cdot P \cdots P = P^n \quad (1.14)$$

We suppose that

$$P(0) = I, \quad (1.15)$$

where I is a unit matrix.

One-dimensional distribution of HMC we write as a one-row matrix

$$p(n) = \left[p_j(n) : j \in S \right], \quad p_j(n) = P(X(n) = j) \quad (1.16)$$

Using the formula for total probability and the Markov property, we obtain

$$p(n) = p(0)P^n, \quad (1.17)$$

$$p(n) = p(n - 1)P. \quad (1.18)$$

A *stationary probability distribution* plays a major role in the theory of Markov chains.

Definition 1.3. A probability distribution

$$\pi = [\pi_i : i \in S] \quad (1.19)$$

satisfying a system of linear equations

$$\sum_{i \in S} \pi_i p_{ij} = \pi_j, \quad j \in S \quad \text{and} \quad \sum_{i \in S} \pi_i = 1 \quad (1.20)$$

is said to be a stationary probability distribution of the Markov chain with transition matrix $P = \left[p_{ij} : i, j \in S \right]$.

A system of linear equations (1.20) in the matrix form is

$$\pi P = \pi, \quad \pi \mathbf{1} = [1], \quad (1.21)$$

where $\mathbf{1}$ is a one-row matrix in which all elements are equal to 1.

Suppose that the initial distribution $p(0)$ is equal to the stationary distribution πP. Note that for $n = 1$ we have $p(1) = p(0)P = \pi P = \pi$. For $n = k$ we suppose $p(k) = \pi$. For $n = k + 1$ we obtain $p(k + 1) = p(k)P = \pi p = \pi$.

Therefore, we jump to the conclusion: if the initial distribution is equal to the stationary distribution, then the one-dimensional distribution of the Markov process $p(n) = [p(n) : i \in S]$ does not depend on n and it is equal to the stationary distribution

$$\pi = p(n) \quad \text{for all } n \in \mathbb{N}_0. \quad (1.22)$$

1.2.2 Classification of states

The properties of *HMC* depend on the shape of the matrix $\mathbf{P}$. For convenience, we can describe the evolution of *HMC* using a graph $G = \{(i, (i, j)) : i, j \in S\}$ corresponding to the matrix P. The vertices $i \in S$ denote the states of *HMC* and the arrows $(i, j) \in S \times S$ correspond to the *positive* probabilities p_{ij}. When $p_{ij} = 0$, the corresponding arrow is omitted. First, we will present the classification of the states of the HMC in terms of properties of the transition probabilities $p_{ij}(n)$.

A state $i \in S$ is *essential*, if

$$\forall_{j \in S} \{\exists_{n \in \mathbb{N}} \, p_{ij}(n) > 0 \Rightarrow \exists_{n \in \mathbb{N}} \quad p_{ji}(m) > 0\}.$$

A state that is not essential is called *inessential*. Hence, the state $i \in S$ is inessential if

$$\exists_{j \in S} \{\exists_{n \in \mathbb{N}} \quad p_{ij}(n) > 0 \wedge \forall_{m \in \mathbb{N}} \, p_{ji}(m) = 0\}.$$

By definition, the state $i \in S$ is inessential if it is possible to escape from it after a finite number of steps (with a positive probability), without ever returning to it.

We say that the state $j \in S$ is *accessible* from the state $i \in S$, $(i \rightarrow j)$ if

$$\exists_{k \in \mathbb{N}_0} \, p_{ij}(k) > 0.$$

We suppose, that $p_{ij}(0) = 1$ for $i = j$ and $p_{ij}(0) = 0$ for $i \neq j$.

From the Chapman-Kolmogorov equation we obtain the following inequality:

$$p_{ij}(n + m) \geqslant p_{ik}(n) p_{kj}(m).$$

From this inequality, it follows that the above-mentioned relation is transitive:

$$(i \rightarrow k) \wedge (k \rightarrow j) \Rightarrow (i \rightarrow j).$$

States $i, j \in S$ *communicate* $(i \leftrightarrow j)$, if state j is accessible from i and i is accessible from j. From this definition it follows that the relation $\leftrightarrow$ is *symmetric* and *reflexive*. It is easy to verify the following implication:

$$[(i \leftrightarrow j) \wedge (j \leftrightarrow k) \Rightarrow (i \leftrightarrow k)]. \tag{1.23}$$

Because the relation $\leftrightarrow$ is *transitive*, it means that $\leftrightarrow$ is the relation of *equivalence*. Hence, the set of the *essential* states separates into a finite or countable number of disjoint sets $S_1, S_2, \ldots$, each of them consisting of the communicating sets but the passage between states belonging to different sets is impossible. The sets $S_1, S_2, \ldots$ are called *classes* or *indecomposable classes* (of the essential communicating sets). A Markov chain is said to be *indecomposable* if its states form a single indecomposable class.

A state $j \in S$ has a period d, if

1. $d > 1$
2. $p_{jj}(n) > 0$ only for $n = m \cdot d$, $m = 1, 2, \ldots$
3. d is the largest number satisfying 2.

Consequently, we can say that the state $j \in S$ has the period d, if the number $d > 1$ is the *greatest common divisor* of all numbers from the set $\{n \in N : \, p_{jj}(n) > 0\}$. Notice that all states of a single indecomposable class have the same period. If $d = 1$, the state j is said to be *aperiodic*.

Example 1.1. For illustration we consider the Markov chain with a set of states $S = \{1, 2, 3, 4, 5\}$ and a transition matrix

$$P = \begin{bmatrix} 0.5 & 0.5 & 0 & 0 & 0 \\ 0 & 0 & 1 & 0 & 0 \\ 0 & 0.4 & 0.6 & 0 & 0 \\ 0 & 0 & 0 & 0 & 1 \\ 0 & 0 & 0 & 1 & 0 \end{bmatrix}$$

The flow graph of this matrix is shown in Figure 1.1. State 1 is *inessential*; however, states 2, 3, 4, and 5 are essential. States 4 and 5 are mutually *accessible*. States 1, 2, and 3 are not accessible from states 4 and 5 and state 1 is not accessible from states 2, 3, 4, and 5. States 2 and 3 *communicate*, and 4 and 5 also communicate. A subset $\{2, 3, \}$ and $\{4, 5\}$ form disjoint classes of essential communicating states. States 4 and 5 are *periodic* with period $d = 2$; however, states 1, 2, and 3 are the aperiodic states.

We shall introduce some main concepts that are important, especially in the case of HMC with a counting state space. A random variable

$$\Delta_A = \min\{n \in \mathbb{N} : X_n \in A\} \tag{1.24}$$

denotes a moment of *the first arrival to subset of states A* by HMC $\{X_n : n \in \mathbb{N}_0\}$ $A \subset S$. Let us add that for $A = \{j\}$, a random variable $\Delta_{\{j\}} = \Delta_j$ denotes a *moment of the first achievement* of state $j \in S$.

Notice the equality of the following events:

$$\{\Delta_A = m\} = \{X_m \in A, X_{m-1} \in A', \dots, X_1 \in A'\} \quad \text{for } m = 2, 3, \dots,$$
$$\{\Delta_A = 1\} = \{X_1 \in A\},$$
$$\{\Delta_A = \infty\} = \bigcap_{k=1}^{\infty}\{X_k \in A'\},$$
$$\{\Delta_A < \infty\} = \bigcup_{k=1}^{\infty}\{X_k \in A\} = \bigcup_{k=1}^{\infty}\{\Delta_A = m\}.$$

The number

$$f_{iA}(m) = P(\Delta_A = m \mid X_0 = i) \tag{1.25}$$

is the probability of the first arrival to the subset A at the moment m, if an initial state is i. Notice that

$$f_{iA}(m) = \begin{cases} P(X_m \in A, X_{m-1} \in A', \dots, X_1 \in A' \mid X_0 = i), & m = 2, 3, \dots \\ P(X_1 \in A \mid X_0 = i), & m = 1. \end{cases} \tag{1.26}$$

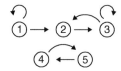

Figure 1.1 Flow graph corresponding to the matrix P.

For $A = \{j\}$ we obtain a probability of the first arrival to the state j at the moment m, if the initial state is i.

$$f_{ij}(m) = \begin{cases} P(X_m = j, X_{m-1} \neq j, \ldots, X_1 \neq j \mid X_0 = i), & m = 2, 3 \ldots \\ P(X_1 = j \mid X_0 = i), & m = 1. \end{cases} \tag{1.27}$$

Let

$$f_{iA} = P(\Delta_A < \infty \mid X_0 = i). \tag{1.28}$$

It is easy to notice that

$$f_{iA} = \sum_{m=1}^{\infty} f_{iA}(m). \tag{1.29}$$

For $A = \{j\}$ we have

$$f_{ij} = P(\{\Delta_j < \infty \mid X_0 = i) = \sum_{m=1}^{\infty} f_{ij}(m). \tag{1.30}$$

This formula provides a following conclusion:

Proposition 1.1. The state $j \in S$ is accessible from the state $i \neq j$ if and only if $f_{ij} > 0$.

We say that a subset of states A is *strongly accessible* from i $(i \overset{1}{\to} A)$ if $f_{iA} = 1$ [32]. A state j is *strongly accessible* from $i \in S$, $(i \overset{1}{\to} j)$ if $f_{ij} = 1$.

A state j is called *recurrent*, if

$$f_{jj} = 1 \tag{1.31}$$

and *nonrecurrent*, if

$$f_{jj} < 1. \tag{1.32}$$

Proposition 1.2. The state j is recurrent if and only if

$$\sum_{n=1}^{\infty} p_{jj}(n) = \infty \tag{1.33}$$

Proposition 1.3. If state j is recurrent and $i \leftrightarrow j$, then state i is also recurrent.
Proof: [47, 88].
Proposition 1.4. If the state j is nonrecurrent, then

$$\sum_{n=1}^{\infty} p_{jj}(n) < \infty. \tag{1.34}$$

The *average time of return* is given by a formula

$$\mu_j = \sum_{n=1}^{\infty} n f_{jj}(n). \tag{1.35}$$

A recurrent state is *positive* if

$$0 < \mu_j < \infty \tag{1.36}$$

and it is *null* if

$$\mu_j = \infty. \tag{1.37}$$

Example 1.2. Consider the Markov chain with transition probability matrix

$$P = \begin{bmatrix} \frac{1}{1\cdot2} & \frac{1}{2\cdot3} & \frac{1}{3\cdot4} & \cdots \\ 1 & 0 & 0 & \cdots \\ 0 & 1 & 0 & \cdots \\ 0 & 0 & 1 & \cdots \\ \cdots\cdots\cdots \end{bmatrix}.$$

A graph corresponding to this matrix is shown in Figure 1.2.

The Markov chain consists of one indecomposable class of states. Notice that

$$f_{11}(n) = p_{1n} \cdot p_{nn-1} \cdots\cdots p_{21} = \frac{1}{n(n+1)} \quad \text{and}$$

$$f_{11} = \sum_{n=1}^{\infty} f_{11}(n) = \sum_{n=1}^{\infty} \frac{1}{n(n+1)} = 1.$$

It means that 1 is a recurrent state. However, the average time of return is

$$\mu_1 = \sum_{n=1}^{\infty} n f_{11}(n) = \sum_{n=1}^{\infty} \frac{1}{n+1} = \infty.$$

It means that this state is null.

Example 1.3. The Markov chain with transition probability matrix

$$P = \begin{bmatrix} \frac{1}{2} & \frac{1}{2^2} & \frac{1}{2^3} & \frac{1}{2^4} & \cdots \\ 1 & 0 & 0 & 0 & \cdots \\ 0 & 1 & 0 & 0 & \cdots \\ 0 & 0 & 1 & 0 & \cdots \\ \cdots\cdots\cdots \end{bmatrix}$$

as in the previous case consists of one indecomposable class of states. As before, we calculate

$$f_{11}(n) = \frac{1}{2^n} \quad \text{and} \quad f_{11} = \sum_{n=1}^{\infty} f_{11}(n) = \sum_{n=1}^{\infty} \frac{1}{2^n} = 1.$$

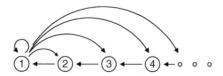

Figure 1.2 Flow graph of the matrix P.

Therefore, state 1 is recurrent. The average time of return is

$$\mu_1 = \sum_{n=1}^{\infty} n \frac{1}{2^n}.$$

Using the Cauchy criterion, we study the convergence of the series. Because

$$\lim_{n \to \infty} \sqrt[n]{\frac{n}{2^n}} = \frac{1}{2},$$

the series converges. So, the recurrent state 1 is positive.

Definition 1.4. The recurrent positive and aperiodic state $i \in S$ is called an ergodic state.

Proposition 1.5. If the state space of a Markov chain is finite, then there exists at least one recurrent state.

Proof: [47].

Theorem 1.2. If i is a recurrent state and the states i and j communicate, then the j is also recurrent.

Proof: [47].

1.2.3 Limiting distribution

One-row matrix

$$p = [p_i : i \in S], \quad \text{where } p_i = \lim_{n \to \infty} p_i(n) \tag{1.38}$$

is said to be limiting distribution of HMC $\{X_n : n \in \mathbb{N}_0\}$.

Theorem 1.3 (Shiryayev [88]). For HMC with a countable states space S, there exists a unique stationary distribution if and only if the space of states S contains exactly one positive recurrent class (of essential communicating states) C.

Theorem 1.4 (Shiryayev [88]). A limit distribution of HMC $\{X(n) : n \in \mathbb{N}_0\}$ with a countable states space S exists if and only if the space S contains exactly one aperiodic positive recurrent class C such that $f_{ij} = 1$ for all $j \in C$ and $i \in S$. Moreover, the limiting distribution is equal to the stationary distribution

$$p = \pi \tag{1.39}$$

and for all $j \in C$ and $i \in S$

$$p_j = \lim_{n \to \infty} p_j(n) = \lim_{n \to \infty} p_{ij}(n) = \frac{1}{\mu_j}. \tag{1.40}$$

A random variable

$$v_j(n) = \delta_{jX(1)} + \cdots + \delta_{jX(n)} \tag{1.41}$$

denotes the "time" being spent in a state j during time $\{1, 2, \ldots, n\}$. Notice that

$$E[v_j(n)|X(0) = i] = E[\delta_{jX(1)}|X(0) = i] + \cdots + E[\delta_{jX(n)}|X(0) = i]$$

and

$$E\left[\delta_{jX(k)}|X(0) = i\right] = 0 \cdot P\{X(k) \neq j|X(0) = i\} + 1 \cdot P\{X(k) = j|(0) = i\}$$
$$= p_{ij}(k).$$

Hence,

$$E\left[v_j(n)|X(0) = i\right] = \sum_{k=1}^{n} p_{ij}(k). \tag{1.42}$$

For any HMC, the limit in the Cesaro sense always exists and

$$q_{ij} = \lim_{n \to \infty} \frac{1}{n} \sum_{k=1}^{n} p_{ij}(k) = \frac{f_{ij}}{\mu_j}. \tag{1.43}$$

If the HMC has one *positive recurrent class* of states, then the numbers q_{ij}, $i, j \in S$ do not depend on an initial state $i \in S$ and they form a stationary distribution

$$q_{ij} = \pi_j, \quad j \in S. \tag{1.44}$$

1.3   Continuous-time homogeneous Markov processes

Let $\{X(t) : t \geqslant 0\}$ be a discrete (finite or counting) state space S Markov process with the piecewise constant and the right-hand side continuous trajectories.

Let $\tau_0 = 0$ denote the start moment of the process and $\tau_1, \tau_2, \ldots$ represent the successive moments of its state changes. A random variable $T_i = \tau_{n+1} - \tau_n|X(\tau_n) = i$, $i \in S$ denotes the time spent in state i when the successor state is unknown. The random variable T_i is called the waiting time in state i [45]. The Chapman-Kolmogorov equation enables proof [17, 47], that the waiting time of a state i always has an exponential distribution with a parameter $\lambda_i > 0$:

$$G_i(t) = P(T_i \leqslant t) = P(\tau_{n+1} - \tau_n \leqslant t \,|\, X(\tau_n) = i) = 1 - e^{-\lambda_i t}, \quad t \geqslant 0, i \in S. \tag{1.45}$$

The inverse implication for the discrete state space process with the piecewise constant and the right-hand side continuous trajectories is also true; if the waiting times for all states $i \in S$ are exponentially distributed then the stochastic process is the Markov process [17, 47].

A square matrix $\boldsymbol{P}(t)$, the elements of which are the functions

$$p_{ij}(t) = P(X(t) = j \,|\, X(0) = i) = P(X(\tau + t) = j|X(\tau)), \quad j \in S, \, t, \tau, t + \tau \in \mathbb{T}, \tag{1.46}$$

is called a transition probabilities matrix. The Chapman-Kolmogorov equation in matrix form is

$$\boldsymbol{P}(s + t) = \boldsymbol{P}(s)\boldsymbol{P}(t). \tag{1.47}$$

For discrete state space homogeneous Markov processes, there always exist the limits [26]

$$\lambda_i = \lim_{h \to 0} \frac{1 - p_{ii}(h)}{h} \leqslant 0, \quad i \in S, \tag{1.48}$$

and

$$\lambda_{ij} = \lim_{h \to 0} \frac{p_{ij}(h)}{h}, \quad i \neq j \in S, \tag{1.49}$$

A number λ_{ij} is called a *transition rate* from the state i to j.
Let

$$\lambda_{ii} = -\lambda_i = \lim_{h \to 0} \frac{p_{ii}(h) - 1}{h}, \quad i \in S. \tag{1.50}$$

A square matrix

$$\Lambda = \left[\lambda_{ij} : i, j \in S \right] \tag{1.51}$$

is said to be a *transition rate matrix*. This matrix has the following properties:

$$\forall_{i \neq j \in S} \lambda_{ij} \geqslant 0, \quad \forall_{i \in S} \lambda_{ii} \leqslant 0, \quad \forall_{i \in S} \sum_{j \in S} \lambda_{ij} = 0. \tag{1.52}$$

From the last identity we get

$$\lambda_{ii} = - \sum_{j \neq i} \lambda_{ij}. \tag{1.53}$$

Subtracting from both sides of this equality a transition probability $p_{ij}(t)$ and dividing by h we obtain

$$\frac{p_{ij}(t + h) - p_{ij}(t)}{h} = \sum_{k \neq j} p_{ik}(t) \frac{p_{kj}(h)}{h} + p_{ij}(t) \frac{p_{jj}(h) - 1}{h}.$$

Passing to a limit with $h \to 0$ we obtain

$$\frac{d p_{ij}(t)}{dt} = \sum_{k \in S} p_{ik}(t) \lambda_{kj}, \quad i, j \in S. \tag{1.54}$$

An initial condition has the form

$$p_{ij}(0) = \delta_{ij} = \begin{cases} 1 & \text{for } j = i \\ 0 & \text{for } j \neq i. \end{cases} \tag{1.55}$$

In matrix form the system of differential equations (1.54) is given by

$$\frac{d P(t)}{dt} = P(t) \Lambda, \quad P(0) = I. \tag{1.56}$$

We construct a system of differential equation for a first-order distribution of the Markov process in a similar way:

$$p(j; t) = p_j(t) = P(X(t) = j), \quad j \in S. \tag{1.57}$$

Using the law of total probability we get

$$p_j(t + h) = \sum_{i \in S} p_i(t) p_{ij}(h), \quad j \in S.$$

Subtracting from both sides of the equality a probability $p_j(t)$, dividing both sides by h, and passing to a limit with $h \to 0$, we obtain

$$p'_j(t) = \sum_{i \in S} p_i(t) \lambda_{ij}, \quad j \in S. \tag{1.58}$$

The initial conditions in this case are

$$p_i(0) = p_i^0, \quad i \in S. \tag{1.59}$$

Usually, those kinds of equations are solved by applying the Laplace transform:

$$L[p_i(t)] = \tilde{p}_i(s) = \int_0^\infty p_i(t) e^{-st} dt.$$

Using a property

$$L\left[p'_i(t)\right] = s \tilde{p}_i(s) - p_i(0), \tag{1.60}$$

by transformation of a system of the differential equations (1.58), we obtain a system of linear algebraic equations with unknown transforms $\tilde{p}_j(s)$, $j \in S$:

$$s \tilde{p}_j(s) - p_j(0) = \sum_{i \in S} \tilde{p}_i(s) \lambda_{ij}, \quad j \in S.$$

This system of equations in matrix form is as follows:

$$(s\boldsymbol{I} - \boldsymbol{\Lambda}^T)\tilde{\boldsymbol{P}}(s) = \boldsymbol{P}(0), \tag{1.61}$$

where

$$\tilde{\boldsymbol{P}}(s) = \left[\tilde{p}_i(s) : i \in S\right]^T, \quad \boldsymbol{P}(0) = [p_i(0) : i \in S]^T, \quad \boldsymbol{\Lambda} = [\lambda_{ij} : i, j \in S].$$

We can investigate a limit behavior of a continuous-time Markov process as $t \to \infty$.

Theorem 1.5. Let $\{X(t) : t \geqslant 0\}$ be a homogeneous Markov process defined by a transition rate matrix $\boldsymbol{\Lambda} = \left[\lambda_{ij} : i, j \in S\right]$. If there exists the limit probabilities

$$\lim_{t \to \infty} p_j(t) = \lim_{t \to \infty} p_{ij}(t) = p_j, \quad j \in S, \tag{1.62}$$

then they satisfy the system of linear equations

$$\sum_{i \in S} p_i \lambda_{ij} = 0, j \in S, \quad \sum_{j \in S} p_j = 1. \tag{1.63}$$

To find the limit distribution of the process we have to solve the above-mentioned system of linear equations.

1.4   Important examples

1.4.1   Poisson process

From the definition of the Poisson process [3, 7, 9, 17, 23, 28, 55, 71] it follows that it is the process with stationary independent increments and

$$P(X(t+h) - X(h) = k) = \frac{(\lambda t)^k}{k!} e^{-\lambda t}, \quad k \in S, \quad \text{for all } t > 0, \ h \geqslant 0$$

Each process with stationary independent increments is a homogeneous Markov process with transition probabilities

$$p_{ij}(t) = P(X(t+h) - X(h) = j - i). \tag{1.64}$$

Hence, the Poisson process is the homogeneous Markov process with the transition probabilities given by

$$p_{ij}(t) = \frac{(\lambda t)^{j-i}}{(j-i)!} e^{-\lambda t}, \quad i, j \in S, \ j \geqslant i \geqslant 0. \tag{1.65}$$

From the definitions (1.48) and (1.49) we will calculate the transition rates of the Poisson process. For $j = i$ we have

$$\lim_{h \to 0} \frac{p_{ii}(h) - 1}{h} = \lim_{h \to 0} \frac{e^{-\lambda h} - 1}{h} = -\lambda.$$

For $j - i \geqslant 1$ we get

$$\lim_{h \to 0} \frac{p_{ij}(h)}{h} = \lim_{h \to 0} \frac{(\lambda h)^{j-i} e^{-\lambda}}{(j-i)! h} = \begin{cases} \lambda & \text{for } j - i = 1 \\ 0 & \text{for } j - i > 1 \end{cases}.$$

Taking into account the properties of a transition rate matrix we get

$$\Lambda = \begin{bmatrix} -\lambda & \lambda & 0 & 0 & 0 & \cdots \\ 0 & -\lambda & \lambda & 0 & 0 & \cdots \\ 0 & 0 & -\lambda & \lambda & 0 & \cdots \\ 0 & 0 & 0 & -\lambda & \lambda & \cdots \\ \vdots & \vdots & \vdots & \vdots & \vdots & \vdots \end{bmatrix}. \tag{1.66}$$

1.4.2   Furry-Yule process

The continuous-time Markov process with a countable state space $S = \{0, 1, 2, \ldots\}$ and a transition rate matrix

$$\Lambda = \begin{bmatrix} -\lambda & \lambda & 0 & 0 & 0 & \cdots \\ 0 & -2\lambda & 2\lambda & 0 & 0 & \cdots \\ 0 & 0 & -3\lambda & 3\lambda & 0 & \cdots \\ 0 & 0 & 0 & -4\lambda & 4\lambda & \cdots \\ \vdots & \vdots & \vdots & \vdots & \vdots & \vdots \end{bmatrix} \tag{1.67}$$

is said to be a Furry-Yule process. The Furry-Yule process is a special case of the so-called birth process.

1.4.3 Finite state space birth and death process

The continuous-time Markov process with finite state space $S = \{0, 1, 2, \ldots, n\}$ and a transition rate matrix

$$
\Lambda = \begin{bmatrix}
-\lambda_0 & \lambda_0 & 0 & 0 & \ldots & 0 & 0 & 0 \\
\mu_1 & -(\mu_1 + \lambda_1) & \lambda_1 & 0 & \ldots & 0 & 0 & 0 \\
0 & \mu_2 & -(\mu_2 + \lambda_2) & \lambda_2 & \ldots & 0 & 0 & 0 \\
\vdots & \vdots & \vdots & & \vdots & \vdots & \vdots & \vdots \\
0 & 0 & 0 & 0 & \ldots & \mu_{n-1} & -(\mu_{n-1} + \lambda_{n-1}) & \lambda_{n-1} \\
0 & 0 & 0 & 0 & \ldots & 0 & \mu_n & -\mu_n
\end{bmatrix}
\tag{1.68}
$$

is called a *finite state space birth and death process*. In this case, a linear system of deferential equations (1.58) takes the form

$$p'_0(t) = -\lambda_0 p_0(t) + \mu_1 p_1(t),$$

$$\ldots \qquad \ldots\ldots\ldots\ldots\ldots$$

$$p'_k(t) = \lambda_{k-1} p_{k-1}(t) - (\mu_k + \lambda_k) p_k(t) + \mu_{k+1} p_{k+1}(t), \quad k = 1, \ldots, n-1, \tag{1.69}$$

$$\ldots \qquad \ldots\ldots\ldots\ldots\ldots\ldots$$

$$p'_n(t) = \lambda_{n-1} p_{n-1}(t) - \mu_n p_n(t).$$

The initial conditions are

$$p_i(0) = p_i^0, \quad i \in S = \{0, 1, \ldots, n\}. \tag{1.70}$$

The limiting distribution

$$p_k = \lim_{t \to \infty} p_k(t) \tag{1.71}$$

can be obtained by solving a system of linear equations

$$-\lambda_0 p_0 + \mu_1 p_1 = 0,$$

$$\ldots\ldots\ldots\ldots\ldots\ldots$$

$$\lambda_{k-1} p_{k-1} - (\mu_k + \lambda_k) p_k + \mu_{k+1} p_{k+1} = 0, \quad k = 1, \ldots, n-1,$$

$$\ldots\ldots\ldots\ldots\ldots\ldots \tag{1.72}$$

$$\lambda_{n-1} p_{n-1} - \mu_n p_n = 0,$$

$$p_0 + p_1 + \cdots + p_n = 1.$$

To solve the system of equations we make the substitution

$$x_k = \mu_k p_k - \lambda_{k-1} p_{k-1}, \quad k = 1, 2, \ldots, n. \tag{1.73}$$

The main part of the system of equations (1.72) takes the form

$$x_1 = 0,$$

$$\cdots\cdots\cdots$$

$$x_k - x_{k+1} = 0, \quad k = 1, 2, \ldots, n - 1, \tag{1.74}$$

$$\cdots\cdots\cdots$$

$$x_n = 0.$$

Hence,

$$x_k = 0 \quad \text{for } k = 1, 2, \ldots, n.$$

Thus,

$$p_k = \frac{\lambda_{k-1}}{\mu_k}, \quad k = 1, 2, \ldots, n. \tag{1.75}$$

We get

$$p_k = \frac{\lambda_0 \lambda_1 \ldots \lambda_{k-1}}{\mu_1 \mu_2 \ldots \mu_k} p_0, \quad k = 1, 2, \ldots, n. \tag{1.76}$$

Using the condition

$$p_0 + p_1 + \cdots + p_n = 1$$

we get

$$p_0 = \frac{1}{1 + \sum_1^n \frac{\lambda_0 \lambda_1 \ldots \lambda_{k-1}}{\mu_1 \mu_2 \ldots \mu_k}}. \tag{1.77}$$

The equalities (1.76) and (1.77) describe the limiting distribution of the birth and death process. This process has numerous applications, particularly in queuing theory and reliability.

1.5 Numerical illustrative examples

Example 1.4 (Markov Alternating Process).
A two state birth and death process is called a *Markov alternating process*. It means that the state space of this process is $S = \{0, 1\}$ and its transition rate matrix is given by

$$\Lambda = \begin{bmatrix} -\lambda_0 & \lambda_0 \\ \mu_1 & -\mu_1 \end{bmatrix}. \tag{1.78}$$

We substitute $\lambda_0 = \mu$ and $\mu_1 = \lambda$. In this case, a linear system of deferential equations (1.69) takes the form

$$\begin{aligned} p_0'(t) &= -\mu p_0(t) + \lambda p_1(t), \\ p_1'(t) &= \mu p_0(t) - \lambda p_1(t). \end{aligned} \tag{1.79}$$

We assume that the initial conditions are

$$p_0(0) = 0, \quad p_1(0) = 1. \tag{1.80}$$

Corresponding to (1.79) and (1.80), the system of linear equations for the Laplace transforms is

$$
\begin{aligned}
s\,\tilde{p}_0(s) &= -\mu\,\tilde{p}_0(s) + \lambda\,\tilde{p}_1(s), \\
s\,\tilde{p}_1(s) - 1 &= \mu\,\tilde{p}_0(s) - \lambda\tilde{p}_1(s).
\end{aligned}
\tag{1.81}
$$

The solution of this equations system is

$$
\begin{aligned}
\tilde{p}_0(s) &= \frac{\lambda}{s(s + \lambda + \mu)}, \\
\tilde{p}_1(s) &= \frac{s + \mu}{s(s + \lambda + \mu)}.
\end{aligned}
\tag{1.82}
$$

Calculating the inverse Laplace transform we obtain the probabilities of states.

$$
\begin{aligned}
p_0(t) &= \frac{\lambda}{\lambda + \mu}\left[1 - e^{-(\lambda+\mu)t}\right], \\
p_1(t) &= \frac{\mu}{\lambda + \mu} + \frac{\mu}{\lambda + \mu}e^{-(\lambda+\mu)t}.
\end{aligned}
\tag{1.83}
$$

The above model can be applied in reliability theory. In this theory, 0 means a state of a technical object failure ("down state") and 1 denotes a working state ("up state"). In the reliability model, the random variable, say η, denoting a duration of the "down state" is equal to the repair time (or renewal time) and it has the exponential distribution with parameter μ. The duration of the "up state" is a random variable, say ζ, that means a lifetime or time to failure and it is governed by the exponential distribution with parameter λ. The functions $p_0(t), t \geqslant 0$ $p_1(t), t \geqslant 0$ given by (1.83) describe probabilities of the "down" and "up" states at the time.

Example 1.5 (Markov Model of Renewable Two-Component Series System).
We will consider the reliability model of a renewable two-component series system under the assumption that the times to failure of both components denoted as ζ_1, ζ_2 are exponentially distributed with parameters λ_1 and λ_2. We also suppose that the renewal (repair) times of components are the random variables η_1, η_2 having the exponential distribution with parameters μ_1, μ_2. We assume that the above-considered random variables and their copies are mutually independent. To construct a random model describing behavior of the system in reliability aspect we start from determining the system states.

1. The system renewal after the failure of the first component (down state)
2. The system renewal after the failure of the second component (down state)
3. Work of the system, both components are up

Notice, that cumulative distribution functions of waiting times for the states 1, 2, 3 and for $t \geqslant 0$ are

$$
\begin{aligned}
P(T_1 \leqslant t) &= P(\eta_1 \leqslant t) = 1 - e^{-\mu_1 t}, \\
P(T_2 \leqslant t) &= P(\eta_2 \leqslant t) = 1 - e^{-\mu_2 t}, \\
P(T_3 \leqslant t) &= P(\min\{\zeta_1, \zeta_2\} \leqslant t) = 1 - e^{-(\lambda_1+\lambda_2)t}.
\end{aligned}
$$

It means that waiting times for all states are exponentially distributed. Therefore, the reliability model of the considering system is a Markov process. To determine this

Markov process we have to define its *transition rate matrix*

$$\boldsymbol{\Lambda} = \left[\lambda_{ij} : i,j \in S = \{1,2,3\} \right].$$

From the model assumptions and definition (1.49), we have

$$\lambda_{12} = \lim_{h \to 0} \frac{p_{12}(h)}{h} = 0,$$

$$\lambda_{13} = \lim_{h \to 0} \frac{p_{13}(h)}{h} = \lim_{h \to 0} \frac{P(\eta_1 \leqslant h) + o(h)}{h} = \lim_{h \to 0} \frac{1 - e^{-\mu_1 h}}{h} = \mu_1,$$

$$\lambda_{21} = \lim_{h \to 0} \frac{p_{21}(h)}{h} = 0,$$

$$\lambda_{23} = \lim_{h \to 0} \frac{p_{23}(h)}{h} = \lim_{h \to 0} \frac{P(\eta_2 \leqslant h) + o(h)}{h} = \lim_{h \to 0} \frac{1 - e^{-\mu_2 h}}{h} = \mu_2,$$

$$\lambda_{31} = \lim_{h \to 0} \frac{p_{31}(h)}{h} = \lim_{h \to 0} \frac{P(\zeta_1 \leqslant h, \zeta_2 > h) + o(h)}{h} = \lim_{h \to 0} \frac{(1 - e^{-\lambda_1 h})e^{-\lambda_2 h}}{h} = \lambda_1,$$

$$\lambda_{32} = \lim_{h \to 0} \frac{p_{32}(h)}{h} = \lim_{h \to 0} \frac{P(\zeta_2 \leqslant h, \zeta_1 > h)) + o(h)}{h} = \lim_{h \to 0} \frac{(1 - e^{-\lambda_2 h})e^{-\lambda_1 h}}{h} = \lambda_2.$$

From (1.78) we have

$$\lambda_{11} = -\mu_1, \quad \lambda_{22} = -\mu_2, \quad \lambda_{33} = -(\lambda_1 + \lambda_2).$$

Finally, the transition rate matrix is

$$\Lambda = \begin{bmatrix} -\mu_1 & \mu_1 & 0 \\ 0 & -\mu_2 & \mu_2 \\ \lambda_1 & \lambda_2 & -(\lambda_1 + \lambda_2) \end{bmatrix}. \tag{1.84}$$

Equation (1.61) for the Laplace transforms has the form

$$\begin{bmatrix} s + \mu_1 & 0 & -\lambda_1 \\ -\mu_1 & s + \mu_2 & -\lambda_2 \\ 0 & -\mu_2 & s + (\lambda_1 + \lambda_2) \end{bmatrix} \begin{bmatrix} \tilde{p}_1(s) \\ \tilde{p}_2(s) \\ \tilde{p}_3(s) \end{bmatrix} = \begin{bmatrix} 0 \\ 0 \\ 1 \end{bmatrix} \tag{1.85}$$

$$(s\boldsymbol{I} - \boldsymbol{\Lambda}^T)\tilde{\boldsymbol{P}}(s) = \boldsymbol{P}(0). \tag{1.86}$$

For

$$\mu_1 = 0.1, \quad \mu_2 = 0.15, \quad \lambda_1 = 0.02, \quad \lambda_2 = 0.01,$$

using the MATHEMATICA computer program we finally obtain the distribution of the process states

$$\begin{aligned} p_1(t) &= 0.1578548 - 0.0259738e^{-0.164495t} - 0.1219208e^{-0.115505t}, \\ p_2(t) &= 0.0526316 - 0.0800327e^{-0.164495t} + 0.0274011e^{-0.115505t}, \\ p_3(t) &= 0.789474 + 0.116007e^{-0.164495t} + 0.0945198e^{-0.115505t}. \end{aligned} \tag{1.87}$$

Semi-Markov process

2

Abstract

This chapter provides the definitions and basic properties related to a discrete state space semi-Markov process (SMP). The SMP is constructed by the so-called Markov renewal process (MRP) that is a special case of the two-dimensional Markov sequence. The MRP is defined by the transition probabilities matrix, called the renewal kernel and an initial distribution, or by other characteristics that are equivalent to the renewal kernel. The counting process corresponding to the SMP allows us to determine the concept of process regularity. The process is said to be regular if the corresponding counting process has a finite number of jumps in a finite period. The chapter also shows the other methods of determining the SMP. The concepts presented are illustrated by some examples. Elements of the SMP statistical estimation are also presented in the chapter. There is considered estimation of the renewal kernel elements by observing one or many sample paths in the time interval, or a given number of the state changes. Basic concepts of the nonhomogeneous semi-Markov processes theory are introduced. The end of the chapter deals with proofs of some theorems.

Keywords: Markov renewal process, Renewal kernel, Semi-Markov process, Statistical estimation of semi-Markov process, Nonhomogeneous semi-Markov process

2.1 Markov renewal processes

The semi-Markov processes (SMPs) were introduced independently and almost simultaneously by Levy [70], Smith [92], and Takács [94] in 1954-1955. The essential developments of SMPs theory were proposed by Cinlar [15], Korolyuk and Turbin [60], and Limnios and Oprisan [72]. We present only SMPs with a discrete state space. A SMP is constructed by the Markov renewal process (MRP), which is defined by the renewal kernel and the initial distribution or by other characteristics that are equivalent to the renewal kernel.

Suppose that $\mathbb{N} = \{1, 2, \ldots\}$, $\mathbb{N}_0 = \{0, 1, 2, \ldots\}$, $\mathbb{R}_+ = [0, \infty)$, and S is a discrete (finite or countable) state space. Let ξ_n be a discrete random variable taking values on S and let ϑ_n be a continuous random variable with values in the set $\mathbb{R}_+$.

Definition 2.1. A two-dimensional sequence of random variables $\{(\xi_n, \vartheta_n) : n \in \mathbb{N}_0\}$ is said to be a MRP if:

1. for all $n \in \mathbb{N}_0$, $j \in S$, $t \in \mathbb{R}_+$

$$P(\xi_{n+1} = j, \vartheta_{n+1} \leqslant t | \xi_n = i, \vartheta_n, \ldots, \xi_0, \vartheta_0)$$

$$= P(\xi_{n+1} = j, \vartheta_{n+1} \leqslant t | \xi_n = i), \quad \text{with prob. 1,} \tag{2.1}$$

2. for all $i, j \in S$,

$$P(\xi_0 = i, \vartheta_0 = 0) = P(\xi_0 = i). \tag{2.2}$$

Semi-Markov Processes: Applications in System Reliability and Maintenance. http://dx.doi.org/10.1016/B978-0-12-800518-7.00002-1

From Definition 2.1, it follows that MRP is a homogeneous two-dimensional Markov chain such that its transition probabilities depend only on the discrete component (they do not depend on the second component).

A matrix

$$Q(t) = \left[Q_{ij}(t) : i, j \in S\right], \tag{2.3}$$

where

$$Q_{ij}(t) = P(\xi_{n+1} = j, \vartheta_{n+1} \leqslant t | \xi_n = i)$$

is called a *renewal matrix*.

A vector $p = [p_i : i \in S]$, where $p_i = P\{\xi_0 = i\}$ defines an initial distribution of the MRP.

It follows from Definition 2.1 that the Markov renewal matrix satisfies the following conditions:

1. The functions

$$Q_{ij}(t),\ t \in \mathbb{R}_+,\ (i,j) \in S \times S$$

   are not decreasing and right-hand continuous.
2. For each pair $(i,j) \in S \times S$, $Q_{ij}(0) = 0$ and $Q_{ij}(t) \leqslant 1$ for $t \in \mathbb{R}_+$.
3. For each $i \in S$, $\lim\limits_{t \to \infty} \sum\limits_{j \in S} Q_{ij}(t) = 1$.

One can prove that a function matrix $Q(t) = \left[Q_{ij}(t) : i, j \in S\right]$ satisfying the above mentioned conditions and a vector $p_0 = \left[p_i^{(0)} : i \in S\right]$ such that

$$\sum_{i \in S} p_i^{(0)} = 1$$

define some MRP.

From definition of the renewal matrix it follows that:

$$P = \left[p_{ij} : i, j \in S\right] \tag{2.4}$$

$$p_{ij} = \lim_{t \to \infty} Q_{ij}(t)$$

is a stochastic matrix. It means that for each pair $(i, j) \in S \times S$

$$p_{ij} \geqslant 0$$

and for each $i \in S$

$$\sum_{j \in S} p_{ij} = 1.$$

It is easy to notice that for each $i \in S$

$$G_i(t) = \sum_{j \in S} Q_{ij}(t) \tag{2.5}$$

is a probability cumulative distribution function (CDF) on $\mathbb{R}_+$.

Definition 2.1 leads to interesting and important conclusions.

Proposition 2.1.

$$P(\vartheta_0 = 0) = 1. \tag{2.6}$$

Proposition 2.2. For a MRP with an initial distribution p_0 and a renewal kernel $Q(t)$, $t \geq 0$ the following equality is satisfied:

$$P(\xi_0 = i_0, \xi_1 = i_1, \vartheta_1 \leq t_1, \dots, \xi_n = i_n, \vartheta_n \leq t_n)$$
$$= p_{i_0} Q_{i_0 i_1}(t_1) Q_{i_1 i_2}(t_2) \dots Q_{i_{n-1} i_n}(t_n). \tag{2.7}$$

For $t_1 \to \infty, \dots, t_n \to \infty$, we obtain

$$P(\xi_0 = i_0, \xi_1 = i_1, \dots, \xi_n = i_n)$$
$$= p_{i_0} p_{i_0 i_1} p_{i_1 i_2} \cdots p_{i_{n-1} i_n}. \tag{2.8}$$

Hence,

Proposition 2.3. A sequence $\{\xi_n : n \in \mathbb{N}_0\}$ is a homogeneous Markov chain with the discrete state space S, defined by the initial distribution $p = [p_{i_0} : i_0 \in S]$ and the transition matrix $P = [p_{ij} : i,j \in S]$, where

$$p_{ij} = \lim_{t \to \infty} Q_{ij}(t). \tag{2.9}$$

Proposition 2.4. The random variables $\vartheta_1, \dots, \vartheta_n$ are conditionally independent if a trajectory of the Markov chain $\{\xi_n : n \in \mathbb{N}_0\}$ is given:

$$P(\vartheta_1 \leq t_1, \vartheta_2 \leq t_2 \dots, \vartheta_n \leq t_n | \xi_0 = i_0, \xi_1 = i_1, \dots, \xi_n = i_n)$$
$$= \prod_{k=1}^{n} P(\vartheta_k \leq t_k | \xi_k = i_k, \xi_{k-1} = i_{k-1}). \tag{2.10}$$

Definition 2.2. The Markov renewal matrix $Q(t) = [Q_{ij}(t) : i,j \in S]$ is called continuous if each row of the matrix contains at least one element having a continuous component in the Lebesgue decomposition of the probability distribution.

Example 2.1. The matrix $Q(t) = [Q_{ij}(t) : i,j \in S]$ with elements

$$Q_{ij}(t) = p_{ij} G_i(t), \quad i \in S,$$

where

$$G_i(t) = c I_{[1,\infty)}(t) + (1 - c) \int_0^t h_i(u) du$$

$c \in (0, 1)$, $p_{ij} \geq 0$, $\sum_{j \in S} p_{ij} = 1$ and $h_i(\cdot)$ is a continuous probability density function, is an example of the continuous Markov renewal matrix.

Example 2.2. The Markov renewal matrix $Q(t) = [Q_{ij}(t) : i,j \in S]$ with elements

$$Q_{ij}(t) = p_{ij} I_{[1,\infty)}(t), \quad i \in S,$$

where $p_{ij} \geq 0$, $\sum_{j \in S} p_{ij} = 1$ is not a continuous Markov renewal matrix.

Moreover, in this book we will assume that the Markov renewal matrix $Q(t) = [Q_{ij}(t) : i,j \in S]$ is continuous.

Let

$$\tau_0 = \vartheta_0, \tag{2.11}$$

$$\tau_n = \vartheta_1 + \vartheta_2 + \cdots + \vartheta_n, n \in \mathbb{N},$$

$$\tau_\infty = \lim_{n \to \infty} \tau_n = \sup\{\tau_n : n \in \mathbb{N}_0\}.$$

Theorem 2.1 (Korolyuk and Turbin [60]). The sequence $\{(\xi_n, \tau_n) : n \in \mathbb{N}_0\}$ is a two-dimensional Markov chain with transition probabilities

$$P(\xi_{n+1} = j, \tau_{n+1} \leqslant t | \xi_n = i, \tau_n = h) = Q_{ij}(t - h), \quad i,j \in S \tag{2.12}$$

and it is also called a MRP.

Proof: [60].

The renewal kernel allows to us calculate a probability distribution of the MRP $\{(\xi_n, \tau_n) : n \in \mathbb{N}_0\}$.

Theorem 2.2 (Korolyuk and Turbin [60] and Grabski [32]). A transition probability

$$R_{ij}^{(n)}(t) = P(\xi_n = j, \tau_n \leqslant t | \xi_0 = i, \tau_0 = 0) \tag{2.13}$$

is given by

$$R_{ij}^{(n)}(t) = \sum_{k_1 \in S} \sum_{k_2 \in S} \cdots \sum_{k_{n-1} \in S} Q_{ik_1} * Q_{k_1 k_2}(t) * \cdots * Q_{k_{n-1} j}(t), \quad n = 2, 3, \ldots. \tag{2.14}$$

Proof: [60].

For convenience we introduce a matrix notation. Let

$$\boldsymbol{R}^{(n)}(t) = \left[R_{ij}^{(n)}(t) : i,j \in S \right]. \tag{2.15}$$

To define a convolution power of the renewal kernel we introduce matrices:

$$\boldsymbol{Q}^{(0)}(t) = \left[Q_{ij}^{(0)}(t) : i,j \in S \right], \quad Q_{ij}^{(0)}(t) = \begin{cases} 1 \text{ for } i = j \\ 0 \text{ for } i \neq j \end{cases}, \quad t \geqslant 0, \tag{2.16}$$

$$\boldsymbol{Q}^{(1)}(t) = \boldsymbol{Q}(t) = \left[Q_{ij}(t) : i,j \in S \right], \quad t \geqslant 0, \tag{2.17}$$

$$\boldsymbol{Q}^{(n)}(t) = \left[Q_{ij}^{(n)}(t) : i,j \in S \right], \tag{2.18}$$

where

$$Q_{ij}^{(n)}(t) := \sum_{k \in S} \int_0^t dQ_{ik}(x) Q_{kj}^{(n-1)}(t - x)$$

$$= \sum_{k \in S} Q_{ik} * Q_{kj}^{(n-1)}(t), \quad t \geqslant 0, n = 2, 3, \ldots.$$

Notice that

$$\boldsymbol{R}^{(n)}(t) = \boldsymbol{Q}^{(n)}(t). \tag{2.19}$$

From the definition of the function $R_{ij}^n(t)$ and from the property of a two-dimensional random variable distribution, we obtain

$$\lim_{t \to \infty} R_{ij}^{(n)}(t) = P(\xi_n = j | \xi_0 = i), \quad i, j \in S, t \geqslant 0, \tag{2.20}$$

$$\sum_{j \in S} R_{ij}^{(n)}(t) = \sum_{j \in S} Q_{ij}^{(n)}(t) = P(\tau_n \leqslant t | \xi_0 = i), \quad i, j \in S, t \geqslant 0. \tag{2.21}$$

Applying the formula for the total probability, we obtain

$$P(\tau_n \leqslant t) = \sum_{i \in S} P\{\tau_n \leqslant t | \xi_0 = i\} P(\xi_0 = i)$$

$$= \sum_{i \in S} \sum_{j \in S} p_i R_{ij}^{(n)}(t) = \sum_{i \in S} \sum_{j \in S} p_i Q_{ij}^{(n)}(t). \tag{2.22}$$

2.2 Definition of discrete state space SMP

We shall present a definition and basic properties of a homogeneous SMP with a countable or finite state space S. The SMP will be determined by the MRP.

Definition 2.3. A stochastic process $\{N(t) : t \geqslant 0\}$ defined by the formula

$$N(t) = \sup\{n \in \mathbb{N}_0 : \tau_n \leqslant t\} \tag{2.23}$$

is called a counting process, corresponding to a random sequence $\{\tau_n : \in \mathbb{N}_0\}$.

A sample path of this process is shown in Figure 2.1.

Definition 2.4. A discrete state space S stochastic process $\{X(t) : t \geqslant 0\}$ with the piecewise constant and the right continuous sample paths given by

$$X(t) = \xi_{N}(t) \tag{2.24}$$

is called a SMP associated with the MRP $\{(\xi_n, \vartheta_n) : n \in \mathbb{N}_0\}$ with the initial distribution $\boldsymbol{p} = [p_i(0) : i \in S]$ and the kernel $\boldsymbol{Q}(t) = [Q_{ij}(t) : i, j \in S], t \geqslant 0$.

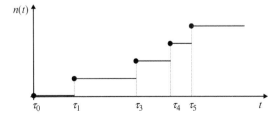

Figure 2.1 A sample path of the counting process.

From Definition 2.24, it follows that:

$$X(t) = \xi_n \quad \text{for } t \in [\tau_n, \tau_{n+1}), \; n \in \mathbb{N}_0. \tag{2.25}$$

Figure 2.2 shows the SMP sample path.

It follows from the definition of the SMP that future states of the process and their sojourn times do not depend on past states and their sojourn times if a present state is known. Let us add that the initial distribution p and the kernel $Q(t)$, $t \geqslant 0$ define completely the SMP. From the definition of SMP, it follows that:

$$X(\tau_n) = \xi_n \quad \text{for } n \in \mathbb{N}_0.$$

This means that a random sequence $\{X(\tau_n) : n \in \mathbb{N}_0\}$ is a *homogeneous Markov chain* with a state space S, defined by the initial distribution $p_0 = \left[p_i^0 : i \in S \right]$ and the stochastic matrix $P = \left[p_{ij} : i, j \in S \right]$, where

$$p_{ij} = \lim_{t \to \infty} Q_{ij}(t). \tag{2.26}$$

The sequence $\{X(\tau_n) : n \in \mathbb{N}_0\}$ is called an *embedded Markov chain* of the SMP $\{X(t) : t \geqslant 0\}$.

2.3   Regularity of SMP

Definition 2.5. A SMP $\{X(t) : t \geqslant 0\}$ is said to be regular if the corresponding counting process $\{N(t) : t \geqslant 0\}$ has a finite number of jumps in a finite period with probability 1:

$$\forall_{t \in \mathbb{R}_+} P(N(t) < \infty) = 1. \tag{2.27}$$

The equality (2.27) is equivalent to a relation

$$\forall_{t \in \mathbb{R}_+} P(N(t) = \infty) = 0. \tag{2.28}$$

Proposition 2.5. A SMP $\{X(t) : t \geqslant 0\}$ is regular if and only if

$$\forall_{t \in \mathbb{R}_+} \lim_{n \to \infty} P(N(t) \geqslant n) = \lim_{n \to \infty} P(\tau_n \leqslant t) = 0. \tag{2.29}$$

Proof: [32].

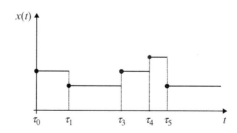

Figure 2.2 A sample path of the semi-Markov process.

Proposition 2.6. If $E[N(t)] < \infty$, then a SMP $\{X(t) : t \geqslant 0\}$ is regular.

Proof: [32].

Theorem 2.3 (Korolyuk and Turbin [60]). Every SMP with a finite state space S is regular.

Proof: [60]

2.4 Other methods of determining the SMP

The SMP was defined by the initial distribution p and renewal kernel $Q(t)$, which determine the Markov renewal process. There are other ways of determining SMP. They are presented, among others, by Korolyuk and Turbin [60] and Limnios and Oprisan [72]. Some definitions of SMP enable its construction. First, we introduce the concepts and symbols that will be necessary for further considerations. For $P(\xi_{n+1} = j, \xi_n = i) > 0$, we define a function

$$F_{ij}(t) = P(\vartheta_{n+1} \leqslant t | \xi_n = i, \xi_{n+1} = j), \quad i,j \in S, \ t \geqslant 0. \tag{2.30}$$

Notice that

$$F_{ij}(t) = P(\vartheta_{n+1} \leqslant t | \xi_{n+1} = j, \xi_n = i) = \frac{P(\vartheta_{n+1} \leqslant t, \xi_{n+1} = j, \xi_n = i)}{P(\xi_{n+1} = j, \xi_n = i)} \tag{2.31}$$

$$= \frac{P(\vartheta_{n+1} \leqslant t, \ \xi_{n+1} = j | \xi_n = i)}{P(\xi_{n+1} = j | \xi_n = i)} = \frac{Q_{ij}(t)}{p_{ij}} \quad \text{for } i,j \in S, \ t \geqslant 0.$$

The function

$$F_{ij}(t) = P(\tau_{n+1} - \tau_n \leqslant t | X(\tau_n) = i, X(\tau_{n+1}) = j) = \frac{Q_{ij}(t)}{p_{ij}} \tag{2.32}$$

is a CDF of some random variable, which is denoted by T_{ij} and it is called a *holding time* [45] in state i, if the next state will be j. From (2.32), we have

$$Q_{ij}(t) = p_{ij}F_{ij}(t). \tag{2.33}$$

The function

$$G_i(t) = P(\tau_{n+1} - \tau_n \leqslant t | X(\tau_n) = i) = P(\vartheta_{n+1} \leqslant t | X(\tau_n) = i) = \sum_{j \in S} Q_{ij}(t) \tag{2.34}$$

is a cumulative probability distribution of a random variable T_i that is called a *waiting time* [45] in state i when a successor state is unknown.

It follows from (2.33) that a SMP with the discrete state space can be defined by the transition probabilities matrix of an embedded Markov chain: $P = [p_{ij} : i,j \in S]$ and the matrix of the holing times CDF $F(t) = [F_{ij}(t) : i,j \in S]$. Therefore, a triple $(p, P, F(t))$ determines the homogeneous SMP with the discrete space S. This method of determining SMP is convenient, for Monte-Carlo simulation of the SMP sample path.

From the Radon-Nikodym theorem it follows that there exist the functions $a_{ij}(x)$, $x \in \mathbb{R}_+$, $i,j \in S$, such that

$$Q_{ij}(t) = \int_0^t a_{ij}(t)\,dG_i(x).$$ (2.35)

Because

$$Q_{ij}(t) = P(\vartheta_{n+1} \leqslant t,\ \xi_{n+1} = j | \xi_n = i)$$

$$= \int_0^t P(\xi_{n+1} = j | \xi_n = i,\ \vartheta_{n+1} = x)dP(\vartheta_{n+1} \leqslant x | \xi_n = i)$$

$$= \int_0^t P(\xi_{n+1} = j | \xi_n = i,\ \vartheta_{n+1} = x)\,dG_i(x)$$

then

$$a_{ij}(x) = P(\xi_{n+1} = j | \xi_n = i,\ \vartheta_{n+1} = x).$$ (2.36)

The function $a_{ij}(x)$, $t \in \mathbb{R}_+$ represents the transition probability from state i to state j under the condition that duration of state i is equal to x. From (2.35), it follows that matrices

$$\boldsymbol{a}(x) = \big[a_{ij}(x) : i,j \in S \big]$$ (2.37)

and

$$\boldsymbol{G}(x) = \big[\delta_{ij}\, G_i(x) : i,j \in S \big]$$ (2.38)

determine the kernel $\boldsymbol{Q}(t) = \big[Q_{ij}(t) : i,j \in S \big]$. Therefore, a triple $(\boldsymbol{p}, \boldsymbol{a}(x), \boldsymbol{G}(x))$ defines the continuous-time SMP with a discrete state space S.

In conclusion, three equivalent ways of determining the SMP are presented in this section:

- by pair $(\boldsymbol{p}, \boldsymbol{Q}(t))$,
- by triple $(\boldsymbol{p}, \boldsymbol{P}, \boldsymbol{F}(t))$,
- by triple $(\boldsymbol{p}, \boldsymbol{a}(x), \boldsymbol{G}(x))$.

It should be added that there exist other ways to define the SMP [60]. Presented here are the ways of defining SMP that seem to be most useful in applications.

2.5 Connection between Semi-Markov and Markov process

A discrete state space and continuous-time SMP is a generalization of that kind of Markov process. The Markov process can be treated as a special case of the SMP.
Theorem 2.4 (Korolyuk and Turbin [60]). Every homogeneous Markov process $\{X(t) : t \geqslant 0\}$ with the discrete space S and the right-continuous trajectories keeping constant values on the half-intervals, given by the transition rate matrix

$\Lambda = \left[\lambda_{ij} : i,j \in S \right]$, $0 < -\lambda_{ii} = \lambda_i < \infty$ is the SMP with the kernel

$$Q(t) = \left[Q_{ij}(t) : i,j \in S \right], \tag{2.39}$$

$$\tag{2.40}$$

where

$$Q_{ij}(t) = p_{ij} \left(1 - e^{-\lambda_i t} \right), \quad t \geqslant 0, \tag{2.41}$$

$$p_{ij} = \frac{\lambda_{ij}}{\lambda_i} \quad \text{for } i \neq j, \ p_{ii} = 0. \tag{2.42}$$

Proof: [24, 32].

From this theorem it follows that the length of interval $[\tau_n, \tau_{n+1})$ given states at instants τ_n and τ_{n+1} is a random variable having an exponential distribution with parameter independent of state at the moment τ_{n+1}:

$$F_{ij}(t) = P(\tau_{n+1} - \tau_n \leqslant t | X(\tau_n) = i, \quad X(\tau_{n+1}) = j) = 1 - e^{-\lambda_i t}, \ t \geqslant 0. \tag{2.43}$$

As we know, the function $F_{ij}(t)$ is a cumulative probability distribution of a holding time in the state i, if the next state is j. Recall that the function

$$G_i(t) = \sum_{j \in S} Q_{ij}(t) = 1 - e^{-\lambda_i t}, \quad t \geqslant 0 \tag{2.44}$$

is a CDF of a waiting time in the state i. For the Markov process, holding times T_{ij}, $i,j \in S$ and waiting times T_i, $j \in S$ have the identical exponential distributions with parameters $\lambda_i = \frac{1}{E(T_i)}$, $i \in S$ that do not depend on state j.

2.6 Illustrative examples

Example 2.3. Presented here as a model is a modification and some extension of a model presented in Example 1.5. We consider the reliability model of a renewable two-component series system under the assumption that the times to failure of booth components denoted as ζ_1, ζ_2 are exponentially distributed with parameters λ_1 and λ_2, but the renewal (repair) times of components are the nonnegative random variables η_1, η_2 with arbitrary distributions defined by CDF $F_{\eta_1}(t)$, $F_{\eta_2}(t)$. We suppose that the above-considered random variables and their copies are mutually independent. The states are defined as identical to those in Example 1.5.

1. The system renewal after the failure of the first component (down state)
2. The system renewal after the failure of the second component (down state)
3. Work of the system, both components are up

A renewal kernel is given by

$$Q(t) = \begin{bmatrix} 0 & 0 & Q_{13}(t) \\ 0 & 0 & Q_{23}(t) \\ Q_{31}(t) & Q_{32}(t) & 0 \end{bmatrix}. \tag{2.45}$$

Using the assumptions, we calculate all elements of this matrix.

$$Q_{13}(t) = F_{\eta_1}(t), \tag{2.46}$$

$$Q_{23}(t) = F_{\eta_2}(t), \tag{2.47}$$

$$Q_{31}(t) = P(\zeta_1 \leqslant t, \ \zeta_1 > \zeta_2) = \iint\limits_{D_{13}} \lambda_1 e^{-\lambda_1 x} \lambda_2 e^{-\lambda_2 y} dx\, dy,$$

where

$$D_{31} = \{(x, y) : x \leqslant t, \ y > x\}.$$

Thus,

$$Q_{31}(t) = \int_0^t \lambda_1 e^{-\lambda_1 x} e^{-\lambda_2 x} dx = \frac{\lambda_1}{\lambda_1 + \lambda_2} \left(1 - e^{-(\lambda_1 + \lambda_2) t}\right). \tag{2.48}$$

In the same way, we obtain

$$Q_{32}(t) = \frac{\lambda_2}{\lambda_1 + \lambda_2} \left(1 - e^{-(\lambda_1 + \lambda_2) t}\right). \tag{2.49}$$

The cumulative distribution of the waiting times for the states 1, 2, and 3 for $t \geqslant 0$ are

$$G_1(t) = P(\eta_1 \leqslant t) = F_{\eta_1}(t),$$

$$G_2(t)) = P(\eta_2 \leqslant t) = F_{\eta_2}(t),$$

$$G_3(t) = Q_{31}(t) + Q_{32}(t) = 1 - e^{(\lambda_1 + \lambda_2) t}.$$

Example 2.4. The transition rate matrix of the Poisson process is defined in the previous chapter by (1.66). From Theorem 2.4, it follows that the Poisson process is a SMP with the state space $S = \{0, 1, 2, \ldots\}$ and the kernel

$$Q(t) = \begin{bmatrix} 0 & Q_{01}(t) & 0 & 0 & 0 & \cdots \\ 0 & 0 & Q_{12}(t) & 0 & 0 & \cdots \\ 0 & 0 & 0 & Q_{23}(t) & 0 & \cdots \\ 0 & 0 & 0 & 0 & Q_{34}(t) & \cdots \\ \vdots & \vdots & \vdots & \vdots & \vdots & \vdots \end{bmatrix}, \tag{2.50}$$

where

$$Q_{i\,i+1}(t) = 1 - e^{-\lambda t}, \quad i = 0, 1, 2, \ldots, \ \lambda > 0. \tag{2.51}$$

Example 2.5. The transition rate matrix of the Furry-Yule process is determined by (1.67). From Theorem 2.4, it follows that the Furry-Yule process is a SMP with the state space $S = \{0, 1, 2, \ldots\}$ and the kernel (2.50), where

$$Q_{i\,i+1}(t) = 1 - e^{-(i+1)\lambda t}, \quad i = 0, 1, 2, \ldots, \ \lambda > 0. \tag{2.52}$$

2.7   Elements of statistical estimation

2.7.1   Observation of SMP sample path

We know that an initial distribution and a renewal kernel define a SMP. Now we want to estimate the elements of the renewal kernel by observing one or many sample paths in the time interval $[0, t]$ or the given number of the state changes (jumps) n. The observation of a SMP sample path in an interval $[0, t]$ is a sequence of number

$$sms(t) = (i_0, i_1, \ldots, i_{n(t)}; x_1, \ldots, x_{n(t)}) \tag{2.53}$$

which is a realization of the random sequence

$$SMS(t) = (\xi_0, \xi_1, \ldots, \xi_{N(t)}; \vartheta_1, \ldots, \vartheta_{N(t)}), \tag{2.54}$$

where $N(t)$ is a random variable the values $n(t)$ of which denotes the number of semi-Markov jumps (state changes) in this time interval. For the semi-Markov process with an initial distribution p_0 and a renewal kernel $Q(t), t \geqslant 0$, the probability distribution of this random sequence [72] is

$$P(N(t) = n, \xi_0 = i_0, \xi_1 = i_1, \vartheta_1 \leqslant t_1, \ldots, \xi_n = i_n, \vartheta_n \leqslant t_n) \tag{2.55}$$
$$= p_{i_0} Q_{i_0 i_1}(t_1) Q_{i_1 i_2}(t_2) \ldots Q_{i_{n-1} i_n}(t_n) \left[1 - G(u_t)\right],$$

where $u_t = t - (t_1 + t_2 + \cdots + t_n)$.

For the given number of jumps n, the observation of the SMP trajectory is a sequence of numbers

$$sms(n) = (i_0, i_1, \ldots, i_n; x_1, \ldots, x_n) \tag{2.56}$$

which is a realization of the random sequence

$$SMS(n) = (\xi_0, \xi_1, \ldots, \xi_n; \vartheta_1, \ldots, \vartheta_n). \tag{2.57}$$

In this case, the probability distribution of the random sequence is given by

$$P(\xi_0 = i_0, \xi_1 = i_1, \vartheta_1 \leqslant t_1, \ldots, \xi_n = i_n, \vartheta_n \leqslant t_n)$$
$$= p_{i_0} Q_{i_0 i_1}(t_1) Q_{i_1 i_2}(t_2) \ldots Q_{i_{n-1} i_n}(t_n). \tag{2.58}$$

The function $q_{ij}(x)$, $x \geqslant 0$ is called the density of $Q_{ij}(x)$, $x \geqslant 0$ if there exists such function $q_{ij}(x)$, $i, j \in S$, $x \geqslant 0$ that

$$Q_{ij}(x) = \int_0^x q_{ij}(u) \mathrm{d}u.$$

If each of the functions $Q_{ij}(x)$, $x \geqslant 0$, $i, j \in S$ has the density $q_{ij}(x)$, $i, j \in S$, $x \geqslant 0$, then the sample density function, corresponding to the distribution (2.55), is

$$f(i_0, i_1, \ldots, i_n; t_1, \ldots, t_n) = p_{i_0} q_{i_0 i_1}(t_1) q_{i_1 i_2}(t_2) \ldots q_{i_{n-1} i_n}(t_n) \left[1 - G(u_t)\right], \tag{2.59}$$

and the density corresponding to the distribution (2.55) is

$$f(i_0, i_1, \ldots, i_n; t_1, \ldots, t_n) = p_{i_0} q_{i_0 i_1}(t_1) q_{i_1 i_2}(t_2) \ldots q_{i_{n-1} i_n}(t_n). \tag{2.60}$$

2.7.2   Empirical estimators

We know (2.33), that $Q_{ij}(x) = p_{ij} F_{ij}(x)$. The maximum likelihood estimator of the transition probability p_{ij} is given by

$$\hat{P}_{ij}(t) = \frac{N_{ij}(t)}{N_i(t)}, \tag{2.61}$$

where

$$N_{ij}(t) = \#\{k \in \{1, 2, \ldots, N(t)\} : \xi_{k-1} = i, \ \xi_k = j\} = \sum_{k=1}^{N(t)} I_{\{\xi_{k-1}=i, \xi_k=j\}}$$

signify the SMP direct number of transitions from the state i to j in a time interval $[0, t]$, and

$$N_i(t) = \#\{k \in \{1, 2, \ldots, N(t)\} : \xi_k = i\} = \sum_{k=1}^{N(t)} I_{\{\xi_k=i\}} = \sum_{j \in S} N_{ij}(t)$$

denotes the number of jumps to the state i in the time interval $[0, t]$. Properties of the estimator $\hat{P}_{ij}$ are presented in Ref. [69]. For the sample (2.54), the empirical estimator of the CDF $F_{ij}(t)$ is given by

$$\hat{F}_{ij}(x, t) = \frac{M_{ij}(x, t)}{N_{ij}(t)}, \tag{2.62}$$

where

$$M_{ij}(x, t) = \sum_{k=1}^{N(t)} I_{\{\xi_{k-1}=i, \xi_k=j, \vartheta_k \leqslant x\}}.$$

The empirical estimator of the renewal kernel element $Q_{ij}(t)$ takes the form

$$\hat{Q}_{ij}(x, t) = \hat{P}_{ij}(t)\hat{F}_{ij}(x, t) = \frac{M_{ij}(x, t)}{N_i(t)}. \tag{2.63}$$

For the semi-Markov sample path (2.57), the corresponding estimators are

$$\hat{P}_{ij} = \frac{N_{ij}}{N_i}, \tag{2.64}$$

where

$$N_{ij} = \#\{k \in \{1, 2, \ldots, n\} : \xi_{k-1} = i, \ \xi_k = j\} = \sum_{k=1}^{n} I_{\{\xi_{k-1}=i, \xi_k=j\}}$$

signify the SMP direct number of transitions from the state i to j in n jumps and

$$N_i = \#\{k \in \{1, 2, \ldots, n\} : \xi_k = i\} = \sum_{k=1}^{n} I_{\{\xi_k=i\}} = \sum_{j \in S} N_{ij}$$

means the number of jumps to the state i. Properties of the estimator $\hat{P}_{ij}$ are presented in Ref. [69].

For the sample (2.57), the empirical estimator of the $F_{ij}(x)$ is

$$\hat{F}_{ij}(x) = \frac{M_{ij}(x)}{N_{ij}} \qquad (2.65)$$

where

$$M_{ij}(x) = \sum_{k=1}^{n} I_{\{\xi_{k-1}=i, \xi_k=j, \vartheta_k \leqslant x\}}.$$

In this case, the empirical estimator of the renewal kernel element $Q_{ij}(t)$ is of the form

$$\hat{Q}_{ij}(x) = \hat{P}_{ij}\hat{F}_{ij}(x) = \frac{M_{ij}(x)}{N_i}. \qquad (2.66)$$

Limnios and Oprisan [72] have proved the following asymptotic properties of the empirical estimator $\hat{Q}_{ij}(x, t)$.

Theorem 2.5 (Limnios and Oprisan [72]). The estimators $\hat{Q}_{ij}(x, t)$ of the $Q_{ij}(x)$, for all $i, j \in S$ satisfy the equality

$$\lim_{t \to \infty} \max_{i,j} \sup_{x \in [0,t)} |\hat{Q}_{ij}(x, t) - Q_{ij}(x)| = 0 \quad \text{a.s.} \qquad (2.67)$$

Theorem 2.6 (Limnios and Oprisan [72] and Ouhbi and Limnios [78]). The estimator $\hat{Q}_{ij}(x, t)$ of the function $Q_{ij}(x)$, for any fixed $x > 0$, has the asymptotically normal distribution if $t \to \infty$,

$$\lim_{t \to \infty} P\left(t^{\frac{1}{2}}\left[\hat{Q}_{ij}(x, t) - Q_{ij}(x)\right] \leqslant u\right) = \frac{1}{\sqrt{2\pi}\,\sigma} \int_{-\infty}^{} u e^{-\frac{y^2}{2\sigma_{ij}^2}} \qquad (2.68)$$

where $\sigma_{ij}^2 = \mu_{ii}\left[Q_{ij}(x)\left[1 - Q_{ij}(x)\right]\right]$ and μ_{ii} denotes mean time of return to the state i.

2.7.3 Nonparametric estimators of kernel elements densities

Now we assume that each of the functions $Q_{ij}(x)$, $i, j \in S$ has the density

$$q_{ij}(x) = p_{ij}f_{ij}(x), \quad i, j \in S, \ x \geqslant 0,$$

where $f_{ij}(x)$ is a probability density function corresponding to CDF $F_{ij}(x)$.

Nonparametric kernel estimator of density

First, we construct the so-called nonparametric kernel estimator of the density function $f_{ij}(x)$. Having the observations of the SMP sample path (2.57), we can construct the sequence

$$\{x_{ij}(m) : m = 1, 2, \ldots, n_{ij}\}, \qquad (2.69)$$

where $x_{ij}(m)$ is a value of the holding time $T_{ij}(m)$. Recall that all random variables $T_{ij}(m) : 1, 2, \ldots, N_{ij}$ are iid and

$$P(T_{ij}(m) \leqslant x) = P(T_{ij} \leqslant x) = P(\vartheta_k \leqslant x | \xi_{k-1} = i, \ \xi_k = j).$$

The real function $K(x)$, which takes the nonnegative values and satisfies the condition

$$\int_{-\infty}^{\infty} K(u)du = 1$$

is called a *statistical kernel*.

A function

$$\hat{f}_{ij}(x) = \frac{1}{n_{ij}h} \sum_{m}^{n_{ij}} K\left(\frac{x - x_{ij}(m)}{h}\right), \quad h > 0 \tag{2.70}$$

is said to be a value of the nonparametric kernel estimator of a density function $f_{ij}(x)$. A number h is called a bandwidth or a smoothing parameter or a window. The choice of the kernel and the choice of the bandwidth is a significant problem of density estimation. One well-known proposal is a Gaussian kernel

$$K(u) = \frac{1}{\sqrt{2\pi}} e^{-\frac{u^2}{2}} I_{(\infty,\infty)}(u) \tag{2.71}$$

A number

$$h = 1.06 \hat{s} n^{-0.2}, \tag{2.72}$$

where $\hat{s}$ is the empirical standard deviation

$$\hat{s} = \sqrt{\frac{1}{n} \sum_{i=1}^{n} (x_i - \bar{x})^2},$$

is a practical estimate of the bandwidth in this case. The value of the nonparametric kernel estimator of the function $q_{ij}(x) : i, j \in S, x \geqslant 0$ is the function

$$\hat{q}_{ij}(x) = \hat{p}_{ij}\hat{f}_{ij}(x), \quad i, j \in S, x \geqslant 0 \tag{2.73}$$

where

$$\hat{p}_{ij} = \frac{n_{ij}}{n_i}, \quad \hat{f}_{ij}(x) = \frac{1}{n_{ij}h} \sum_{m}^{n_{ij}} K\left(\frac{x - x_{ij}(m)}{h}\right), \, h > 0.$$

Ciesielski nonparametric estimator of density

Ciesielski nonparametric density estimator is based on the concept of *the spline function*. The definition and properties of the estimator are presented by Ciesielski [14]. Other certain properties of the estimator are examined and announced by Krzykowski [68]. The value of this estimator depends on the value $x = (x_1, x_2, \ldots, x_n)$ of the simple sample $X = (X_1, X_2, \ldots, X_n)$ and a natural number r, which is called an order of the spline function. We obtain smoothing parameter h using the formula [68]

$$h = \sqrt{\frac{6}{r n (n-1)} \sum_{j=1}^{n} (x_j - \bar{x})^2},$$

where $\bar{x} = \frac{\sum_{j=1}^{n} x_j}{n}$ is a mean of the sample realization $(x_1, x_2, \ldots, x_n)$.

We denote:

- $$s_0 = \left[\frac{x_{\min}}{h} - v\right] - r, \quad s_m = \left[\frac{x_{\max}}{h} - v\right] + 1$$

  where $v = 0$, if r is an even number, and $v = \frac{1}{2}$, if r is an odd number.

- $$N_j(x) = r \sum_{i=j}^{r} \frac{(-1)^{r-i}}{i!(r-i)!} \left((s+v+i) - \frac{x}{h}\right)^{r-1}, \quad j = 1, 2, \ldots, r$$

- $$N_{s,h}^{(r)}(x) = \begin{cases} N_1(x) & \text{dla } x \in A_1 = [(s+v)h, (s+v+1)h) \\ N_2(x) & \text{dla } x \in A_2 = [(s+v+1)h, (s+v+2)h) \\ \vdots & \vdots \quad \vdots \\ N_j(x) & \text{dla } x \in A_j = [(s+v+j-1)h, (s+v+j)h) \\ \vdots & \vdots \quad \vdots \\ N_r(x) & \text{dla } x \in A_r = [(s+v+r-1)h, (s+v+r)h) \\ 0 & \text{dla } x < (s+v)h \quad \text{albo} \quad x \geqslant (s+v+r)h \end{cases}$$

- $$a_{s,h} = \frac{1}{n} \sum_{j=1}^{n} \frac{1}{h} N_{s,h}^{(r)}(x_j), \quad s = s_0, s_0 + 1, \ldots, s_m$$

The value of the Ciesielski nonparametric density estimator is given by

$$\hat{f}_{ij}(x) = \sum_{s=s_0}^{s_m} a_{s,h} N_{s,h}^{(r)}(x), \quad x \in \mathbb{R}$$

and the function

$$\hat{q}_{ij}(x) = \hat{p}_{ij} \hat{f}_{ij}(x), \quad i,j \in S, \ x \geqslant 0 \tag{2.74}$$

is the corresponding nonparametric spline estimator of the density $q_{ij}(x) : i,j \in S, \ x \geqslant 0$.

We should add that there is another method of the semi-Markov kernel estimation. We can use one of the parametric methods of the densities $f_{ij}(x)$, $i,j \in S, x \geqslant 0$ estimation.

2.8 Nonhomogeneous Semi-Markov process

The nonhomogeneous semi-Markov process (NHSMP) was introduced independently by Iosifescu-Manu [46] and Hoem [42]. The results of Iosifescu-Manu were generalized by Jensen and De Dominicisis [48]. Theory of discrete time NHSMP was developed by Vassiliou and Papadopoulou [96] and Papadopoulou and Vassiliou [80].

Definition 2.6. A two-dimensional Markov chain $\{(\xi_n, \tau_n) : n \in \mathbb{N}_0\}$ with transition probabilities

$$Q_{ij}(t,x) = P(\xi_{N(t)+1} = j, \ \tau_{N(t)+1} - \tau_{N(t)} \leqslant x | \xi_{N(t)} = i, \ \tau_{N(t)} = t),$$
$$i,j \in S, \ x,t \in \mathbb{R}_+ \tag{2.75}$$

is called nonhomogeneous Markov renewal process (NHMRP).

Definition 2.7. A stochastic process $\{X(t) : t \geqslant 0\}$ with the piecewise constant and the right continuous sample paths, which is given by

$$X(t) = \xi_{N(t)} \tag{2.76}$$

is called a NHSMP associated with NHMRP $\{(\xi_n, \vartheta_n) : n \in \mathbb{N}_0\}$ determined by the initial distribution $\boldsymbol{p} = [p_i(0) : i \in S]$ and the kernel $\boldsymbol{Q}(t, x) = [Q_{ij}(t, x) : i, j \in S]$, $t \geqslant 0$.

Recall that $N(t) = \sup\{n \in \mathbb{N}_0 : \tau_n \leqslant t\}$ denotes a number of the state changes in a time interval $[0, t]$ and $\{N(t) : t \in \mathbb{R}_+\}$ is a counting process. Recall also that $\tau_{N(t)+1} - \tau_{N(t)} = \vartheta_{N(t)+1}$.

Definition 2.8. The functions

$$p_{ij}(t) = P(\xi_{N(t)+1} = j | \xi_{N(t)} = i, \tau_{N(t)} = t) = \lim_{x \to \infty} Q_{ij}(t, x), \quad t \in \mathbb{R}_+, \ i, j \in S \tag{2.77}$$

are called the transition probabilities of the embedded nonhomogeneous Markov chain $\{\xi_n : n \in \mathbb{N}_0\}$.

Those functions form a square matrix $\boldsymbol{p}(t) = [p_{ij}(t) : i, j \in S]$.

Similar to the case of a homogeneous SMP, we can introduce a CDF of a holding time $\{T_{ij}(t) : t \in \mathbb{R}_+\}$, $i, j \in S$. The CDF is given by

$$F_{ij}(t, x) = P(\vartheta_{N(t)+1} \leqslant x | \xi_{N(t)+1} = j, \ \xi_{N(t)} = i, \ \tau_{N(t)} = t)$$

$$= P(T_{ij}(t) \leqslant x), \quad i, j \in S, \ x, t \in \mathbb{R}_+. \tag{2.78}$$

It is easy to show that

$$Q_{ij}(t, x) = p_{ij}(t) F_{ij}(t, x). \tag{2.79}$$

The CDF of a waiting time $\{T_i(t) : t \in \mathbb{R}_+\}$ in a state i is given by the formula

$$G_i(t, x) = \sum_{j \in S} Q_{ij}(t, x). \tag{2.80}$$

It means that

$$G_i(t, x) = P(T_i(t) \leqslant x) = P(\vartheta_{N(t)+1} \leqslant x | \xi_{N(t)} = i, \ \tau_{N(t)} = t). \tag{2.81}$$

The interval transition probabilities

$$P_{ij}(t, s) = P(X(s) = j | X(\tau_{N(t)}) = i, \tau_{N(t)} = t), \quad 0 \leqslant t < s, \ i, j \in S \tag{2.82}$$

are some of the important characteristics of the NHSMP. Assume that $i \neq j$. The NHSMP that starts from a state i at the moment $\tau_{N(t)} = t$ will be in state j at the moment $s > t > 0$ if in an instant $\tau_{N(t)+1}$ the process will pass to a state $k \in S$, and in a time interval $(\tau_{N(t)+1}, s]$ there takes place at least one change of the state from the state k to j. Using a memoryless property of a SMP in the instant $\tau_{N(t)+1}$ and the theorem of the total probability we have

$$P_{ij}(t, s) = P(X(s) = j | X(\tau_{N(t)}) = i, \tau_{N(t)} = t)$$

$$= \sum_{k \in S} \int_0^u P(X(s) = j | X(\tau_{N(t)+1}) = k, \vartheta_{N(t)+1} = u))$$

$$P(X(\tau_{N(t)+1}) = k, \vartheta_{N(t)+1} \in du | X(0) = i)$$

$$= \sum_{k \in S} \int_0^s P_{kj}(t + u, s - u) Q_{ik}(t, du).$$

Therefore,

$$P_{ij}(t, s) = \sum_{k \in S} \int_0^s P_{kj}(t + u, s - u) Q_{ik}(t, du), \quad i, j \in S, i \neq j, 0 \leqslant t < s.$$

Assume now that $i = j$. The process starting from the state $i \in S$ at the moment $\tau_{N(t)} = t$ will have value $i \in S$ and in the instant $s > t \geqslant 0$ will also have the same value, if the event $\{\vartheta_{N(t)+1} > s\}$ occurs. Because $P(\vartheta_{N(t)+1} \leqslant s | X(\tau_{N(t)}) = i, \tau_{N(t)} = t) = G_i(t, s)$ then

$$P(\vartheta_{N(t)+1} > s | X(\tau_{N(t)}) = i, \tau_{N(t)} = t) = 1 - G_i(t, s).$$

Hence, for any $i \in S$,

$$P_{ii}(t, s) = 1 - G_i(t, s) + \sum_{k \in S} \int_0^s P_{ki}(t + u, s - u) Q_{ik}(t, du), \quad i \in S, 0 \leqslant t < s.$$

Therefore, we obtain the following system of integral equations:

$$P_{ij}(t, s) = \delta_{ij} [1 - G_i(t, s)]$$

$$+ \sum_{k \in S} \int_0^s P_{kj}(t + u, s - u) Q_{ik}(t, du), \quad i, j \in S, 0 \leqslant t < s, \qquad (2.83)$$

with an initial condition

$$P_{ij}(t, 0) = \begin{cases} 1 & \text{if } i = j \\ 0 & \text{if } i \neq j \end{cases} \qquad (2.84)$$

Often, the system of integral equations is called the *evolution system of equations*. The NHSMP has been applied to problems relating to life insurance, medicine as well as issues of reliability and maintenance.

Characteristics and parameters of SMP

3

Abstract

The cumulative distribution functions of the first passage time from the given states to a subset of states and expected values and second moments corresponding to them are considered in this chapter. The equations for these quantities are presented. Moreover, the concept of interval transition probabilities is discussed and the corresponding equations are also derived. These equations allowed obtaining the interval transition probabilities for the alternating process and also for the Poisson and Furry-Yule processes. Furthermore, the reliability and maintainability characteristics and parameters in semi-Markov models are considered in the chapter. At the end of the chapter, numerical illustrative examples and proofs of some theorems are shown.

Keywords: First passage time of semi-Markov process, Interval transition probability, Reliability characteristics and parameters, Maintainability characteristics and parameters

3.1 First passage time to subset of states

From the previous chapter, it follows that a semi-Markov process (SMP) is defined by a renewal kernel and the initial distribution of states or other equivalent parameters. Those quantities contain full information about the process and they allow us to find many characteristics and parameters of the process, which we can translate on the reliability characteristics in the semi-Markov reliability model.

Let $\{X(t) : t \geqslant 0\}$ be the continuous-time SMP with a discrete state space S and a kernel $\boldsymbol{Q}(t)$, $t \geqslant 0$. A value of random variable

$$\Delta_A = \min\{n \in \mathbb{N} : X(\tau_n) \in A\} \tag{3.1}$$

denotes a discrete time (a number of state changes) of *a first arrival* to the set of states $A \subset S$ of the embedded Markov chain, $\{X(\tau_n) : n \in \mathbb{N}_0\}$. A number

$$f_{iA}(m) = P(\Delta_A = m | X(0) = i), \quad m = 1, 2, \ldots \tag{3.2}$$

is a conditional probability of the *first arrival* to subset A at time m, if the initial state is i. Notice that

$$f_{iA}(m) = \begin{cases} P(X(m) \in A, X(m-1) \notin A, \ldots, X(1) \notin A | X(0) = i), & m = 2, 3, \ldots \\ P(X(1) \in A | X(0) = i), & m = 1 \end{cases}$$

$$f_{iA} = P(\Delta_A < \infty | X(0) = i) = 1 - P(\Delta_A = \infty | X(0) = i)$$

Semi-Markov Processes: Applications in System Reliability and Maintenance. http://dx.doi.org/10.1016/B978-0-12-800518-7.00003-X

$$= 1 - P\left(\bigcap_{k=1}^{\infty} X(k) \in A' | X(0) = i\}\right) = P\left(\bigcup_{k=1}^{\infty}\{\Delta_A = m | X(0) = i\right)$$

$$= \sum_{m=1}^{\infty} f_{iA}(m).$$

A value of a random variable

$$\Theta_A = \tau_{\Delta_A} \tag{3.3}$$

denotes a first passage time to the subset A or the time of *a first arrival* at the set of states $A \subset S$ of SMP $\{X(t) : t \geqslant 0\}$. A function

$$\Phi_{iA}(t) = P(\Theta_A \leqslant t | X(0) = i), \quad t \geqslant 0 \tag{3.4}$$

is the cumulative distribution function (CDF) of a random variable Θ_{iA} denoting the first passage time from the state $i \in A'$ to the subset A. Thus,

$$\Phi_{iA}(t) = P(\Theta_{iA} \leqslant t). \tag{3.5}$$

We will present some theorems concerning distributions and parameters of the random variables Θ_{iA}. First, we define a function

$$\Psi_{iA}(t, n) = P(\tau_n \leqslant t, \ \Delta_A = n | X(0) = i), \quad n \in \mathbb{N}, \ t \in \mathbb{R}_+. \tag{3.6}$$

Note that

$$\Psi_{iA}(t, n) = \begin{cases} P(\tau_n \leqslant t, X(\tau_n) \in A, X(\tau_{n-1}) \in A', \ldots, X(\tau_1) \in A' | X(0) = i) \\ \qquad\qquad\qquad\qquad\qquad\qquad\qquad \text{for } n = 2, 3, \ldots \\ P(\tau_1 \leqslant t, X(\tau_1) \in A | X(0) = i) \\ \qquad\qquad\qquad\qquad\qquad\qquad\qquad\qquad \text{for } n = 1. \end{cases} \tag{3.7}$$

A value of the function $\Psi_{iA}(t, n)$ denotes a first passage time from a state $i \in A'$ to a subset A during time no greater than t, as a result of an nth jump of the SMP $\{X(t) : t \in \mathbb{R}_+\}$. This function has the following properties:

1.

$$\forall_{t \in \mathbb{R}_+} \Psi_{iA}(t, n) \leqslant f_{iA}(n), \quad n \in \mathbb{N}. \tag{3.8}$$

2.

$$\forall_{n \in \mathbb{R}_+} \lim_{t \to \infty} \Psi_{iA}(t, n) = f_{iA}(n). \tag{3.9}$$

3. If the process $\{X(t) : t \in \mathbb{R}_+\}$ is regular, then

$$\forall_{t \in \mathbb{R}_+} \lim_{n \to \infty} \Psi_{iA}(t, n) = 0. \tag{3.10}$$

4. If $i \xrightarrow{1} A$, then

$$\lim_{n \to \infty} \Psi_{iA}(t, n) = 0. \tag{3.11}$$

Lemma 3.1. The functions $\Psi_{iA}, i \in A'$ satisfy equations

$$\Psi_{iA}(t, 1) = \sum_{j \in A} Q_{ij}(t), \tag{3.12}$$

$$\Psi_{iA}(t, n) = \sum_{k \in A'} \int_0^t \Psi_{kA}(t - x, n - 1) dQ_{ik}(x), \quad n = 2, 3, \ldots. \tag{3.13}$$

Moreover,

$$\Psi_{iA}(t, n) = \sum_{j \in A} \sum_{r_1, r_2, \ldots, r_{n-1} \in A'} Q_{ir} * Q_{r_1 r_2} * \cdots * Q_{r_{n-1} j}(t). \tag{3.14}$$

Proof: [32].
Note that

$$\Phi_{iA}(t) = \sum_{n=1}^{\infty} \Psi_{iA}(t, n). \tag{3.15}$$

Theorem 3.1 (Korolyuk and Turbin [60], Silvestrov [90], and Grabski [32]). For the regular SMPs such that,

$$f_{iA} = P(\Delta_A < \infty | X(0) = i) = 1, \quad i \in A', \tag{3.16}$$

the distributions $\Phi_{iA}(t)$, $i \in A'$ are proper and they are the only solutions of the equations system

$$\Phi_{iA}(t) = \sum_{j \in A} Q_{ij}(t) + \sum_{k \in S} \int_0^t \Phi_{kA}(t - x) dQ_{ik}(x), \quad i \in A'. \tag{3.17}$$

Proof: [32, 60, 90].
Applying a Laplace-Stieltjes (L-S) transformation for the system of integral equations (3.17) we obtain the linear system of equations for (L-S) transforms

$$\tilde{\varphi}_{iA}(s) = \sum_{j \in A} \tilde{q}_{ij}(s) + \sum_{k \in A'} \tilde{q}_{ik}(s) \tilde{\varphi}_{kA}(s), \tag{3.18}$$

where

$$\tilde{\varphi}_{iA}(s) = \int_0^{\infty} e^{-st} d\Phi_{iA}(t) \tag{3.19}$$

are L-S transforms of the unknown CDF of the random variables Θ_{iA}, $i \in A'$ and

$$\tilde{q}_{ij}(s) = \int_0^{\infty} e^{-st} dQ_{ij}(t) \tag{3.20}$$

are L-S transforms of the given functions $Q_{ij}(t)$, $i, j \in S$. That linear system of equations is equivalent to the matrix equation

$$\left(I - \tilde{q}_{A'}(s) \right) \tilde{\varphi}_{A'}(s) = \tilde{b}(s), \tag{3.21}$$

where

$$I = \left[\delta_{ij} : i, j \in A' \right] \tag{3.22}$$

is the unit matrix,

$$\tilde{q}_{A'}(s) = \left[\tilde{q}_{ij}(s) : i, j \in A' \right] \tag{3.23}$$

is the square submatrix of the L-S transforms of the matrix $\tilde{q}(s)$, while

$$\tilde{\varphi}_{A'}(s) = \left[\tilde{\varphi}_{iA}(s) : i \in A' \right]^T, \quad \boldsymbol{b}(s) = \left[\sum_{j \in A} \tilde{q}_{ij}(s) : i \in A' \right]^T \tag{3.24}$$

are one-column matrices of the corresponding L-S transforms.

The linear system of equations (3.18) for the L-S transforms allows us to obtain the linear system of equations for the moments of random variables Θ_{iA}, $i \in A'$.

Theorem 3.2 (Korolyuk and Turbin [60], Silvestrov [90], and Grabski [32]). If

- assumptions of Theorem 3.1 are satisfied
- $\exists_{c>0} \ \forall_{i,j \in S} \ 0 < E(T_{ij}) \leqslant c$,
- $\forall_{i \in A} \mu_{iA} = \sum_{n=1}^{\infty} n f_{iA}(n) < \infty$,

then there exist expectations $E(\Theta_{iA})$, $i \in A'$ and they are unique solutions of the linear equations system, which have the following matrix form:

$$(\boldsymbol{I} - \boldsymbol{P}_{A'})\overline{\boldsymbol{\Theta}}_{A'} = \overline{\boldsymbol{T}}_{A'}, \tag{3.25}$$

where

$$\boldsymbol{P}_{A'} = \left[p_{ij} : i, j \in A' \right], \quad \overline{\boldsymbol{\Theta}}_{A'} = \left[E(\Theta_{iA}) : i \in A' \right]^T, \quad \overline{\boldsymbol{T}}_{A'} = \left[E(T_i) : i \in A' \right]$$

and $\boldsymbol{I}$ is the unit matrix.

Proof: [32, 60, 90].

Theorem 3.3 (Korolyuk and Turbin [60], Silvestrov [90], and Grabski [32]). If

- assumptions of Theorem 3.1 are fulfilled,
- $\exists_{d>0} \ \forall_{i,j \in S} \ 0 < E(T_{ij}^2) \leqslant d$,
- $\forall_{i \in A} \mu_{iA}^2 = \sum_{n=1}^{\infty} n^2 f_{iA}(n) < \infty$,

then there exist second moments $E(\Theta_{iA}^2)$, $i \in A'$ and they are unique solutions of the linear system equations, which have the following matrix form:

$$(\boldsymbol{I} - \boldsymbol{P}_{A'})\overline{\boldsymbol{\Theta}}_{A'}^2 = \boldsymbol{B}_A, \tag{3.26}$$

where

$$\boldsymbol{P}_{A'} = \left[p_{ij} : i, j \in A' \right], \quad \overline{\boldsymbol{\Theta}}_{A'} = \left[E(\Theta_{iA}^2) : i \in A' \right]^T,$$

$$\boldsymbol{B}_A = \left[b_{iA} : i \in A' \right]^T, \quad b_{iA} = E(T_i^2) + 2 \sum_{k \in A'} p_{ik} E(T_{ik}) E(\Theta_{kA}).$$

Proof: [32, 60, 90].

A first return time to a given state $j \in S$ is an important and useful quantity.

Definition 3.1. A function

$$\Phi_{jj}(t) = P(\tau_{\Delta_j} \leqslant t | X(0) = j), \quad j \in S \tag{3.27}$$

where

$$\Delta_j = \min\{n \in \mathbb{N} : X(\tau_n) = j\}$$

is called a CDF of the first return time to the state j of the SMP $\{X(t) : t \geqslant 0\}$. The random variable having distribution (3.26) is denoted by Θ_{jj}.

Theorem 3.4. Let $\{X(t) : t \geqslant 0\}$ be the regular SMP with the kernel $Q(t) = [Q_{ij}(t) : i, j \in S]$. If for each $i \in S$, $f_{ij} = 1$, then there exists only one, proper CDF of the first return time to the state j and it is given by the equality

$$\Phi_{jj}(t) = Q_{jj}(t) + \sum_{k \in S - \{j\}} \int_0^t \Phi_{kj}(t-x) dQ_{jk}(x). \tag{3.28}$$

Applying L-S transformation, we obtain

$$\tilde{\varphi}_{jj}(s) = \tilde{q}_{jj}(s) + \sum_{k \in S - \{j\}} \tilde{q}_{jk}(s)\tilde{\varphi}_{kj}(s). \tag{3.29}$$

We take the L-S transforms $\tilde{\varphi}_{kj}(s)$ for $k \neq j$ from (3.18).

Theorem 3.5. If

- assumptions of Theorem 3.1 are fulfilled,
- $\exists_{d>0} \forall_{i,j \in S} \ 0 < E(T_{ij}^2) \leqslant d$,
- $\forall_{i \in S} \ \mu_{ij}^2 = \sum_{n=1}^{\infty} n^2 f_{ij}(n) < \infty$,

then there exist expectations $E(\Theta_{jj})$ and second moments $E(\Theta_{jj}^2)$ and

$$E(\Theta_{jj}) = E(T_j) + \sum_{k \in S - \{j\}} p_{jk} E(\Theta_{kj}), \tag{3.30}$$

$$E(\Theta_{jj}^2) = E(T_j^2) + \sum_{k \in S - \{j\}} p_{jk} E(\Theta_{kj}^2) + 2 \sum_{k \in S - \{j\}} p_{jk} E(T_{jk}) E(\Theta_{kj}). \tag{3.31}$$

Example 3.1. Consider a SMP with the state space $S = \{1, 2, 3\}$ defined by the kernel

$$Q(t) = \begin{bmatrix} 0 & 0.2(1 - e^{-0.1t}) & 0.8(1 - e^{-0.1t}) \\ 1 - e^{-t} & 0 & 0 \\ 1 - (1 + 0.2t)e^{-0.2t} & 0 & 0 \end{bmatrix}, \quad t \geqslant 0.$$

We will calculate PDF (probability density function) of the first passage times from state $i \in A' = \{1, 2\}$ to state 3. The L-S transform of the kernel is

$$\tilde{q}(s) = \begin{bmatrix} 0 & \frac{0.02}{s+0.1} & \frac{0.08}{s+0.1} \\ \frac{1}{s+1} & 0 & 0 \\ \frac{0.04}{(s+0.2)^2} & 0 & 0 \end{bmatrix}.$$

Matrices of transforms from (3.21) are as follows:

$$\tilde{q}_{A'}(s) = \begin{bmatrix} 0 & \frac{0.02}{s+0.1} \\ \frac{1}{s+1} & 0 \end{bmatrix}, \quad \tilde{\phi}_{A'}(s) = \begin{bmatrix} \tilde{\phi}_{13}(s) \\ \tilde{\phi}_{23}(s) \end{bmatrix}, \quad \tilde{b}_{A'} = \begin{bmatrix} \frac{0.08}{s+0.1} \\ 0 \end{bmatrix}. \tag{3.32}$$

In this case, (3.21) is of the form

$$\begin{bmatrix} 1 & -\frac{0.02}{s+0.1} \\ -\frac{1}{s+1} & 1 \end{bmatrix} \begin{bmatrix} \tilde{\varphi}_{13}(s) \\ \tilde{\varphi}_{23}(s) \end{bmatrix} = \begin{bmatrix} \frac{0.08}{s+0.1} \\ 0 \end{bmatrix}. \tag{3.33}$$

The solution of this equation is

$$\tilde{\varphi}_{13}(s) = \frac{0.0781599}{s + 0.0783009} + \frac{0.00184008}{s + 1.0217}, \tag{3.34}$$

$$\tilde{\varphi}_{23}(s) = \frac{0.0847998}{s + 0.0783009} - \frac{0.0847998}{s + 1.0217}. \tag{3.35}$$

We obtain the PDFs of random variables Θ_{13}, Θ_{23} as the inverse transforms of the functions $\tilde{\varphi}_{13}(s)$, $\tilde{\varphi}_{23}(s)$.

$$\varphi_{13}(t) = 0.0781599e^{-0.0783009t} + 0.00184008e^{-1.0217t}, \quad t \geqslant 0, \tag{3.36}$$

$$\varphi_{23}(t) = 0.0847998e^{-0.0783009t} - 0.08479980e^{-1.0217t}, \quad t \geqslant 0. \tag{3.37}$$

Plots of those functions appear in Figures 3.1 and 3.2.

The first and second moments of considered random variables can be found from the formula

$$E(\Theta_{i3}^k) = \int_0^\infty t^k \phi(t) dt, \quad i \in A' = 1, 2. \tag{3.38}$$

Variances of these random variables are given by

$$V(\Theta_{i3}) = E(\Theta_{i3}^2) - [E(\Theta_{i3})]^2, \quad i \in A' = 1, 2. \tag{3.39}$$

The first and second moments of the considered random variables may also be found from the equations of Theorems 3.2 and 3.3. In our case, the transition matrix of the embedded Markov chain is

$$P = \begin{bmatrix} 0 & 0.2 & 0.8 \\ 1 & 0 & 0 \\ 1 & 0 & 0 \end{bmatrix}.$$

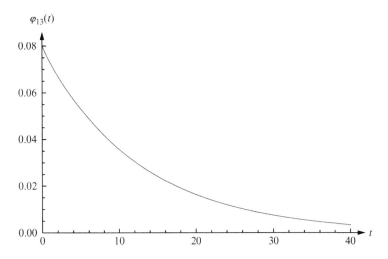

Figure 3.1 The density function $\phi_{13}(t)$.

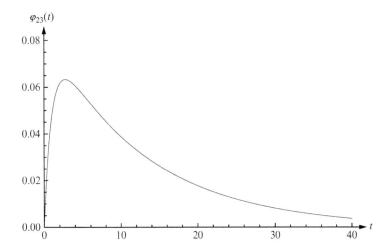

Figure 3.2 The density function $\phi_{23}(t)$.

CDFs of waiting times T_i, $i \in S$ and CDFs of holding times T_{ij}, $i, j \in S$, are

$$G_1(t) = P(T_1 \leqslant t) = 1 - e^{-0.1t}, \quad t \geqslant 0,$$
$$G_2(t) = P(T_2 \leqslant t) = 1 - e^{-t}, \quad t \geqslant 0,$$
$$G_3(t) = P(T_3 \leqslant t) = 1 - (1 + 0.2t)e^{-0.2t}, \quad t \geqslant 0,$$
$$F_{12}(t) = P\{T_{12} \leqslant t\} = 1 - e^{-0.1t}, \quad t \geqslant 0,$$
$$F_{13}(t) = P\{T_{13} \leqslant t\} = 1 - e^{-0.1t}, \quad t \geqslant 0,$$
$$F_{21}(t) = P\{T_{21} \leqslant t\} = 1 - e^{-t}, \quad t \geqslant 0,$$
$$F_{31}(t) = P\{T_{31} \leqslant t\} = 1 - (1 + 0.2t)e^{-0.2t}, \quad t \geqslant 0.$$

Hence,

$$E(T_1) = 10, \quad E(T_2) = 1, \quad E(T_3) = 10,$$
$$E(T_1^2) = 200, \quad E(T_2^2) = 2, \quad E(T_3^2) = 150.$$

$\triangle$

3.2 Interval transition probabilities

The conditional probability

$$P_{ij}(t) = P(X(t) = i | X(0) = i), \quad i, j \in S, \ t \geqslant 0 \tag{3.40}$$

is called *the interval transition probability from the state i to the state j* in the interval $[0, t]$. The number $P_{ij}(t)$ denotes the probability that the SMP $\{X(t) : t \geqslant 0\}$ will occupy the state j at an instant t if it starts from the state i at moment 0. Feller [21], introduced equations that allow us to calculate the interval transition probability by means of the process kernel. We will explain a way of deriving the equations. Assume that $i \neq j$. The SMP that starts from the state i at the moment 0 will be in the state j at the moment $t > 0$ if in an instant τ_1 the process passes to the state $k \in S$, and in a time interval $(\tau_1, t]$ at least one change of the state from state k to j takes place. These considerations and the memoryless property of a SMP in the instant τ_1 lead us to the equation

$$P_{ij}(t) = P(X(t) = j | X(0) = i)$$

$$= \sum_{k \in S} \int_0^t P(X(t) = j | X(\tau_1) = k) P(X(\tau_1) = k, \tau_1 \in dx | X(0) = i)$$

$$= \sum_{k \in S} \int_0^t P_{kj}(t - x) dQ_{ik}(x), \quad i, j \in S, \ i \neq j.$$

Assume now that $i = j$. The SMP starting from the state $i \in S$ at the moment 0 will have a value $i \in S$ and in the instant $t > 0$ it will have the same value also if at the moment τ_1 the event $\{\tau_1 > t\}$ occurs. Because $P(\tau_1 \leqslant t | X(0) = i) = G_i(t)$, then $P(\tau_1 > t | X(0) = i) = 1 - G_i(t)$. Therefore, for $i \in S$

$$P_{ii}(t) = 1 - G_i(t) + \sum_{k \in S} \int_0^t P_{ki}(t - x) dQ_{ik}(x).$$

Finally, we obtain the following system of integral equations:

$$P_{ij}(t) = \delta_{ij} [1 - G_i(t)] + \sum_{k \in S} \int_0^t P_{kj}(t - x) dQ_{ik}(x), \quad i, j \in S. \tag{3.41}$$

Only one solution of the system exists if the SMP $\{X(t) : t \geqslant 0\}$ is regular and for all $i, j \in S, i \to j$.

We can obtain the solution of that system of equations applying Laplace transformation. Let

$$\tilde{P}_{ij}(s) = \int_0^\infty e^{-st} P_{ij}(t)dt, \quad \tilde{q}_{ik}(s) = \int_0^\infty e^{-st} dQ_{ik}(t), \quad \tilde{G}_i(s) = \int_0^\infty e^{-st} G_i(t)dt.$$

(3.42)

Using properties of Laplace transformation we pass to the linear system of equations, where the complex functions $\tilde{P}_{ij}(s)$, $i,j \in S$ are unknown Laplace transforms. Finally, we have

$$\tilde{P}_{ij}(s) = \delta_{ij}\left[\frac{1}{s} - \tilde{G}_i(s)\right] + \sum_{k \in S} \tilde{q}_{ik}(s)\tilde{P}_{kj}(s), \quad i,j \in S.$$

(3.43)

If we place this system of equations in matrix form, we get

$$\tilde{P}(s) = \left(\frac{1}{s}I - \tilde{G}(s)\right) + \tilde{q}(s)\tilde{P}(s),$$

(3.44)

where

$$\tilde{P}(s) = \left[\tilde{P}_{ij}(s) : i,j \in S\right], \quad \tilde{G}(s) = \left[\delta_{ij}\tilde{G}_i(s) : i,j \in S\right],$$
$$\tilde{q}(s) = \left[\tilde{q}_{ij}(s) : i,j \in S\right], \quad I = \left[\delta_{ij} : i,j \in S\right].$$

Hence,

$$\tilde{P}(s) = (I - \tilde{q}(s))^{-1}\left(\frac{1}{s}I - \tilde{G}(s)\right).$$

(3.45)

3.2.1 Interval transition probabilities for alternating process

Let $\{X(t) : t \in \mathbb{R}_+\}$ be the SMP with the state space $S = \{0, 1\}$ and the kernel

$$Q(t) = \begin{bmatrix} 0 & G_0(t) \\ G_1(t) & 0 \end{bmatrix}$$

(3.46)

where $G_0(t)$ and $G_1(t)$ are the CDFs with nonnegative support. This process is called the *alternating stochastic process*. For the process, the matrices from (3.45) are

$$\left(\frac{1}{s}I - \tilde{G}(s)\right) = \begin{bmatrix} \frac{1}{s} - \tilde{G}_0(s) & 0 \\ 0 & \frac{1}{s} - \tilde{G}_1(s) \end{bmatrix}, \quad \tilde{q}(s) = \begin{bmatrix} 0 & \tilde{g}_0(s) \\ \tilde{g}_1(s) & 0 \end{bmatrix}.$$

(3.47)

Now, the solution is

$$\tilde{P}(s) = \frac{1}{1 - \tilde{g}_0(s)\tilde{g}_1(s)} \begin{bmatrix} \frac{1}{s} - \tilde{G}_0(s) & \tilde{g}_0(s)\left[\frac{1}{s} - \tilde{G}_1(s)\right] \\ \tilde{g}_1(s)\left[\frac{1}{s} - \tilde{G}_0(s)\right] & \frac{1}{s} - \tilde{G}_1(s) \end{bmatrix}.$$

(3.48)

Example 3.2. We assume that the waiting times of the states process are exponentially distributed: $G_0(t) = 1 - e^{-\mu t}$, $t \geq 0$, $\mu > 0$ and $G_1(t) = 1 - e^{-\lambda t}$, $t \geq 0, \lambda > 0$.

In that case, $\{X(t) : t \in \mathbb{R}_+\}$ is a two-states Markov process. The corresponding Laplace transforms are

$$\tilde{G}_0(s) = \frac{1}{s} - \frac{1}{s + \mu}, \quad \tilde{G}_1(s) = \frac{1}{s} - \frac{1}{s + \lambda}.$$

The matrix of the Laplace transforms of the interval transition probabilities (3.48) takes the form

$$\tilde{P}(s) = \begin{bmatrix} \frac{s+\lambda}{s(s+\lambda+\mu)} & \frac{\mu}{(s(s+\lambda+\mu))} \\ \frac{\lambda}{s(s+\lambda+\mu)} & \frac{s+\mu}{s(s+\lambda+\mu)} \end{bmatrix}. \tag{3.49}$$

The inverse Laplace transformation of elements of this matrix leads to the interval transition probabilities

$$P_{00}(t) = \frac{\lambda}{\lambda+\mu} + \frac{\mu}{\lambda+\mu}e^{-(\lambda+\mu)t}, \quad P_{01}(t) = \frac{\mu}{\lambda+\mu} - \frac{\mu}{\lambda+\mu}e^{-(\lambda+\mu)t},$$

$$P_{10}(t) = \frac{\lambda}{\lambda+\mu} + \frac{\lambda}{\lambda+\mu}e^{-(\lambda+\mu)t}, \quad P_{11}(t) = \frac{\mu}{\lambda+\mu} - \frac{\lambda}{\lambda+\mu}e^{-(\lambda+\mu)t}. \tag{3.50}$$

This way we obtain a well-known result.   △

Example 3.3. We assume that

$$G_0(t) = 1 - (1+t)e^{-t}, \quad G_1(t) = 1 - e^{-0.1t}, \quad t \geqslant 0.$$

Using the Laplace transformation, we calculate the interval transition probabilities of the process. The matrix of the Laplace transform of the kernel is

$$\tilde{q}(s) = \begin{bmatrix} 0 & \frac{1}{(s+1)^2} \\ \frac{0.1}{s+0.1} & 0 \end{bmatrix}. \tag{3.51}$$

The matrix of the Laplace transforms of the waiting times CDF is

$$\tilde{G}(s) = \begin{bmatrix} \frac{1}{s(s+1)^2} & 0 \\ 0 & \frac{0.1}{s(s+0.1)} \end{bmatrix}. \tag{3.52}$$

Using (3.47) we obtain

$$\tilde{P}(s) = \begin{bmatrix} \frac{(0.1+s)(2+s)}{s(1.2+2.1s+s^2)} & \frac{1}{s(1.2+2.1s+s^2)} \\ \frac{0.1(2+s)}{s(1.2+2.1s+s^2)} & \frac{(1+s)^2}{s(1.2+2.1s+s^2)} \end{bmatrix}. \tag{3.53}$$

Finally, we have

$$\tilde{P}_{00}(s) = \frac{0.16667}{s} + \frac{1.75 + 0.83333s}{1.2 + 2.1s + s^2},$$

$$\tilde{P}_{01}(s) = \frac{0.83333}{s} - \frac{1.75 + 0.83333s}{1.2 + 2.1s + s^2},$$

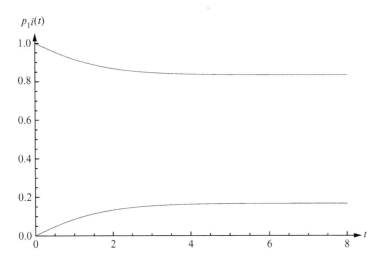

Figure 3.3 The interval transition probabilities $P_{10}(t)$ and $P_{11}(t)$.

$$\tilde{P}_{10}(s) = \frac{0.16667}{s} - \frac{0.25 + 0.16667s}{1.2 + 2.1s + s^2},$$

$$\tilde{P}_{11}(s) = \frac{0.83333}{s} + \frac{0.25 + 0.16667s}{1.2 + 2.1s + s^2}.$$

Hence,

$$P_{00}(t) = 0.16667 + e^{-1.05t}[0.83333\cos(0.31225t) + 2.80225\sin(0.31225t)],$$

$$P_{01}(t) = 0.83333 - e^{-1.05t}[0.83333\cos(0.31225t) + 2.80225\sin(0.31225t)],$$

$$P_{10}(t) = 0.16667 - e^{-1.05t}[0.16667\cos(0.31225t) + 0.24019\sin(0.31225t)],$$

$$P_{11}(t) = 0.83333 + e^{-1.05t}[0.16667\cos(0.31225t) + 0.24019\sin(0.31225t)].$$

The interval transition probabilities $P_{10}(t)$ and $P_{11}(t)$ are plotted in Figure 3.3. △

3.2.2 Interval transition probabilities for Poisson process

As we know (Example 2.4), the Poisson process is a SMP $\{X(t) : t \geqslant 0\}$ with the countable state space $S = \{0, 1, 2, \ldots\}$, defined by the initial distribution

$$\boldsymbol{p}(0) = \begin{bmatrix} 1 & 0 & 0 & \ldots \end{bmatrix} \tag{3.54}$$

and the renewal kernel

$$\boldsymbol{Q}(t) = \begin{bmatrix} 0 & G_0(t) & 0 & 0 & 0 & \ldots \\ 0 & 0 & G_1(t) & 0 & 0 & \ldots \\ 0 & 0 & 0 & G_2(t) & 0 & \ldots \\ 0 & 0 & 0 & 0 & G_3(t) & 0 \\ \cdot & \cdot & \cdot & \cdot & \cdot & \cdot \end{bmatrix}, \tag{3.55}$$

where $G_i(t), i = 0, 1, \ldots$ are identical exponential CDF of waiting times T_i:

$$G_i(t) = P(T_i \leqslant t) = 1 - e^{-\lambda t}, \quad t \geqslant 0, \lambda > 0. \tag{3.56}$$

The matrix of the L-S transform of the process kernel is

$$\tilde{q}(s) = \begin{bmatrix} 0 & \frac{\lambda}{s+\lambda} & 0 & 0 & 0 & \cdots \\ 0 & 0 & \frac{\lambda}{s+\lambda} & 0 & 0 & \cdots \\ 0 & 0 & 0 & \frac{\lambda}{s+\lambda} & 0 & \cdots \\ 0 & 0 & 0 & 0 & \frac{\lambda}{s+\lambda} & 0 \\ & & \cdots\cdots\cdots & & \end{bmatrix}. \tag{3.57}$$

Because $\tilde{p}_{ij}(s) = 0$ for $i > j$, then the linear system of equation (3.43) takes the form

$$\tilde{P}_{ii}(s) = \frac{1}{s+\lambda}, \quad j \in \mathbb{N}_0, \tag{3.58}$$

$$\tilde{P}_{ij}(s) = \frac{\lambda}{s(s+\lambda)}\tilde{P}_{i+1j}(s), \quad i,j \in S, \ 0 \leqslant i+1 \leqslant j.$$

Hence,

$$\tilde{P}_{ii+1}(s) = \frac{\lambda}{s+\lambda}\tilde{P}_{i+1i+1}(s) = \frac{\lambda}{(s+\lambda)^2}, \quad i \in S.$$

Finally, we get

$$\tilde{P}_{ii+k}(s) = \frac{\lambda^k}{(s+\lambda)^{k+1}}, \quad i \in S, \ k \in \mathbb{N}.$$

We obtain the interval transition probabilities by calculating the inverse Laplace transforms.

$$P_{ii}(t) = L^{-1}\left[\tilde{P}_{ii}(s)\right] = L^{-1}\left[\frac{1}{s+\lambda}\right] = e^{-\lambda t}, \quad t \geqslant 0, j \in \mathbb{N}_0, \tag{3.59}$$

$$P_{ii+k}(t) = L^{-1}\left[\frac{\lambda^k}{(s+\lambda)^{k+1}}\right] = \frac{(\lambda t)^k e^{-\lambda t}}{k!}, \quad t \geqslant 0, i \in \mathbb{N}_0, k \in \mathbb{N}. \tag{3.60}$$

3.2.3 Interval transition probabilities for Furry-Yule process

The *Furry-Yule* process (Example 2.5) is the SMP $\{X(t) : t \geqslant 0\}$ with the countable state space $S = \{0, 1, 2, \ldots\}$, given by the initial distribution

$$p(0) = \begin{bmatrix} 1 & 0 & 0 & \cdots \end{bmatrix} \tag{3.61}$$

and the renewal kernel (3.55), where

$$G_i(t) = 1 - e^{-(i+1)\lambda t}, \quad t \geqslant 0, j \in \mathbb{N}_0, \tag{3.62}$$

The matrix of the L-S transform of the Furry-Yule process kernel is

$$\tilde{q}(s) = \begin{bmatrix} 0 & \frac{\lambda}{s+\lambda} & 0 & 0 & 0 & \cdots \\ 0 & 0 & \frac{2\lambda}{s+2\lambda} & 0 & 0 & \cdots \\ 0 & 0 & 0 & \frac{3\lambda}{s+3\lambda} & 0 & \cdots \\ 0 & 0 & 0 & 0 & \frac{4\lambda}{s+4\lambda} & 0 \\ & & & \cdots\cdots\cdots\cdots \end{bmatrix}. \tag{3.63}$$

Because $\tilde{P}_{ij}(s) = 0$ for $i > j$, then the linear system of equations (3.18) takes the form:

$$\tilde{P}_{ii}(s) = \frac{1}{s + (i+1)\lambda}, \quad i \in \mathbb{N}_0,$$

$$\tilde{P}_{ij}(s) = \frac{(i+1)\lambda}{s + (i+1)\lambda}\tilde{P}_{i+1j}(s), \quad i,j \in S, \ 0 \leqslant i+1 \leqslant j.$$

Hence,

$$\tilde{P}_{ii+1}(s) = \frac{(i+1)\lambda}{s+(i+1)\lambda}, \quad \tilde{P}_{i+1i+1}(s) = \frac{(i+1)\lambda}{(s+(i+1)\lambda)(s+(i+2)\lambda)}, \quad i \in S.$$

By induction, we obtain

$$\tilde{P}_{ii+k}(s) = \frac{(i+1)(i+2)\ldots(i+k)\lambda^k}{(s+(i+1)\lambda)\ldots(s+(i+k+1))\lambda)}, \quad i \in S.$$

Consequently, we have

$$\tilde{P}_{ii+k}(s) = \binom{i+k}{k} \frac{\lambda^k k!}{(s+(i+1)\,\lambda)(s+(i+1)\lambda+\lambda)\ldots(s+(i+1)\lambda+k\lambda)}.$$

Remember the following properties of the Laplace transformation

$$L\left[e^{-at}f(t)\right] = \tilde{f}(s+a), \quad L\left[(1-e^{-\lambda t})^k\right] = \frac{\lambda^k k!}{s(s+\lambda)(s+2\lambda)\ldots(s+k\lambda)},$$

for $a = (i+1)\lambda$, we have

$$\tilde{P}_{ii+k}(s) = L\left[\binom{i+k}{k} e^{-(i+1)\lambda t}(1-e^{-\lambda t})^k\right].$$

Therefore, for $i, k \in \mathbb{N}_0$ we obtain the formula

$$P_{ii+k}(t) = \binom{i+k}{k} e^{-(i+1)\lambda t}(1-e^{-\lambda t})^k. \tag{3.64}$$

For $k = 0$, we get

$$P_{ii}(t) = e^{-(i+1)\lambda t}, \quad t \geqslant 0, \ i \in \mathbb{N}_0,$$

while for $i = 0$, we have

$$P_{0k}(t) = e^{-\lambda t}(1-e^{-\lambda t})^k. \tag{3.65}$$

Notice that for the initial distribution (3.61)

$$P_{0k}(t) = P_k(t) = P(X(t) =), \quad k \in S.$$

Therefore, the one-dimensional distribution of the Furry-Yule process is

$$P_k(t) = e^{-\lambda t}(1 - e^{-\lambda t})^k, \quad k \in S = \mathbb{N}_0. \tag{3.66}$$

We obtain an expectation of the process applying the geometric transformation

$$E[X(t)] = \phi'(z)|_{z=1}, \quad \text{where } \phi(z) = \sum_{k \in \mathbb{N}_0} P_k(t) z^k, \quad |z| \leqslant 1. \tag{3.67}$$

For the considering process, we have

$$\phi(z) = \sum_{k \in \mathbb{N}_0} e^{-\lambda t} \left(1 - e^{-\lambda t}\right)^k z^k = \frac{e^{-\lambda t}}{1 - \left(1 - e^{-\lambda t}\right) z} \tag{3.68}$$

and

$$E[X(t)] = \phi'(z)|_{z=1} = e^{\lambda t} - 1. \tag{3.69}$$

3.3 The limiting probabilities

In many cases, the interval transitions probabilities $P_{ij}(t)$, $t \geqslant 0$ and the states probabilities

$$P_j(t) = P(X(t) = j), \quad t \geqslant 0, j \in S \tag{3.70}$$

approach the constant values for large t. Let

$$P_j = \lim_{t \to \infty} P_j(t) = \lim_{t \to \infty} P(X(t) = j), \quad j \in S, \tag{3.71}$$

$$P_{ij} = \lim_{t \to \infty} P_{ij}(t) = \lim_{t \to \infty} P(X(t) = j | X(0) = i), \quad j \in S. \tag{3.72}$$

As a conclusion from theorems presented by Korolyuk and Turbin [60], we introduce the following theorem

Theorem 3.6 (Korolyuk and Turbin [60]). Let $\{X(t) : t \geqslant 0\}$ be the regular SMP with the discrete state S and the kernel $\mathbf{Q}(t) = \left[Q_{ij}(t) : i, j \in S\right]$. If

• the embedded Markov chain $\{X(\tau_n) : n \in \mathbb{N}_0\}$ of the SMP $\{X(t) : t \in \mathbb{R}_+\}$ contains one positive recurrent class C, such that for

$$\forall_{i \in S, j \in C} f_{ij} = 1,$$

•

$$\exists_{a>0} \forall_{i \in S} 0 < E(T_i) = \int_0^\infty [1 - G_i(t)] \, dt \leqslant a,$$

then there exist the limiting probabilities of $P_{ij}(t)$, $i, j \in S$ and $P_j(t)$, $j \in S$ as $t \to \infty$. Moreover

$$P_{ij} = \lim_{t \to \infty} P_{ij}(t) = P_j = \lim_{t \to \infty} P_j(t) = \frac{\pi_j E(T_j)}{\sum_{k \in S} \pi_k E(T_k)}, \tag{3.73}$$

where π_i, $i \in S$ form the stationary distribution of the embedded Markov chain $\{X(\tau_n) : n \in \mathbb{N}_0\}$.

P r o o f: [60].

Recall that the stationary distribution of $\{X(\tau_n) : n \in \mathbb{N}_0\}$ is the unique solution of the linear system of equations

$$\sum_{i \in S} \pi_i p_{ij} = \pi_j, \; j \in S, \quad \sum_{j \in S} \pi_j = 1, \tag{3.74}$$

where

$$p_{ij} = \lim_{t \to \infty} Q_{ij}(t).$$

As we have noticed, a calculation of the limiting state probabilities is very easy. From Theorem 3.6 it follows that we have to calculate the expectation of the waiting time of an SMP in each state and we have to solve the linear system of equations (3.74).

Example 3.4. Let $\{X(t) : i \in \mathbb{R}_+\}$ be an SMP with the state space $S = \{1, 2, 3\}$ and the kernel

$$Q(t) = \begin{bmatrix} 0 & 0.4\left[1 - (1 + 0.2t)e^{-0.2t}\right] & 0.6\left[1 - (1 + 0.2t)e^{-0.2t}\right] \\ 1 - e^{-0.2t} & 0 & 0 \\ 1 - e^{-0.5t} & 0 & 0 \end{bmatrix}. \tag{3.75}$$

In this case,

$$\begin{aligned} G_1(t) &= 1 - (1 + 0, 2t)e^{-0.2t}, \quad t \geq 0, \\ G_2(t) &= 1 - e^{-0.2t}, \quad t \geq 0, \\ G_3(t) &= 1 - e^{-0.5t}, \quad t \geq 0. \end{aligned}$$

and

$$E(T_1) = 10, \quad E(T_2) = 5, \quad E(T_3) = 2.$$

The transition matrix of the embedded Markov chain is

$$P = \begin{bmatrix} 0 & 0.4 & 0.6 \\ 1 & 0 & 0 \\ 1 & 0 & 0 \end{bmatrix}. \tag{3.76}$$

This process satisfies all assumptions of the introduced theorem. The linear system of equations for the stationary distribution of the embedded Markov chain takes the form

$$\pi_2 + \pi_3 = \pi_1, \quad 0.4\pi_1 = \pi_2, \quad 0.6\pi_1 = \pi_3, \quad \pi_1 + \pi_2 + \pi_3 = 1.$$

Hence,

$$\pi_1 = 0.5, \quad \pi_2 = 0.2, \quad \pi_3 = 0.3.$$

Using the equality (3.73), we obtain the limiting distribution of the SMP presented above,

$$P_1 = \frac{25}{33}, \quad P_2 = \frac{5}{33}, \quad P_3 = \frac{3}{33}.$$

$\triangle$

Example 3.5. Let $\{X(t) : i \in \mathbb{R}_+\}$ be the alternating process. Recall that this process is SMP with the state space $S = \{0, 1\}$ and the kernel

$$Q(t) = \begin{bmatrix} 0 & G_0(t) \\ G_1(t) & 0 \end{bmatrix}. \tag{3.77}$$

Suppose that expectations of the waiting times are positive, finite numbers

$$0 < E(T_i) = \int_0^\infty [1 - G_i(t)] \, dt < \infty \quad i \in S.$$

The transition matrix of the embedded Markov chain is

$$P = \begin{bmatrix} 0 & 1 \\ 1 & 0 \end{bmatrix}. \tag{3.78}$$

The stationary distribution of the embedded Markov chain is

$$\pi_1 = \pi_0, \quad \pi_0 = \pi_1, \quad \pi_0 + \pi_1 = 1.$$

Hence,

$$\pi_1 = 0.5, \quad \pi_2 = 0.5.$$

From the equality (3.67), we get the limiting distribution of this SMP

$$P_0 = \frac{E(T_0)}{E(T_0) + E(T_1)}, \quad P_1 = \frac{E(T_1)}{E(T_0) + E(T_1)}.$$

$\triangle$

Example 3.6. Suppose that $\{X(t) : i \in \mathbb{R}_+\}$ is a SMP on the countable state space $S = \{1, 2, \ldots\}$, which is determined by the kernel

$$Q(t) = \begin{bmatrix} \frac{1}{2}G_1(t) & \frac{1}{4}G(t) & \frac{1}{8}G(t) & \frac{1}{16}G(t) & \cdots \\ G(t) & 0 & 0 & 0 & \cdots \\ 0 & G(t) & 0 & 0 & \cdots \\ 0 & 0 & G(t) & 0 & \cdots \\ & & \cdots\cdots\cdots\cdots & & \end{bmatrix}, \tag{3.79}$$

where

$$G_1(t) = 1 - e^{-\alpha t}, \quad t \geqslant 0, \ \alpha > 0$$

and

$$G(t) = 1 - (1 + \lambda t)e^{-\lambda t}, \quad t \geqslant 0, \ \lambda > 0.$$

In this case,

$$E(T_1) = \frac{1}{\alpha}, \quad E(T_k) = \frac{2}{\lambda} \quad \text{for } k = 2, \ldots.$$

The transition matrix of the embedded Markov chain is identical to the matrix from Example 1.3:

$$P = \begin{bmatrix} \frac{1}{2} & \frac{1}{2^2} & \frac{1}{2^3} & \frac{1}{2^4} & \cdots \\ 1 & 0 & 0 & 0 & \cdots \\ 0 & 1 & 0 & 0 & \cdots \\ 0 & 0 & 1 & 0 & \cdots \\ & & \cdots\cdots\cdots & & \end{bmatrix}$$

In Chapter 1 we have proved that state 1 is recurrent and positive. Moreover, all states of the embedded Markov chain form one positive recurrent class. Thus, all assumptions of Theorem 3.6 are satisfied. In this case, the system of equation (3.74) takes the form

$$\frac{1}{2}\pi_1 + \pi_2 = \pi_1, \quad \frac{1}{4}\pi_1 + \pi_3 = \pi_2, \quad \frac{1}{8}\pi_1 + \pi_4 = \pi_3, \ldots, \tag{3.80}$$

$$\pi_1 + \pi_2 + \pi_3 + \cdots = 1,$$

and the solution is

$$\pi_k = \frac{1}{2^k}, \quad k = 1, 2, \ldots.$$

Using equality (3.73) of Theorem 3.6, we obtain the limiting distribution of the process

$$P_1 = \frac{\lambda}{\lambda + 2\alpha}, \quad P_k = \frac{1}{2^{k-1}}\frac{2\alpha}{\lambda + 2\alpha}, \quad k = 2, 3, \ldots.$$

$$\triangle$$

The interval transition probabilities $P_{ij}(t)$, $i, j \in S$ and the CDFs of the first passage times $\Phi_{ij}(t)$, $i, j \in S$ are connected by equations [45]

$$P_{ij}(t) = \int_0^t P_{jj}(t - x)\mathrm{d}\Phi_{ij}(x) \quad \text{for } i \neq j, \; i, j \in S \tag{3.81}$$

$$P_{jj}(t) = 1 - G_j(t) + \int_0^t P_{jj}(t - x)\mathrm{d}\Phi_{jj}(x) \quad \text{for } j \in S \tag{3.82}$$

The corresponding equations for the Laplace transforms are

$$\tilde{P}_{ij}(s) = \tilde{P}_{jj}(s)\tilde{\varphi}_{ij}(s) \quad \text{for } i \neq j, i, j \in S, \tag{3.83}$$

$$\tilde{P}_{jj}(s) = \frac{1}{s} - \tilde{G}_j(s) + \tilde{P}_{jj}(s)\tilde{\varphi}_{jj}(s) \quad \text{for } j \in S, \tag{3.84}$$

where

$$\tilde{P}_{ij}(s) = \int_0^\infty e^{-st} P_{ij}(t)dt, \quad \tilde{\varphi}_{ik}(s) = \int_0^\infty e^{-st} d\Phi_{ik}(t), \quad \tilde{G}_i(s) = \int_0^\infty e^{-st} G_i(t)dt.$$

(3.85)

From (3.84) we can find the transform $\tilde{\varphi}_{jj}(s)$:

$$\tilde{\varphi}_{jj}(s) = \frac{s\tilde{P}_{jj}(s) + s\tilde{G}_j(s) - 1}{s\tilde{P}_{jj}(s)}.$$

(3.86)

We know that an expected value of a return time Θ_{jj} can be calculated as

$$E(\Theta_{jj}) = -\tilde{\varphi}'_{jj}(s)|_{s=0}.$$

(3.87)

The derivative of the function $\tilde{\varphi}_{jj}(s)$ at the point 0 we calculate from definition.

$$\frac{d\tilde{\varphi}_{jj}(s)}{ds}|_{s=0} = \lim_{s \to 0} \frac{\tilde{\varphi}_{jj}(s) - \tilde{\varphi}_{jj}(0)}{s} = \lim_{s \to 0} \frac{\frac{s\tilde{P}_{jj}(s) + s\tilde{G}_j(s) - 1}{s\tilde{P}_{jj}(s)} - 1}{s}$$

$$= -\lim_{s \to 0} \frac{\frac{s\tilde{G}(s)-1}{s}}{s\tilde{P}_{jj}(s)} = \frac{-\lim_{s \to 0} \frac{s\tilde{G}(s)-1}{s}}{\lim_{s \to 0} s\tilde{P}_{jj}(s)}.$$

(3.88)

If there exists a PDF $g(t) = G'(t)$, $t > 0$ and $G(0) = 0$, then

$$s\tilde{G}(s) = L[G'(t)] = L[g(t)] = \tilde{g}(s).$$

Hence,

$$-\lim_{s \to 0} \frac{s\tilde{G}(s) - 1}{s} = -\lim_{s \to 0} \frac{\tilde{g}(s) - 1}{s} = -\frac{d\tilde{g}(s)}{ds}|_{s=0} = E(T_j).$$

(3.89)

From properties of the Laplace transformation we have

$$\lim_{s \to 0} s\tilde{P}_{jj}(s) = \lim_{t \to \infty} P_{jj}(t) = P_j.$$

(3.90)

Finally, we obtain an important equality

$$E(\Theta_{jj}) = \frac{E(T_j)}{P_j},$$

(3.91)

where $E(T_j)$ represents the expected value of the waiting time in a state j, and P_j denotes the limiting probability of the state j.

3.4 Reliability and maintainability characteristics

Reliability and maintainability characteristics and parameters in semi-Markov models are presented in many books: [6, 7, 11, 24, 32, 60, 72, 90, 93]. Assume that the evolution of a system's reliability is described by a finite state space S SMP $\{X(t) : t \geqslant$

0}. Elements of a set S represent the reliability states of the system. Let S_+ consists of the functioning states (up states) and S_- contains all the failed states (down states). The subset S_+ and S_- form a partition of S, i.e., $S = S_+ \cup S_-$ and $S_+ \cap S_- = \emptyset$.

3.4.1 Reliability function and parameters of the system

Suppose that $i \in S_+$ is an initial state of the process. The conditional reliability function is defined by

$$R_i(t) = P(\forall u \in [0, t], \quad X(u) \in S_+ | X(0) = i), \ i \in S_+. \tag{3.92}$$

Note that from the Chapman-Kolmogorov equation of a two-dimensional Markov chain $\{X(\tau_n), \tau_n : n \in \mathbb{N}_0\}$, we obtain

$$
\begin{aligned}
R_i(t) &= P(\forall u \in [0, t], \ X(u) \in S_+ | X(0) = i) \\
&= P(\forall u \in [0, t], \ X(u) \in S_+, \ \tau_1 > t | X(0) = i) \\
&\quad + P(\forall u \in [0, t], \ X(u) \in S_+, \ \tau_1 \leqslant t | X(0) = i) \\
&= [1 - P(\vartheta_1 > t)] I_{S_+}(i) \\
&\quad + \sum_{j \in S+} \int_0^t P(\forall u \in [0, t], \ X(t - \tau_1) \in S_+ | \tau_1 \\
&= u, X(\tau_1) = j)P(X(\tau_1) = j, \ \tau_1 \in du | X(0) = i) \\
&= [1 - G_i(t)] I_{S_+}(i) + \sum_{j \in S+} \int_0^t R_j(t - u) dQ_{ij}(u).
\end{aligned}
$$

Finally, we have

$$R_i(t) = 1 - G_i(t) + \sum_{j \in S+} \int_0^t R_j(t - u) dQ_{ij}(u), \quad i \in S_+. \tag{3.93}$$

Passing to the Laplace transform, we get

$$\tilde{R}_i(s) = \frac{1}{s} - \tilde{G}_i(s) + \sum_{j \in S+} \tilde{q}_{ij}(s)\tilde{R}_j(s), \quad i \in S_+, \tag{3.94}$$

where

$$\tilde{R}_j(s) = \int_0^\infty e^{-st} R_j(t) dt.$$

The matrix form of the equation system is

$$\left(I - \tilde{q}_{S_+}(s)\right) \tilde{R}(s) = \tilde{W}_{S_+}(s), \tag{3.95}$$

where

$$\tilde{R}(s) = \left[\tilde{R}_i(s) : i \in S_+\right]^{\mathrm{T}}, \quad \tilde{W}_{S_+}(s) = \left[\frac{1}{s} - \tilde{G}_i(s) : i \in S_+\right]^{\mathrm{T}} \tag{3.96}$$

are one-column matrices and

$$\tilde{\boldsymbol{q}}_{S_+}(s) = \left[\tilde{q}_{ij}(s) : i,j \in S_+\right], \quad \boldsymbol{I} = \left[\delta_{ij} : i,j \in S_+\right] \tag{3.97}$$

are square matrices. Note that

$$\tilde{G}_i(s) = \frac{1}{s} \sum_{j \in S_+} \tilde{q}_{ij}(s). \tag{3.98}$$

Elements of the matrix $\tilde{\boldsymbol{R}}(s)$ are the Laplace transforms of the conditional reliability functions. We obtain the reliability functions $R_i(t)$, $i \in S_+$ by inverting the Laplace transforms $\tilde{R}_i(s)$, $i \in S_+$.

If $A = S_-$, and $A' = S_+$, then the first passage time from a state $i \in A'$ to a subset A denotes a time to failure or lifetime of the system under condition that i is an initial state. Therefore, the random variable Θ_{S_-} is time to failure of the system. Notice that

$$R_i(t) = P(\forall u \in [0, t], \ X(u) \in S_+ | X(0) = i) = P(\Theta_{S_-} > t | X(0) = i)$$
$$= 1 - P(\Theta_{S_-} \leqslant t | X(0) = i) = 1 - \Phi_{iS_-}(t) \quad i \in S_+. \tag{3.99}$$

A conditional mean time to failure is the conditional expectation

$$\bar{\Theta}_{iS_-} = E(\Theta_{S_-} | X(0) = i) = \int_0^\infty R_i(t)\mathrm{d}t, \quad i \in S_+. \tag{3.100}$$

A corresponding second moment is

$$\bar{\Theta}^2_{iS_-} = E(\Theta_{S_-} | X(0) = i) = 2 \int_0^\infty t R_i(t)\mathrm{d}t, \quad i \in S_+. \tag{3.101}$$

We can calculate those reliability parameters of the system in another way. According to equality (3.99) we can solve (3.25) and (3.26), substituting $A = S_-$, and $A' = S_+$.

The unconditional reliability function of the system is defined as

$$R(t) = P(\forall u \in [0, t], \ X(u) \in S_+). \tag{3.102}$$

Note that

$$R(t) = P(\forall u \in [0, t], \ X(u) \in S_+) = \sum_{i \in S+} p_i R_i(t), \tag{3.103}$$

as

$$P(X(0) = i) = 0 \quad \text{for } i \in S_-.$$

3.4.2 Pointwise availability

The availability at time t is determined as

$$A(t) = P(X(t) \in S_+). \tag{3.104}$$

Notice that

$$A(t) = P(X(t) \in S_+)$$
$$= P(X(t) \in S_+, X(0) \in S)$$
$$= \sum_{i \in S} \sum_{j \in S_+} P(X(t) = j | X(0) = i) \, P(X(0) = i)$$
$$= \sum_{i \in S} \sum_{j \in S_+} p_i \, P_{ij}(t). \tag{3.105}$$

From the equality it follows that we can calculate the pointwise availability having an initial distribution $p_i = P(X(0) = i)$, $i \in S$ and the interval transition probabilities $P_{ij}(t)$, $i \in S$, $j \in S_+$.

If there exist the limit probabilities

$$P_j = \lim_{t \to \infty} P_{ij}(t), \quad j \in S_+,$$

then the limiting (steady-state) availability

$$A = \lim_{t \to \infty} A(t) \tag{3.106}$$

is given by

$$A = \sum_{j \in S_+} P_j. \tag{3.107}$$

3.4.3 Maintainability function and parameters of the system

In a similar way we obtain the maintainability function and parameters of the system, symmetrically to the reliability case. Now we assume that $i \in S_-$ is an initial state of the process. The conditional maintainability function is determined by

$$M_i(t) = P(\forall u \in [0, t], \quad X(u) \in S_- | X(0) = i), \quad i \in S_-. \tag{3.108}$$

These functions fulfill the equations system

$$M_i(t) = 1 - G_i(t) + \sum_{j \in S_-} \int_0^t M_j(t - u) \mathrm{d}Q_{ij}(u), \quad i \in S_-, \tag{3.109}$$

which is equivalent to the system of equations for the Laplace transform

$$\tilde{M}_i(s) = \frac{1}{s} - \tilde{G}_i(s) + \sum_{j \in S_-} \tilde{q}_{ij}(s) \tilde{M}_j(s), \quad i \in S_-, \tag{3.110}$$

where

$$\tilde{M}_j(s) = \int_0^\infty \mathrm{e}^{-st} M_j(t) \mathrm{d}t.$$

We obtain the maintainability functions $M_i(t)$, $i \in S_-$ by inverting the Laplace transforms $\tilde{M}_i(s)$, $i \in S_-$ being a solution of the above system of equations.

If $A = S_+$, and $A' = S_-$, then the first passage time from a state $i \in A'$ to a subset A denotes a *time to repair* or a *maintenance time* of the system under the condition that $i \in S_-$ is an initial state. Therefore, the random variable Θ_{S_+} denotes the maintenance time of the system. Notice that

$$M_i(t) = P(\forall u \in [0, t], X(u) \in S_- | X(0) = i) = P(\Theta_{S_+} > t | X(0) = i)$$
$$= 1 - P(\Theta_{S_-} \leqslant t | X(0) = i) = 1 - \Phi_{iS_+}(t) \quad i \in S_-. \tag{3.111}$$

The conditional mean maintenance time and the corresponding second moment are

$$\bar{\Theta}_{iS_+} = E(\Theta_{S_+} | X(0) = i) = \int_0^\infty M_i(t)dt, \quad i \in S_-. \tag{3.112}$$

$$\bar{\Theta}_{iS_+}^2 = E(\Theta_{S_+}^2 | X(0) = i) = 2 \int_0^\infty t M_i(t)dt, \quad i \in S_-. \tag{3.113}$$

We can also calculate those parameters of the system in another way. According to (3.99), we can solve (3.25) and (3.26) substituting $A = S_+$, and $A' = S_-$.

The unconditional maintainability function of the system is defined as

$$M(t) = P(\forall u \in [0, t], X(u) \in S_-). \tag{3.114}$$

Note that

$$M(t) = P(\forall u \in [0, t], X(u) \in S_-) = \sum_{i \in S-} p_i M_i(t). \tag{3.115}$$

3.5   Numerical illustrative example

3.5.1   Description and assumptions

The object (device) works by performing the two types of tasks 1 and 2. The duration of task k is a nonnegative random variable ξ_k, $k = 1, 2$ governed by a CDF $F_{\xi_k}(x)$, $x \geqslant 0$, $k = 1, 2$. The working object may be damaged. Time to failure of the object executing a task k is a nonnegative random variable ζ_k, $k = 1, 2$ with PDF $f_{\zeta_k}(x)$, $x \geqslant 0$, $k = 1, 2$. A repair (renewal, maintenance) time of the object performing task k is a nonnegative random variable η_k, $k = 1, 2$, governed by PDF $f_{\eta_k}(x)$, $x \geqslant 0$, $k = 1, 2$. After each operation or maintenance, the object waits for the next task. In this case a waiting time is a nonnegative random variable γ, having PDF $f_\gamma(x)$, $x \geqslant 0$, $k = 1, 2$ wherein it is task 1 with a probability p or task 2 with a probability $q = 1 - p$. Furthermore, we assume that all random variables and their copies are independent and they have the finite second moments.

3.5.2 Model construction

We start the construction of the model with the determination of the operation process states.

1. State of waiting for the tasks performing
2. Performing of task 1
3. Performing of task 2
4. Repair after failure during executing of task 1
5. Repair after failure during executing of task 2

Possible state changes of the process appears as Figure 3.4.

A model of a object operation is a SMP with a state space $S = \{1, 2, 3, 4, 5\}$ and a kernel

$$Q(t) = \begin{bmatrix} 0 & Q_{12}(t) & Q_{13}(t) & 0 & 0 \\ Q_{21}(t) & 0 & 0 & Q_{24}(t) & 0 \\ Q_{31}(t) & 0 & 0 & 0 & Q_{35}(t) \\ Q_{41}(t) & 0 & 0 & 0 & 0 \\ Q_{51}(t) & 0 & 0 & 0 & 0 \end{bmatrix}, \tag{3.116}$$

The model is constructed if all kernel elements are determined. According to assumptions, we calculate elements of the matrix $Q(t)$. In detail, we explain calculation of an element $Q_{21}(t)$. Transition from a state 2 to 1 in time no greater than t takes place if the duration of task 1, denoted as ξ_1, is less or equal t and time to failure is greater then ξ_1. Hence,

$$Q_{21}(t) = P(\xi_1 \leqslant t, \ \zeta_1 > \xi_1) = \iint\limits_{D_{21}} \mathrm{d}F_{\xi_1}(x)\mathrm{d}F_{\zeta_1}(y),$$

where

$$D_{21} = \{(x, y) : 0 \leqslant x \leqslant t, \ y > x\}.$$

Therefore,

$$Q_{21}(t) = \int_0^t \mathrm{d}F_{\xi_1}(x) \int_x^\infty \mathrm{d}F_{\zeta_1}(x) = \int_0^t \left[1 - F_{\zeta_1}(x)\right] \mathrm{d}F_{\xi_1}(x),$$

where

$$F_{\zeta_1}(x) = \int_0^t f_{\zeta_1}(x)\mathrm{d}x.$$

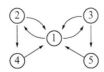

Figure 3.4 Possible state changes of the process.

We calculate the rest of elements similar way. Finally, we obtain:

$$Q_{12}(t) = p F_\gamma(t), \qquad\qquad Q_{13}(t) = q F_\gamma(t),$$

$$Q_{21}(t) = \int_0^t \left[1 - F_{\zeta_1}(x)\right] dF_{\xi_1}(x), \quad Q_{24}(t) = \int_0^t \left[1 - F_{\xi_1}(x)\right] dF_{\zeta_1}(x),$$

$$Q_{31}(t) = \int_0^t \left[1 - F_{\zeta_2}(x)\right] dF_{\xi_2}(x), \quad Q_{35}(t) = \int_0^t \left[1 - F_{\xi_2}(x)\right] dF_{\zeta_2}(x),$$

$$Q_{41}(t) = F_{\eta_1}(t), \qquad\qquad Q_{51}(t) = F_{\eta_2}(t).$$

(3.117)

The transition probability matrix of the embedded Markov chain $\{X(\tau_n) : n \in \mathbb{N}_0\}$ is of the form

$$P = \begin{bmatrix} 0 & p & q & 0 & 0 \\ p_{21} & 0 & 0 & p_{24} & 0 \\ p_{31} & 0 & 0 & 0 & p_{35} \\ 1 & 0 & 0 & 0 & 0 \\ 1 & 0 & 0 & 0 & 0 \end{bmatrix},$$

(3.118)

where

$$p_{21} = \int_0^\infty \left[1 - F_{\zeta_1}(x)\right] dF_{\xi_1}(x), \quad p_{24} = \int_0^\infty \left[1 - F_{\xi_1}(x)\right] dF_{\zeta_1}(x) = 1 - p_{21},$$

$$p_{31} = \int_0^\infty \left[1 - F_{\zeta_2}(x)\right] dF_{\xi_2}(x), \quad p_{35} = \int_0^\infty \left[1 - F_{\xi_2}(x)\right] dF_{\zeta_2}(x) = 1 - p_{31}.$$

The L-S transform of the kernel (3.116) is given by

$$\tilde{q}(s) = \begin{bmatrix} 0 & \tilde{q}_{12}(s) & \tilde{q}_{13}(s) & 0 & 0 \\ \tilde{q}_{21}(s) & 0 & 0 & \tilde{q}_{24}(s) & 0 \\ \tilde{q}_{31}(s) & 0 & 0 & 0 & \tilde{q}_{35}(s) \\ \tilde{q}_{41}(s) & 0 & 0 & 0 & 0 \\ \tilde{q}_{51}(s) & 0 & 0 & 0 & 0 \end{bmatrix}.$$

(3.119)

3.5.3 Reliability characteristics and parameters

In the model a subset $S_+ = \{1, 2, 3\}$ consists of the functioning states, and a subset $S_- = \{4, 5\}$ contains all failed states of the object operation process. In this case, the matrix equation (3.95) for the Laplace transform of the conditional reliability function takes the form

$$\begin{bmatrix} 1 & -\tilde{q}_{12}(s) & -\tilde{q}_{13}(s) \\ -\tilde{q}_{21}(s) & 1 & 0 \\ -\tilde{q}_{31}(s) & 0 & 1 \end{bmatrix} \begin{bmatrix} \tilde{R}_1(s) \\ \tilde{R}_2(s) \\ \tilde{R}_3(s) \end{bmatrix} = \begin{bmatrix} \frac{1}{s}(1 - \tilde{q}_{12}(s) - \tilde{q}_{13}(s)) \\ \frac{1}{s}(1 - \tilde{q}_{21}(s) - \tilde{q}_{24}(s)) \\ \frac{1}{s}(1 - \tilde{q}_{31}(s) - \tilde{q}_{35}(s)) \end{bmatrix}.$$

(3.120)

Solving this equation, we obtain

$$\tilde{R}_1(s) = \frac{\tilde{A}_1(s)}{\tilde{D}(s)}, \quad \tilde{R}_2(s) = \frac{\tilde{A}_2(s)}{\tilde{D}(s)}, \quad \tilde{R}_3(s) = \frac{\tilde{A}_3(s)}{\tilde{D}(s)}, \tag{3.121}$$

where

$$\tilde{A}_1(s) = -1 + \tilde{q}_{12}(s)\tilde{q}_{21}(s) + \tilde{q}_{12}(s)\tilde{q}_{24}(s) + \tilde{q}_{13}(s)\tilde{q}_{31}(s) + \tilde{q}_{13}(s)\tilde{q}_{35}(s),$$

$$\tilde{A}_2(s) = -1 + \tilde{q}_{24}(s) + \tilde{q}_{12}(s)\tilde{q}_{21}(s) + \tilde{q}_{13}(s)\tilde{q}_{31}(s) - \tilde{q}_{13}(s)\tilde{q}_{24}(s)\tilde{q}_{31}(s)$$
$$+ \tilde{q}_{13}(s)\tilde{q}_{21}(s)\tilde{q}_{35}(s),$$

$$\tilde{A}_3(s) = -1 + \tilde{q}_{35}(s) + \tilde{q}_{12}(s)\tilde{q}_{21}(s) + \tilde{q}_{13}(s)\tilde{q}_{31}(s) + \tilde{q}_{12}(s)\tilde{q}_{24}(s)\tilde{q}_{31}(s)$$
$$- \tilde{q}_{12}(s)\tilde{q}_{21}(s)\tilde{q}_{35}(s),$$

$$\tilde{D}(s) = s\left(-1 + \tilde{q}_{12}(s)\tilde{q}_{21}(s) + \tilde{q}_{13}(s)\tilde{q}_{31}(s)\right).$$

We obtain the reliability functions $R_i(t)$, $i \in S_+$ by inverting the Laplace transforms $\tilde{R}_i(s)$, $i \in S_+$.

The numbers $\bar{\Theta}_{iS_-} = E(\Theta_{iS_-})$, $i \in S_+$ signify conditional mean times to failure. They are unique solutions of the linear equations system (3.25), which in this case have the following matrix form

$$(I - P_{S_+})\overline{\Theta}_{S_+} = \overline{T}_{S_+}, \tag{3.122}$$

$$P_{S_+} = \left[p_{ij} : i,j \in S_+\right], \quad \overline{\Theta}_{S_+} = \left[E(\Theta_{iS_-}) : i \in S_+\right]^{\mathrm{T}}, \quad \overline{T}_{S_+} = [E(T_i) : i \in S_+]^{\mathrm{T}}.$$

and I is the unit matrix. Hence,

$$\begin{bmatrix} 1 & -p & -\tilde{q} \\ -p_{21} & 1 & 0 \\ -p_{31} & 0 & 1 \end{bmatrix} \begin{bmatrix} E(\Theta_{1S_-}) \\ E(\Theta_{2S_-}) \\ E(\Theta_{3S_-}) \end{bmatrix} = \begin{bmatrix} E(T_1) \\ E(T_2) \\ E(T_3) \end{bmatrix}, \tag{3.123}$$

$$\bar{g}_1 = E(T_1) = E(\gamma), \quad \bar{g}_2 = E(T_2) = E(\min\{\xi_1, \zeta_1\}),$$
$$\bar{g}_3 = E(T_3) = E(\min\{\xi_2, \zeta_2\}).$$

Solving the equation, we get

$$E(\Theta_{1S_-}) = \frac{\bar{g}_1 + p\bar{g}_2 + q\bar{g}_3}{1 - pp_{21} - qp_{31}},$$

$$E(\Theta_{2S_-}) = \frac{p_{21}\bar{g}_1 + (1 - qp_{31})\bar{g}_2 + qp_{21}\bar{g}_3}{1 - pp_{21} - qp_{31}}, \tag{3.124}$$

$$E(\Theta_{3S_-}) = \frac{p_{31}\bar{g}_1 + pp_{31}\bar{g}_2 + (1 - pp_{21})\bar{g}_3}{1 - pp_{21} - qp_{31}}.$$

3.5.4 Numerical illustrative example

Now we assume that the random variables presented in description have the specific probability distributions.

We suppose:

- PDF of the task k, $k = 1, 2$ duration:

$$f_{\xi_k}(t) = (\alpha_k)^2 \, t \, e^{-\alpha_k t}, \quad t \geqslant 0, \; k = 1, 2,$$

- PDF of the time to failure of the object executing a task k:

$$f_{\zeta_k}(t) = \lambda_k \, e^{-\lambda_k t}, \quad t \geqslant 0, \; k = 1, 2,$$

- PDF of the repair time of the object performing task k:

$$f_{\eta_k}(t) = (\mu_k)^2 \, t \, e^{-\mu_k t}, \quad t \geqslant 0, \; k = 1, 2,$$

- PDF of the waiting time for the any task

$$f_{\gamma}(t) = \beta \, e^{-\beta t}, \quad t \geqslant 0.$$

First, we have to calculate all elements of the process kernel. From equalities (3.117) for $t \geqslant 0$, we get:

$$Q_{12}(t) = p \, (1 - e^{-\beta t}), \quad Q_{13}(t) = q \, (1 - e^{-\beta t}),$$

$$Q_{21}(t) = \frac{\alpha_1^2}{(\alpha_1 + \lambda_1)^2} \left[1 - (1 + (\alpha_1 + \lambda_1) t) \, e^{-(\alpha_1 + \lambda_1) t} \right],$$

$$Q_{24}(t) = \frac{\lambda_1}{(\alpha_1 + \lambda_1)^2} \left[(\alpha_1 + \lambda_1) \left(1 - e^{-(\alpha_1 + \lambda_1) t} \right) \right.$$
$$\left. + \alpha_1 \left(1 - (1 + (\alpha_1 + \lambda_1) t) \, e^{-(\alpha_1 + \lambda_1) t} \right) \right],$$

$$Q_{31}(t) = \frac{\alpha_2^2}{(\alpha_2 + \lambda_2)^2} \left[1 - (1 + (\alpha_2 + \lambda_2) t) \, e^{-(\alpha_2 + \lambda_2) t} \right],$$

$$Q_{35}(t) = \frac{\lambda_2}{(\alpha_2 + \lambda_2)^2} \left[(\alpha_2 + \lambda_2) \left(1 - e^{-(\alpha_2 + \lambda_2) t} \right) \right.$$
$$\left. + \alpha_2 \left(1 - (1 + (\alpha_2 + \lambda_2) t) \, e^{-(\alpha_2 + \lambda_2) t} \right) \right],$$

$$Q_{41}(t) = 1 - (1 + \mu_1 t) \, e^{-\mu_1 t}, \quad Q_{51}(t) = 1 - (1 + \mu_2 t) \, e^{-\mu_2 t}. \tag{3.125}$$

Passing to limit with $t \to \infty$ in the above functions, we obtain elements of the matrix (3.118).

$$
\begin{aligned}
& p_{12} = p, && p_{13} = q, \\
& p_{21} = \frac{\alpha_1^2}{(\alpha_1 + \lambda_1)^2}, && p_{24} = \frac{\lambda_1^2 + 2\alpha_1 \lambda_1}{(\alpha_1 + \lambda_1)^2}, \\
& p_{31} = \frac{\alpha_2^2}{(\alpha_2 + \lambda_2)^2}, && p_{35} = \frac{\lambda_2^2 + 2\alpha_2 \lambda_2}{(\alpha_2 + \lambda_2)^2}, \\
& p_{41} = 1, && p_{51} = 1.
\end{aligned}
\tag{3.126}
$$

The CDF of waiting times in states $i \in S$ and their expected values and second moments are

$$G_1(t) = 1 - e^{-\beta t}, \qquad E(T_1) = \frac{1}{\beta}, \qquad E(T_1^2) = \frac{2}{\beta^2},$$

$$G_2(t) = 1 - (1 + \alpha_1 t)\, e^{-(\alpha_1 + \lambda_1)t}, \quad E(T_2) = \frac{2\alpha_1 + \lambda_1}{(\alpha_1 + \lambda_1)^2}, \quad E(T_2^2) = \frac{2(3\alpha_1 + \lambda_1)}{(\alpha_1 + \lambda_1)^3},$$

$$G_3(t) = 1 - (1 + \alpha_2 t)\, e^{-(\alpha_2 + \lambda_2)t}, \quad E(T_3) = \frac{2\alpha_2 + \lambda_2}{(\alpha_2 + \lambda_2)^2}, \quad E(T_3^2) = \frac{2(3\alpha_2 + \lambda_2)}{(\alpha_2 + \lambda_2)^3},$$

$$G_4(t) = 1 - (1 + \mu_1 t)\, e^{-\mu_1 t}, \qquad E(T_4) = \frac{2}{\mu_1}, \qquad E(T_4^2) = \frac{6}{\mu_1^2},$$

$$G_5(t) = 1 - (1 + \mu_2 t)\, e^{-\mu_2 t} \qquad E(T_5) = \frac{2}{\mu_2}, \qquad E(T_5^2) = \frac{6}{\mu_2^2}.$$

$$(3.127)$$

The elements of the matrix (3.119) are

$$\tilde{q}_{12}(s) = p\, \frac{\beta}{(s + \beta)}, \qquad \tilde{q}_{13}(s) = q\, \frac{\beta}{(s + \beta)},$$

$$\tilde{q}_{21}(s) = \frac{\alpha_1^2}{(s + \alpha_1 + \lambda_1)^2}, \qquad \tilde{q}_{24}(s) = \frac{\lambda_1\,(s + 2\,\alpha_1 + \lambda_1)}{(s + \alpha_1 + \lambda_1)^2},$$

$$\tilde{q}_{31}(s) = \frac{\alpha_2^2}{(s + \alpha_2 + \lambda_2)^2}, \qquad \tilde{q}_{35}(s) = \frac{\lambda_2\,(s + 2\,\alpha_2 + \lambda_2)}{(s + \alpha_2 + \lambda_2)^2},$$

$$\tilde{q}_{41}(s) = \frac{\mu_1}{s + \mu_1}, \qquad \tilde{q}_{51}(s) = \frac{\mu_2}{s + \mu_2}.$$

$$(3.128)$$

We assume the following values of the above-presented parameters:

$$p = 0.6, \quad q = 0.4,$$

$$\beta = 0.4, \quad \alpha_1 = 0.12 \quad \lambda_1 = 0.01, \quad \alpha_2 = 0.25, \quad \lambda_2 = 0.005; \quad \left[\tfrac{1}{h}\right],$$

$$\mu_1 = 0.15, \quad \mu_2 = 0.2; \quad \left[\tfrac{1}{h}\right].$$

$$(3.129)$$

A procedure written in the MATHEMATICA computer program enables us to obtain the reliability characteristics and parameters in the considered model. The numbers $\bar{\Theta}_{iS_-} = E(\Theta_{iS_-})$, $i \in S_+$, denoting conditional mean times to failure, are

$$\bar{\Theta}_{1S_-} = 138.86, \quad \bar{\Theta}_{2S_-} = 133.12, \quad \bar{\Theta}_{1S_-} = 141.24; \quad [h].$$

The corresponding conditional standard deviations are

$$D(\bar{\Theta}_{1S_-}) = 136.52, \quad D(\Theta_{2S_-}) = 136.27, \quad D(\Theta_{1S_-}) = 136.70; \quad [h].$$

Using (3.121) and (3.128) for parameters (3.129) we get the Laplace transform of the conditional reliability functions. Finally, applying the procedure InverseLaplaceTransform$[R_i, s, t]$ in the MATHEMATICA program, we get the conditional reliability functions of the considered system. The formulas are relatively

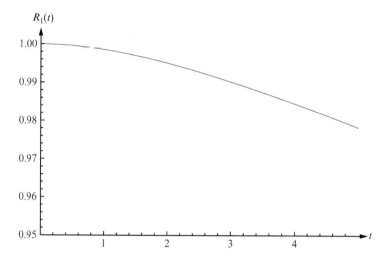

Figure 3.5 Conditional reliability function $R_1(t)$ on interval $[0, 5]$ $[h]$.

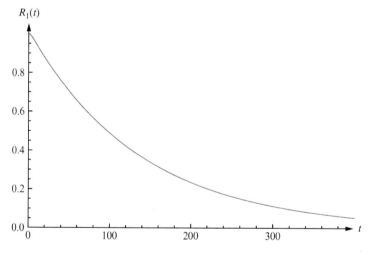

Figure 3.6 Conditional reliability function $R_1(t)$ on interval $[0, 400]$.

simple, but very long. So, we present only figures representing the reliability function $R_1(t)$, $t \geqslant 0$. Figure 3.5 shows the reliability function on interval $[0, 5]$ and Figure 3.6 shows the same function on interval $[0, 400]$.

Applying Theorem 3.6, we obtain the limiting distribution of the process states.

$$P_1 = 0.158, \quad P_2 = 0.560, \quad P_3 = 0.192, \quad P_4 = 0.075 \quad P_5 = 0.015. \tag{3.130}$$

As we know (3.107), the limiting (steady-state) availability is given by

$$A = \sum_{j \in S_+} P_j.$$

Hence,

$$A = P_1 + P_2 + P_3 = 0.910. \tag{3.131}$$

This section shows that calculating many reliability parameters is relatively simple. To obtain the limiting distribution of the process states, the steady-state availability, the conditional means time to failure, and the corresponding second moments we have to solve the appropriate system of linear equations. It should be added that characteristics and parameters of SMPs and Markov processes were discussed in many other books and papers [3, 6, 7,12, 13, 15, 16,17, 19, 25, 45, 53, 54, 87, 90].

Perturbed Semi-Markov processes

Abstract

The chapter covers the application of the perturbed semi-Markov processes (SMPs) in reliability problems. Many kinds of the perturbed SMPs exist. Some of the simplest types are considered in this chapter. All concepts of the perturbed SMPs are explained in the same simple example. Moreover, an exemplary approximation of the system reliability function with an illustrative numerical example is presented in this chapter. The last section is devoted to the state space aggregation method.

Keywords: Perturbed semi-Markov processes, Approximation, Approximate reliability function, State space aggregation method

4.1 Introduction

It is well-known that in the case of complex semi-Markov models, calculating the exact probability distribution of the first passage time to the subset of states is usually very difficult. Then, it seems that the only way is to find the approximate probability distribution of that random variable. It is possible by using the results from the theory of semi-Markov processes (SMPs) perturbations. The perturbed SMPs are defined in different way by different authors. This theory has a rich literature. The most significant and original results include the books of Korolyuk and Turbin [60, 61], Korolyuk and Limnios [65], and Gyllenberg and Silvestrov [40]. We can find many interesting results concerning perturbed SMPs in papers by Gertsbakh [25], Pavlov and Ushakov [83], Shpak [89], Gyllenberg and Silvestrov [39], Domsta and Grabski [18], and many more. In this chapter we present only a few of the simplest types of perturbed SMPs. The first of them is presented by Shpak, the second one is introduced by Pavlov and Ushakov and was presented by Gertsbakh, and the third one is defined by Koryoluk and Turbin.

4.2 Shpak concept

A perturbed SMP in the paper by Shpak [89] is called an associated process to the SMP. We introduce our version of the Shpak definition. Let $A' = \{1, 2, \ldots, N\}, A = \{0\}$ and $S = A \cup A'$. Suppose that $\{X^0(t) : t \geq 0\}$ is the SMP on the state space A and the kernel $\mathbf{Q}^0(t) = \left[Q^0_{ij}(t) : i, j \in A' \right]$.

Semi-Markov Processes: Applications in System Reliability and Maintenance. http://dx.doi.org/10.1016/B978-0-12-800518-7.00004-1

Definition 4.1. The SMP $\{X(t) : t \geqslant 0\}$ with the state space S is said to be the perturbed process with respect to the process $\{X^0(t) : t \geqslant 0\}$ if the kernel $\boldsymbol{Q}(t) = \left[Q_{ij}(t) : i, j \in S \right]$ of the process $\{X(t) : t \geqslant 0\}$ is given by

$$Q_{ij}(t) = \begin{cases} \int_0^t [1 - F_i(x)] \, dQ_{ij}^0(x) & \text{dla } i, j \in A' \\ 0 & \text{dla } i \in A, \ j \in A' \end{cases}, \tag{4.1}$$

$$Q_{i0}(t) = \int_0^t \left[1 - G_i^0(x) \right] dF_i(x), \quad i \in A', \tag{4.2}$$

where $G_i^0(t) = \sum_{j \in A'} Q_{ij}^0(t)$, while the functions $F_i(t) = P(Z_i \leqslant t)$, $i \in A'$ represent the cumulative density functions (CDFs) of the nonnegative random variables Z_i, $i \in A'$ with the positive finite expected values.

The random variable

$$\Theta_{i0} = \inf\{t : X(t) = 0 | X(0) = i\}, \quad i \in A' = \{1, \ldots, N\}$$

denotes the first passage time from a state $i \in A'$ to state 0. Let

$$m_i^0 = \int_0^\infty \left[1 - G_i^0(t) \right] dt, \quad i \in A', \tag{4.3}$$

where

$$G_i^0(t) = \sum_{j \in A'} Q_{ij}^0(t). \tag{4.4}$$

The number m_i^0 is an expectation of the waiting time of the state $i \in A'$ for the SMP $\{X^0(t) : t \geqslant 0\}$. The number

$$\varepsilon_i = p_{i0} = \lim_{t \to \infty} Q_{i0}(t) \tag{4.5}$$

is the transition probability of the embedded Markov chain $\{X(\tau_n) : n \in \mathbb{N}_0\}$ from the state $i \in A'$ to state 0. Let $\pi^0 = \left[\pi_1^0, \ldots, \pi_N^0 \right]$ denote the stationary distribution of the embedded Markov chain and let

$$\varepsilon = \sum_{i \in A'} \pi_i^0 \varepsilon_i. \tag{4.6}$$

Theorem 4.1. If the embedded Markov chain $\{X(\tau_n) : n \in \mathbb{N}_0\}$ has the stationary distribution $\pi^0 = \left[\pi_1^0, \ldots, \pi_N^0 \right]$, $\varepsilon > 0$ and the distributions $G_i^0(t)$, $i = 1, \ldots, N$ have the finite and positive expectations, then

$$\lim_{\varepsilon \to 0} P\{\varepsilon \Theta_{i0} > t\} = \exp\left[-\lambda t \right], \quad t \geqslant 0, \tag{4.7}$$

where

$$\lambda = \frac{1}{\sum_{i \in A'} \pi_i^0 m_i^0}.$$

Proof: [89].

Note first that the asymptotic distribution of the random variable Θ_{i0} does not depend on the state of $i \in A'$.

If the subset A' is the set of "up" states, while $A = \{0\}$ is a state of the object failure, then a thesis of Theorem 4.1 allows us to find the approximate reliability function

$$R(t) = P\{\Theta_{iA} > t\} = P\{\varepsilon\Theta_{iA} > \varepsilon t\} \approx \exp[-\lambda \, \varepsilon t], \quad t \geqslant 0, \tag{4.8}$$

when ε is small.

Example 4.1. In many real cases we can treat the duration of the functioning periods and repair or replacement periods of a system as the random variables. We can describe the operation process of the system by a two-states stochastic process. In practice, the danger of an emergency can occur in a functioning period as well as in a repair period of the system. This event can stop or break the operation process. Our aim is to construct the stochastic model describing the process of operation with the perturbations of the system reliability. This model will allow us to obtain the reliability characteristics of the system.

We assume that the duration of the functioning periods are the independent copies of the positive random variable ξ with CDF

$$F_\xi(t) = P(\xi \leqslant t).$$

The duration of the renewal periods (repair or replacement periods) are the independent copies of the positive random variable η with CDF

$$F_\eta(t) = P(\eta \leqslant t).$$

During both a functioning period and a repair period the system can incur damage. The time to failure during a functioning period is a positive random variable ζ_1 with a probability density function $f_{\zeta_1}(x)$, $x \geqslant 0$. The lifetime during a functioning renewal period is a positive random variable ζ_2 with a probability density function $f_{\zeta_2}(x)$, $x \geqslant 0$. The failure causes stopping the operation process.

To consider a stochastic model of the above-mentioned operation process, we start with introducing the following states of the system:

$s_1 \leftrightarrow 1$: work of the system
$s_2 \leftrightarrow 2$: renewal of the system
$s_0 \leftrightarrow 0$: failure of the system

Possible states changes of the system are shown in Figure 4.1.

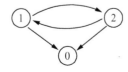

Figure 4.1 Possible states changes of the system.

Let $\{X(t) : t \geqslant 0\}$ be a stochastic process describing the process of state changes of the object. Denote moments of the state changes of the process by $\tau_0 = 0, \tau_1, \tau_2, \ldots$. This process takes constant values on half intervals $[\tau_0 \, \tau_1), [\tau_1, \tau_2), \ldots$. From description and assumptions it follows that $\{X(t) : t \geqslant 0\}$ is the SMP with the state space $S = \{0, 1, 2\}$ and the kernel

$$
\boldsymbol{Q}(t) = \begin{bmatrix} Q_{00}(t) & 0 & 0 \\ Q_{10}(t) & 0 & Q_{12}(t) \\ Q_{20}(t) & Q_{21}(t) & 0 \end{bmatrix},
\tag{4.9}
$$

where

$$
Q_{10}(t) = P(\zeta \leqslant t, \zeta \leqslant \xi) = \int_0^t \left[1 - F_\xi(x)\right] f_{\zeta_1}(x) \mathrm{d}x,
\tag{4.10}
$$

$$
Q_{12}(t) = P(\xi \leqslant t, \ \zeta > \xi) = \int_0^t \left[1 - F_{\zeta_1}(x)\right] \mathrm{d}F_\xi(x),
\tag{4.11}
$$

$$
Q_{20}(t) = P(\zeta_2 \leqslant t, \ \zeta_2 \leqslant \eta) = \int_0^t \left[1 - F_\eta(x)\right] f_{\zeta_2}(x) \, \mathrm{d}x,
\tag{4.12}
$$

$$
Q_{21}(t) = P(\eta \leqslant t, \ \zeta_2 > \eta) = \int_0^t \left[1 - F_{\zeta_2}(x)\right] \mathrm{d}F_\eta(x),
\tag{4.13}
$$

and $Q_{00} = G_0(t)$ is an arbitrary PDF concentrated on the $\mathbb{R}_+$. We can treat this SMP as a perturbed process with respect to a SMP $\{X^0(t) : t \geqslant 0\}$ with the state space $A' = \{1, 2\}$ defined by the kernel

$$
\boldsymbol{Q}^0(t) = \begin{bmatrix} 0 & Q^0_{12}(t) \\ Q^0_{21}(t) & 0 \end{bmatrix},
\tag{4.14}
$$

where

$$
Q^0_{12}(t) = F_\xi(t), \quad Q^0_{21}(t) = F_\eta(t).
$$

A time to failure ζ_1 plays a role of the random variable Z_1, and the random variable ζ_2 is identical to the random variable Z_2. The quantities from the theorem presented herein are

$$
\pi_1^0 = 0.5, \quad \pi_2^0 = 0.5, \quad m_1^0 = E(\xi), \quad m_2^0 = E(\eta),
\tag{4.15}
$$

$$
\varepsilon_1 = \int_0^\infty \left[1 - F_\xi(u)\right] f_{\zeta_1}(u) \mathrm{d}u,
$$

$$
\varepsilon_2 = \int_0^\infty \left[1 - F_\eta(u)\right] f_{\zeta_2}(u) \mathrm{d}u, \varepsilon = 0.5(\varepsilon_1 + \varepsilon_2).
\tag{4.16}
$$

Thus, the reliability function of the object is expressed by approximation

$$
R(t) \approx \exp\left[-\frac{\int_0^\infty \left[1 - F_\xi(u)\right] f_{\zeta_1}(u) \mathrm{d}u + \int_0^\infty \left[1 - F_\eta(u)\right] f_{\zeta_2}(u) \mathrm{d}u}{E(\xi) + E(\eta)} t \right].
\tag{4.17}
$$

This formula can be used only when ε is small. This number is small if probability of the failure in the functioning interval of length ξ, and probability of the failure in the renewal interval of length η are both small. $\triangle$

4.3 Pavlov and Ushakov concept

Let $A' = S - A$ be a finite subset of states and A be at most countable subset of S. Suppose that $\{X(t) : t \geqslant 0\}$ is the SMP with the state space $S = A \cup A'$ and the kernel $Q(t) = [Q_{ij}(t) : i, j \in S]$, the elements of which have the form of $Q_{ij}(t) = p_{ij}F_{ij}(t)$. Recall that p_{ij} is the transition probability from the state $i \in S$ to the state $j \in S$ of the embedded Markov chain of the SMP $\{X(t) : t \geqslant 0\}$, while $F_{ij}(t)$ denotes the CDF of a holding time $T_{ij}, i, j \in S$. Assume

$$\varepsilon_i = \sum_{j \in A} p_{ij} \tag{4.18}$$

and

$$p_{ij}^0 = \frac{p_{ij}}{1 - \varepsilon_i}, \quad i, j \in A'. \tag{4.19}$$

Notice that $\sum_{j \in A'} p_{ij}^0 = 1$.

Definition 4.2. A SMP $\{X(t) : t \geqslant 0\}$ with the discrete state space S defined by the renewal kernel $Q(t) = [p_{ij}F_{ij}(t) : i, j \in S]$ is called the perturbed process with respect to a SMP $\{X^0(t) : t \geqslant 0\}$ with the state space A' and defined by the kernel $Q^0(t) = [p_{ij}^0 F_{ij}(t) : i, j \in A']$.

We will present the theorem that is some version of the theorem introduced by Gertsbakh [25]. The random variable

$$\Theta_{iA} = \Theta_{iA} = \inf\{t : X(t) \in A | X(0) = i\}, \quad i \in A', \tag{4.20}$$

represents the first passage time from state $i \in A'$ to the subset A. The number

$$m_i = \int_0^\infty [1 - G_i(t)] \, dt, \quad i \in A', \tag{4.21}$$

where

$$G_i(t) = \sum_{j \in S} Q_{ij}(t)$$

is an expectation of the waiting time in the state i for the SMP $\{X(t) : t \geqslant 0\}$. The number

$$m_i^0 = \int_0^\infty \left[1 - G_i^0(t)\right] dt, \quad i \in A', \quad \text{where } G_i^0(t) = \sum_{j \in A'} Q_{ij}^0(t) \tag{4.22}$$

is an expected value of the waiting time in state i for the process $\{X^0(t) : t \geqslant 0\}$. Denote by $\pi^0 = \left[\pi_i^0 : i \in A'\right]$ the stationary distribution of the embedded Markov chain in the SMP $\{X^0(t) : t \geqslant 0\}$. Let

$$\varepsilon = \sum_{i \in A'} \pi_i^0 \varepsilon_i \quad \text{and} \quad m^0 = \sum_{i \in A'} \pi_i^0 m_i^0. \tag{4.23}$$

We are interested in the limiting distribution of the random variable $\Theta_{iA}, i \in A'$.

Theorem 4.2. If the embedded Markov chain defined by the matrix of transition probabilities $P = \left[p_{ij} : i, j \in S\right]$ satisfies conditions

-
$$f_{iA} = P(\Delta_A < \infty | X(0) = i) = 1, \quad i \in A', \tag{4.24}$$

-
$$\exists_{c > 0} \forall_{i,j \in S}, \quad 0 < E(T_{ij}) \leqslant c, \tag{4.25}$$

-
$$\forall_{i \in A} \mu_{iA} = \sum_{n=1}^{\infty} n f_{iA}(n) < \infty, \tag{4.26}$$

then

$$\lim_{\varepsilon \to 0} P(\varepsilon \Theta_{iA} > x) = e^{-\frac{x}{m^0}}, \tag{4.27}$$

where $\pi^0 = \left[\pi_i : i \in A'\right]$ is the unique solution of the linear system of equations

$$\pi^0 = \pi^0 P^0, \quad \pi^0 \mathbf{1} = 1. \tag{4.28}$$

From that theorem it follows that for small ε we obtain the following approximate formula:

$$P(\Theta_{iA} > t) \approx \exp\left[-\frac{\sum\limits_{i \in A'} \pi_i^0 \varepsilon_i}{\sum\limits_{i \in A'} \pi_i^0 m_i^0} t\right], \quad t \geqslant 0. \tag{4.29}$$

Example 4.2. We apply the above-mentioned theorem for the model presented in Example 4.1. As we know, $\{X(t) : t \geqslant 0\}$ is a SMP with the state space $S = \{0, 1, 2\}$ and the kernel given by (4.9). We can treat this SMP as the perturbed process with respect to the SMP $\{X^0(t) : t \geqslant 0\}$ with the state space $A' = \{1, 2\}$ and the kernel

$$Q^0(t) = \begin{bmatrix} 0 & Q_{12}^0(t) \\ Q_{21}^0(t) & 0 \end{bmatrix}, \tag{4.30}$$

where

$$Q_{12}^0(t) = p_{12}^0 F_{12}(t), \quad Q_{21}^0(t) = p_{21}^0 F_{21}(t).$$

As $A = \{0\}$, then

$$\varepsilon_1 = p_{10} = Q_{10}(\infty) = \int_0^\infty \left[1 - F_\xi(x)\right] f_{\zeta_1}(x)\,dx,$$

$$\varepsilon_2 = \int_0^\infty \left[1 - F_\eta(u)\right] f_{\zeta_2}(u)\,du. \tag{4.31}$$

Note that

$$p_{12}^0 = \frac{p_{12}}{1 - \varepsilon_1} = 1 \quad \text{and} \quad p_{21}^0 = \frac{p_{21}}{1 - \varepsilon_2} = 1.$$

From the equality $F_{ij}(t) = \frac{Q_{ij}(t)}{p_{ij}}$, we have

$$Q_{12}^0(t) = F_{12}(t) = \frac{\int_0^t \left[1 - F_{\zeta_1}(x)\right] dF_\xi(x)}{\int_0^\infty \left[1 - F_{\zeta_1}(x)\right] dF_\xi(x)},$$

$$Q_{21}^0(t) = F_{21}(t) = \frac{\int_0^t \left[1 - F_{\zeta_2}(x)\right] dF_\eta(x)}{\int_0^\infty \left[1 - F_{\zeta_2}(x)\right] dF_\xi(x)} \tag{4.32}$$

The transition matrix of the embedded Markov chain of the SMP $\{X^0(t) : t \geqslant 0\}$ takes the form

$$P^0 = \begin{bmatrix} 0 & 1 \\ 1 & 0 \end{bmatrix}. \tag{4.33}$$

The solution of the system of equations

$$\begin{bmatrix} \pi_1^0, \pi_2^0 \end{bmatrix} \begin{bmatrix} 0 & 1 \\ 1 & 0 \end{bmatrix} = \begin{bmatrix} \pi_1^0, \pi_2^0 \end{bmatrix}, \tag{4.34}$$

$$\pi_1^0 + \pi_2^0 = 1 \tag{4.35}$$

is the stationary distribution of the above-mentioned embedded Markov chain:

$$\pi^0 = [0.5,\ 0.5].$$

Using (4.23), we obtain parameters

$$\varepsilon = 0.5(\varepsilon_1 + \varepsilon_2), \tag{4.36}$$

$$m^0 = 0.5(m_1^0 + m_2^0), \tag{4.37}$$

where

$$m_1^0 = \int_0^\infty t\,dQ_{12}^0(t) = \frac{\int_0^\infty t\left[1 - F_{\zeta_1}(t)\right] dF_\xi(t)}{\int_0^\infty \left[1 - F_{\zeta_1}(x)\right] dF_\xi(x)},$$

$$m_2^0 = \int_0^\infty t\,dQ_{12}^0(t) = \frac{\int_0^\infty t\left[1 - F_{\zeta_2}(t)\right] dF_\eta(t)}{\int_0^\infty \left[1 - F_{\zeta_2}(x)\right] dF_\eta(x)}. \tag{4.38}$$

For small ε, we get an approximation of the reliability function

$$R(t) = P(\Theta_{iA} > t) = P(\varepsilon\Theta_{iA} > \varepsilon t)$$

$$\approx \exp\left[-\frac{\varepsilon}{m^0}t\right] = \exp\left[-\frac{\varepsilon_1 + \varepsilon_2}{m_1^0 + m_2^0}t\right], \quad t \geqslant 0. \tag{4.39}$$

Notice that for small ε_1 and ε_2

$$m_1^0 \approx \int_0^\infty t\mathrm{d}F_\xi(x) = E(\xi), \quad m_2^0 \approx \int_0^\infty t\mathrm{d}F_\eta(x) = E(\eta)$$

and we get the approximation of the reliability function identical to the function given by (4.17). $\triangle$

4.4 Korolyuk and Turbin concept

Now we present the concept of the simplest type of the perturbed process coming from the monograph of Korolyuk and Turbin [60].

Suppose that A' is at most a countable set and let $A = \{0\}$. Let $\{X^\varepsilon(t) : t \geqslant 0\}$ be a SMP with the state space $S = A \cup A'$ and the kernel

$$\boldsymbol{Q}^\varepsilon(t) = \left[Q_{ij}^\varepsilon(t) : i, j \in S\right], \tag{4.40}$$

the elements of which have the form of

$$Q_{ij}^\varepsilon(t) = p_{ij}^\varepsilon F_{ij}(t), \tag{4.41}$$

where

$$p_{ij}^\varepsilon = \begin{cases} p_{ij}^0 - \varepsilon b_{ij} & \text{for } i,j \in A' \\ \varepsilon q_i & \text{for } i \in A', j = 0 \\ 1 & \text{for } i = j = 0 \end{cases} \tag{4.42}$$

and

$$q_i = \sum_{j \in A'} b_{ij}, \quad i \in A'. \tag{4.43}$$

Definition 4.3. The SMP $\{X(t) : t \geqslant 0\}$ with the state space $S = A \cup A'$ and the kernel $\boldsymbol{Q}^\varepsilon(t) = \left[Q_{ij}^\varepsilon(t) : i,j \in S\right]$ given by (4.40)–(4.43) is called perturbed with respect to the SMP $\{X^0(t) : t \geqslant 0\}$ with the state space A' and the kernel $\boldsymbol{Q}^0(t) = \left[p_{ij}^0 F_{ij}(t) : i,j \in A'\right]$.

Let $\pi^0 = \left[\pi_i^0 : i \in A'\right]$ denote a stationary distribution of the embedded Markov chain $\{X^0(\tau_n) : n \in \mathbb{N}_0\}$. As we know, this distribution is the only solution of the system of equations

$$\pi^0 = \pi^0 \boldsymbol{P}^0, \quad \pi^0 \boldsymbol{1} = 1. \tag{4.44}$$

Let

$$q = \sum_{i \in A'} \pi_i^0 \, q_i. \tag{4.45}$$

and as in the previous theorem

$$m^0 = \sum_{i \in A'} \pi_i^0 \, m_i^0. \tag{4.46}$$

Theorem 4.3. If

- the embedded Markov chain $\{X^0(\tau_n) : n \in \mathbb{N}_0\}$ with the matrix of transition probabilities $P^0 = \left[p_{ij}^0 : i, j \in A' \right]$ consists of exactly one positive, strongly recurrent class A' such that $f_{ij} = 1$ for all $i, j \in A'$,
- $\exists_{C < \infty} \forall_{i \in A'} \, m_i^0 \leqslant C$,
- $\exists_{i \in A'} \, p_{i0}^\varepsilon > 0$,

then

$$\lim_{\varepsilon \to 0} P\{\varepsilon \Theta_{i0} > x\} = e^{-\frac{q}{m^0} x}. \tag{4.47}$$

Proof: [60].

From this theorem it follows that for $\varepsilon > 0$ we get the approximate formula

$$P(\Theta_{iA} > t) = P(\varepsilon \Theta_{iA} > \varepsilon t) \approx e^{-\frac{\varepsilon q}{m^0} t}. \tag{4.48}$$

Example 4.3. Let us consider the same model. Now, the SMP describing the reliability evolution of the object is denoted by $\{X^\varepsilon(t) : t \geqslant 0\}$ and its kernel is given by (4.9). A transition matrix of the corresponding embedded Markov chain is

$$P^\varepsilon = \begin{bmatrix} 1 & 0 & 0 \\ p_{10} & 0 & p_{12} \\ p_{20} & p_{21} & 0 \end{bmatrix}, \tag{4.49}$$

where

$$p_{10} = \int_0^\infty \left[1 - F_\xi(x) \right] f_{\zeta_1}(x) dx, \quad p_{12} = \int_0^\infty \left[1 - F_{\zeta_1}(x) \right] dF_\xi(x), \tag{4.50}$$

$$p_{20} = \int_0^\infty \left[1 - F_\eta(x) \right] f_{\zeta_2}(x) dx, \quad p_{21} = \int_0^\infty \left[1 - F_{\zeta_2}(x) \right] dF_\eta(x). \tag{4.51}$$

The matrix of CDF of the holding times T_{ij}, $i, j \in S$ is

$$F^\varepsilon(t) = \begin{bmatrix} F_{00}(t) & 0 & 0 \\ F_{10}(t) & 0 & F_{12}(t) \\ F_{20}(t) & F_{21}(t) & 0 \end{bmatrix}, \tag{4.52}$$

where

$$F_{10}(t) = \frac{\int_0^t \left[1 - F_\xi(x)\right] f_{\zeta_1}(x) dx}{\int_0^\infty \left[1 - F_\xi(x)\right] f_{\zeta_1}(x) dx}, \quad F_{12}(t) = \frac{\int_0^t \left[1 - F_{\zeta_1}(x)\right] dF_\xi(x)}{\int_0^\infty \left[1 - F_{\zeta_1}(x)\right] dF_\xi(x)},$$

$$F_{20}(t) = \frac{\int_0^t \left[1 - F_\eta(x)\right] f_{\zeta_2}(x) dx}{\int_0^\infty \left[1 - F_\eta(x)\right] f_{\zeta_2}(x) dx}, \quad F_{12}(t) = \frac{\int_0^t \left[1 - F_{\zeta_2}(x)\right] dF_\xi(x)}{\int_0^\infty \left[1 - F_{\zeta_2}(x)\right] dF_\xi(x)},$$

and

$$F_{00} = G_0(t).$$

The kernel of the process $\{X^0(t) : t \geq 0\}$ with the state space $A' = \{1, 2\}$ is

$$Q^0(t) = \begin{bmatrix} 0 & Q_{12}^0(t) \\ Q_{21}^0(t) & 0 \end{bmatrix}, \tag{4.53}$$

where

$$Q_{12}^0(t) = F_{12}(t), \quad Q_{21}^0(t) = F_{21}(t).$$

The probability transition matrix of the embedded Markov chain in process $\{X^0(t) : t \geq 0\}$ is

$$P^0 = \begin{bmatrix} 0 & 1 \\ 1 & 0 \end{bmatrix}, \tag{4.54}$$

and the stationary distribution is $\pi^0 = [0.5, 0.5]$. From (4.42), we get

$$\varepsilon b_{ij} = p_{ij}^0 - p_{ij}^\varepsilon, \quad i, j \in A'. \tag{4.55}$$

Thus,

$$\varepsilon b_{11} = 0, \quad \varepsilon b_{12} = P(\zeta_1 \leq \xi) = \int_0^\infty \left[1 - F_\xi(x)\right] f_{\zeta_1}(x) dx, \tag{4.56}$$

$$\varepsilon b_{21} = P(\zeta_2 \leq \eta) = \int_0^\infty \left[1 - F_\eta(x)\right] f_{\zeta_2}(x) dx, \quad \varepsilon b_{22} = 0 \tag{4.57}$$

and

$$\varepsilon q_1 = \int_0^\infty \left[1 - F_\xi(x)\right] f_{\zeta_1}(x) dx, \quad \varepsilon q_2 = \int_0^\infty \left[1 - F_\eta(x)\right] f_{\zeta_2}(x) dx. \tag{4.58}$$

Therefore, from (4.45) we get

$$\varepsilon q = 0.5(q_1 + q_2). \tag{4.59}$$

From (4.45), we have

$$m^0 = 0.5(m_1^0 + m_2^0),$$

where m_1^0 and m_2^0 are given by (4.40). Finally, we obtain the same approximation of the reliability function as in the previous case. $\triangle$

4.5 Exemplary approximation of the system reliability function

Let us consider the example presented in Section 3.5. To obtain the approximate reliability function, we use the concept of the semi-Markov perturbed process introduced by Pavlov and Ushakov. The model of the object operation is the SMP $\{X(t) : t \geqslant 0\}$ with the five elements state space $S = \{1, 2, 3, 4, 5\}$ where the subset $A' = S_+ = \{1, 2, 3\}$ consists of the functioning states and $A = S_- = \{4, 5\}$ contains all failed states of the object. The kernel of the process is given by (3.116). We know that the kernel can be expressed as $Q(t) = \left[p_{ij}F_{ij}(t) : i, j \in S \right]$. We have to construct the SMP $\{X^0(t) : t \geqslant 0\}$ with the state space $A' = \{1, 2, 3\}$ and the kernel

$$Q^0(t) = \left[p_{ij}^0 F_{ij}(t) : i, j \in A' \right], \tag{4.60}$$

where

$$p_{ij}^0 = \frac{p_{ij}}{1 - \varepsilon_i}, \quad i, j \in A', \quad \text{and} \quad \varepsilon_i = \sum_{j \in A} p_{ij}, \quad i \in A'. \tag{4.61}$$

So, we are obliged to find all elements of the matrix

$$Q^0(t) = \begin{bmatrix} 0 & Q_{12}^0(t) & Q_{13}^0(t) \\ Q_{21}^0(t) & 0 & 0 \\ Q_{31}^0(t) & 0 & 0 \end{bmatrix}. \tag{4.62}$$

The transition matrix of the embedded Markov chain of SMP generated by this kernel is

$$P^0 = \begin{bmatrix} 0 & p & q \\ 1 & 0 & 0 \\ 1 & 0 & 0 \end{bmatrix}. \tag{4.63}$$

To find a stationary distribution of the embedded Markov chain we have to solve the system of equations

$$\left[\pi_1^0, \pi_2^0, \pi_3^0 \right] \begin{bmatrix} 0 & p & q \\ 1 & 0 & 0 \\ 1 & 0 & 0 \end{bmatrix} = \left[\pi_1^0, \pi_2^0, \pi_3^0 \right], \tag{4.64}$$

$$\pi_1^0 + \pi_2^0 + \pi_3^0 = 1. \tag{4.65}$$

The solution is

$$\pi_1^0 = 0.5, \quad \pi_2^0 = 0.5p, \quad \pi_3^0 = 0.5q. \tag{4.66}$$

The PDF's functions $F_{ij}(t)$, $i,j \in A'$ are

$$F_{12}(t) = F_\gamma(t), \quad F_{13}(t) = F_\gamma(t), \tag{4.67}$$

$$F_{21}(t) = \frac{\int_0^t \left[1 - F_{\zeta_1}(x)\right] dF_{\xi_1}(x)}{\int_0^\infty \left[1 - F_{\zeta_1}(x)\right] dF_{\xi_1}(x)}, \quad F_{31}(t) = \frac{\int_0^t \left[1 - F_{\zeta_2}(x)\right] dF_{\xi_2}(x)}{\int_0^\infty \left[1 - F_{\zeta_2}(x)\right] dF_{\xi_2}(x)}. \tag{4.68}$$

The elements of the matrix (4.62) take the form

$$Q_{12}^0(t) = pF_\gamma(t), \quad Q_{13}^0(t) = qF_\gamma(t), \quad Q_{21}^0(t) = F_{21}(t), \quad Q_{31}^0(t) = F_{31}(t). \tag{4.69}$$

From (4.61), we get

$$\varepsilon_1 = 0, \quad \varepsilon_2 = p_{24} = \int_0^\infty \left[1 - F_{\xi_1}(x)\right] dF_{\zeta_1}(x),$$

$$\varepsilon_3 = p_{35} = \int_0^\infty \left[1 - F_{\xi_2}(x)\right] dF_{\zeta_2}(x). \tag{4.70}$$

The parameters ε and m^0 given by (4.23) are

$$\varepsilon = 0.5\varepsilon_2 + 0.5\varepsilon_3 = 0.5 \left(\int_0^\infty \left[1 - F_{\xi_1}(x)\right] dF_{\zeta_1}(x) + \int_0^\infty \left[1 - F_{\xi_2}(x)\right] dF_{\zeta_2}(x) \right) \tag{4.71}$$

and

$$m^0 = 0.5(m_1^0 + p\,m_2^0 + q\,m_3^0), \tag{4.72}$$

where

$$m_1^0 = E(\gamma), \quad m_2^0 = \frac{\int_0^\infty x\left[1 - F_{\zeta_1}(x)\right] dF_{\xi_1}(x)}{\int_0^\infty \left[1 - F_{\zeta_1}(x)\right] dF_{\xi_1}(x)}, \quad m_3^0 = \frac{\int_0^\infty x\left[1 - F_{\zeta_2}(x)\right] dF_{\xi_2}(x)}{\int_0^\infty \left[1 - F_{\zeta_2}(x)\right] dF_{\xi_2}(x)}. \tag{4.73}$$

From Theorem 4.2, it follows that for small ε we get the approximate reliability function

$$R(t) = P(\Theta_{iA} > t) = P(\varepsilon\Theta_{iA} > \varepsilon t) = P(\varepsilon\Theta_{iA} > \varepsilon t) \approx \exp\left[-\frac{\varepsilon}{m^0}t\right], \quad t \geqslant 0. \tag{4.74}$$

where ε is determined by (4.71) and m^0 is given by (4.72). Note that for small ε_2 and ε_3

$$m_2^0 \approx \int_0^\infty t\,dF_{\xi_1}(x) = E(\xi_1), \quad m_3^0 \approx \int_0^\infty t\,dF_{\xi_2}(x) = E(\xi_2)$$

and the approximate reliability function takes the form

$$R(t) = P(\Theta_{iA} > t)$$

$$\approx \exp\left[-\frac{\int_0^\infty \left[1 - F_{\xi_1}(x)\right] dF_{\zeta_1}(x) + \int_0^\infty \left[1 - F_{\xi_2}(x)\right] dF_{\zeta_2}(x)}{E(\gamma) + pE(\xi_1) + qE(\xi_2)} t\right], \quad t \geqslant 0.$$

(4.75)

4.5.1 Numerical illustrative example

As the numerical illustrative example, we will calculate the approximate reliability function for the model considered in Section 3.5. First, we calculate the necessary characteristics and parameters. The PDF's functions $F_{ij}(t)$, $i, j \in A'$ are

$$F_{12}(t) = F_\gamma(t), \quad F_{13}(t) = F_\gamma(t),$$

(4.76)

$$F_{21}(t) = 1 - (1 + (\alpha_1 + \lambda_1)t) e^{-(\alpha_1 + \lambda_1)t},$$

$$F_{31}(t) = 1 - (1 + (\alpha_2 + \lambda_2)t) e^{-(\alpha_2 + \lambda_2)t}.$$

(4.77)

The elements of the matrix (4.62) take the form

$$Q_{12}^0(t) = pF_\gamma(t), \quad Q_{12}^0(t) = qF_\gamma(t), \quad Q_{21}^0(t) = F_{21}(t), \quad Q_{31}^0(t) = F_{31}(t).$$

(4.78)

From (4.61), we get

$$\varepsilon_1 = 0, \quad \varepsilon_2 = p_{24} = \frac{\lambda_1^2 + 2\alpha_1\lambda_1}{(\alpha_1 + \lambda_1)^2}, \quad \varepsilon_3 = p_{35} = \frac{\lambda_2^2 + 2\alpha_2\lambda_2}{(\alpha_2 + \lambda_2)^2}.$$

(4.79)

The parameters ε given by (4.71) and m^0 given by (4.72) have the form of

$$\varepsilon = 0.5(p\varepsilon_2 + \varepsilon_3 q)$$

(4.80)

and

$$m^0 = 0.5(m_1^0 + pm_2^0 + qm_3^0)$$

(4.81)

where

$$m_1^0 = \frac{1}{\beta}, \quad m_2^0 = \frac{2\alpha_1 + \lambda_1}{(\alpha_1 + \lambda_1)^2}, \quad m_3^0 = \frac{2\alpha_2 + \lambda_2}{(\alpha_2 + \lambda_2)^2}.$$

(4.82)

Thus, for small ε the approximate reliability function is given by

$$R(t) = P(\Theta_{iA} > t) \approx \exp\left[-\frac{p\varepsilon_2 + \varepsilon_3 q}{m_1^0 + pm_2^0 + qm_3^0} t\right], \quad t \geqslant 0.$$

(4.83)

We assume the same values of parameters as in Section 3.5:

$$p = 0.6, \quad q = 0.4,$$

$$\beta = 0.4, \quad \alpha_1 = 0.12, \quad \lambda_1 = 0.01, \quad \alpha_2 = 0.25, \quad \lambda_2 = 0.005; \quad \left[\frac{1}{h}\right].$$

(4.84)

From (4.80) and (4.81), we calculate the parameter

$$\lambda = \frac{\varepsilon}{m^0}$$

of the exponential distribution. The value of this parameter is $\lambda = 0.00826$. The mean time to failure is given by

$$E(\Theta_A) = \frac{1}{\lambda}$$

and its value is $E(\Theta_A) = 121.02$. Finally, the approximate reliability function in this case is

$$R(t) \approx \exp(-0.00826t), \quad t \geqslant 0.$$

Comparing the above-presented results with results from Section 3.5, we can see that the difference between the conditional expectations of the time to failure and the approximate expectation of that one are small; similarly, the differences between values of the conditional reliability function and its approximation are very small. Calculating the approximate reliability function of the system is much simpler than the exact calculation. However, we must remember that we can apply the theorems from the perturbation theory only if, in considered cases, the assumptions of those theorems are satisfied.

4.6 State space aggregation method

The theory of the SMP perturbation is closely related to the *state space aggregation method*. The aggregation method is also called a merging or lumping or consolidation method. In this section we present a theorem in the simple case of the state space aggregation method. This theorem is a kind of conclusion from the theorem presented by Korolyuk and Turbin [60] (see Chapter 7, p. 151). Its assumptions come from the paper [64]. The theorem contains the conditions under which the SMP with a discrete state space S can be approximated by a continuous-time Markov process with a finite state space $E = \{1, 2, \ldots, m\}$, where the states represent the numbers of disjoint subset $S_1, S_2, \ldots, S_m$ which are a partition of the state space S:

$$S = S_1 \cup S_2 \cup, \ldots, \cup S_m, \quad \text{and} \quad S_i \cap S_j = 0 \quad \text{for } i \neq j.$$

Let $\{X^\varepsilon(t) : t \geqslant 0\}$ be a SMP with a discrete state space S defined by the kernel $Q^\varepsilon(t) = \left[Q_{ij}^\varepsilon(t) : i, j \in S\right]$ depending on parameter $\varepsilon > 0$.

Theorem 4.4. Let the following assumptions be satisfied:

- The elements of the kernel depend on the small parameter as follows ε

$$Q_{ij}^{\varepsilon}(t) = p_{ij}^{\varepsilon} F_{ij}(t) \quad i, j \in S, \tag{4.85}$$

$$p_{ij}^{\varepsilon} = \begin{cases} p_{ij}^{(k)} - \varepsilon q_{ij}^{(k)} & i, j \in S_k \\ \varepsilon q_{ij}^{(k)} & i \in S_k, \ j \notin S_k \end{cases}, \tag{4.86}$$

where

$$\sum_{j \in S_k} p_{ij}^{(k)} = 1, \quad i \in S_k, \ k = 1, 2, \ldots, m.$$

- The Markov chains determined by the transition probability matrices $\boldsymbol{P}^{(k)} = \left[p_{ij}^{(k)} : i, j \in S_k \right]$ are ergodic with stationary probabilities $\pi_i^{(k)}$, $i \in S_k$, $k = 1, 2, \ldots, m$.

Then, the SMP $\{X^{\varepsilon}(t) : t \geqslant 0\}$ for small ε can be approximated by the Markov process $\{Y(t) : t \geqslant 0\}$ with the state space $E = \{1, 2, \ldots, m\}$, where states represent the classes of states $S_1, S_2, \ldots, S_m$. This process is equivalent to the SMP defined by the kernel

$$\bar{\boldsymbol{Q}}(t) = \left[\bar{p}_{kr} \left(1 - e^{-\lambda_k t} \right) : k, r \in E \right], \tag{4.87}$$

where

$$\bar{p}_{kr} = \frac{\bar{\alpha}_{kr}}{\bar{m}_k}, \qquad \lambda_k = \frac{1}{\bar{m}_k}, \tag{4.88}$$

$$\bar{\alpha}_{kr} = \sum_{i \in S_k} \pi_i^{(k)} \sum_{j \in S_r} b_{ij}, \quad b_{ij}^{(k)} = \varepsilon q_{ij}^{(k)}, \quad \bar{m}_k = \sum_{i \in S_k} \pi_i^{(k)} m_i, \quad m_i = E(T_i).$$

From this theorem it follows that for small ε instead of a semi-Markov model with a state space $S = S_1 \cup S_2 \cup, \ldots, \cup S_m$ there can be considered the approximate Markov model with a set of states $E = \{1, 2, \ldots, m\}$. The Markov process with states representing the aggregated states of the SMP is much easier for calculations and analysis as a reliability model.

It should be noted that significant generalizations of the above-presented result can be found in Refs. [60, 61, 65].

4.7 Remarks on advanced perturbed Semi-Markov processes

The concepts of perturbed SMPs presented in previous sections belong to the simplest ones. In books by Korolyuk and Turbin [60, 61], Korolyuk and Limnios [65], Korolyuk and Swishchuk [63], and Gyllenberg and Silvestrov [40] and also in the paper [91] more advanced concepts of the perturbed SMPs are presented. In Ref. [60], different types of perturbed discrete state SMPs are considered and phase merging (phase aggregation) of those kinds of processes are also presented. In Refs. [61, 63, 65], one can find some applications of the perturbation theory to the limit theorems for characteristics of SMPs with a general state space. Furthermore, the phase merging of the general SMP is developed in those monographs.

The book by Gyllenberg and Silvestrov [40] is devoted to methods of the asymptotic analysis of nonlinearly perturbed stochastic processes based on some types of asymptotic expansions for perturbed renewal equation for regenerative processes, SMPs, and Markov chains. In this book, moreover, we can find applications for the analysis of quasi-stationary phenomena in nonlinearly perturbed queueing systems, population dynamics and epidemic models, and risk processes.

Stochastic processes associated with the SM process

Abstract

Random processes determined by the characteristics of the semi-Markov process are considered in this chapter. First is a renewal process generated by return times of a given state. The systems of equations for the distribution and expectation of them have been derived. The limit theorem for the process is formulated by adopting a theorem of the renewal theory. The limiting properties of the alternating process and integral functionals of the semi-Markov process are presented in this chapter. The chapter contains illustrative examples.

Keywords: Renewal process, Alternating process, Cumulative process, Integral functionals of the semi-Markov process

5.1   The renewal process generated by return times

Suppose that $\{X(t) : t \geqslant 0\}$ is a semi-Markov process determined by the kernel $Q(t)$, $t \geqslant 0$, where S is discrete state space and let $\{X(\tau_n) : n \in \mathbb{N}_0\}$ be the embedded Markov chain of this process. Assume that a state $j \in S$ is strongly recurrent and strongly accessible from the state $i \in S$. Consider the sequence of random variables

$$\Theta_{ij}^{(1)}, \Theta_{jj}^{(2)}, \Theta_{jj}^{(3)}, \ldots, \tag{5.1}$$

where $\Theta_{ij}^{(1)} = \Theta_{ij}$ represents a first passage time from the state i to the state j while random variables $\Theta_{jj}^{(n)}$, $n = 2, 3, \ldots$ form the sequence of return times to the state j. The function

$$\Phi_{ij}(t) = P(\Theta_{ij}^{(1)} \leqslant t), \quad t \geqslant 0 \tag{5.2}$$

is the CDF of the random variable $\Theta_{ij}^{(1)}$ while

$$\Phi_{jj}(t) = P(\Theta_{jj}^{(n)} \leqslant t) = P(\Theta_{jj} \leqslant t), \quad t \geqslant 0$$

is the CDF of the random variables $\Theta_{jj}^{(n)}, \ldots, n = 2, 3, \ldots$. It can be proved that the random variables

$$\Theta_{ij}^{(1)}, \Theta_{jj}^{(2)}, \ldots, \Theta_{jj}^{(n)}, \quad n = 2, 3, \ldots$$

are mutually independent.

Semi-Markov Processes: Applications in System Reliability and Maintenance. http://dx.doi.org/10.1016/B978-0-12-800518-7.00005-3

Let

$$S_{ij}^{(n)} = \Theta_{ij}^{(1)} + \Theta_{jj}^{(2)} + \cdots + \Theta_{jj}^{(n)}, \quad n = 2, 3, \ldots \tag{5.3}$$

and

$$V_{ij}(t) = \sum_{n=1}^{\infty} I_{[0,t]}(S_{ij}^{(n)}), \quad t \geqslant 0, \tag{5.4}$$

where

$$I_{[0,t]}(S_{ij}^{(n)}) = \begin{cases} 1 & \text{if } S_{ij}^{(n)} \in [0,t] \\ 0 & \text{if } S_{ij}^{(n)} \notin [0,t] \end{cases}.$$

The value of a random variable $V_{ij}(t)$ denotes the number of visits of the SM process in the state $j \in S$ on an interval $[0,t]$ if the initial state is $i \in S$. The stochastic process $\{V_{ij}(t) : t \in \mathbb{R}_+\}$ is called *the renewal process generated by the first passage time and return times of the SM process.*

5.1.1 Characteristics and parameters

Let

$$W_{ij}(t,n) = P(V_{ij}(t) = n), \quad n \in \mathbb{N}_0 \tag{5.5}$$

be the probability distribution of the process $\{V_{ij}(t) : t \in \mathbb{R}_+\}$. From the equality of events

$$\{V_{ij}(t) = 0\} = \{\Theta_{ij} > t\}$$

we get

$$W_{ij}(t,0) = P(V_{ij}(t) = 0) = P(\Theta_{ij} > t) = 1 - \Phi_{ij}(t). \tag{5.6}$$

The SM process, which starts from the state i, will take the value j n times in an interval $[0,t]$, if at the moment $x \in [0,t]$ a first passage to state j takes place, and in the interval $(x,t]$ the SM process achieves state j $n-1$ times. This remark, together with the properties of the SM process, lead to equations

$$W_{ij}(t,n) = \int_0^t W_{jj}(t-x, n-1) \mathrm{d}\Phi_{ij}(x), \quad n \in \mathbb{N}, \; i,j \in S. \tag{5.7}$$

For $n \in \mathbb{N}_0$, we can write both those equations as

$$W_{ij}(t,n) = \delta_{n0}\left[1 - \Phi_{ij}(t)\right] + [1 - \delta_{n0}] \int_0^t W_{jj}(t-x, n-1) \mathrm{d}\Phi_{ij}(x). \tag{5.8}$$

Those equations were presented by R. Howard [45]. Under the assumption of Theorem 3.4 the system of equations has a unique solution. We can obtain the solution using a transformation [45] given by

$$\tilde{W}_{ij}(s, z) = \sum_{n=0}^{\infty} z^n \int_0^{\infty} W_{ij}(t, m) e^{-st} dt. \tag{5.9}$$

Applying this transformation to (5.8), we get the system of equations for transforms

$$\tilde{W}_{ij}(s, z) = \frac{1 - \tilde{\varphi}_{ij}(s)}{s} + z \tilde{\varphi}_{ij}(s) \tilde{W}_{jj}(s, z), \quad i, j \in S. \tag{5.10}$$

Particularly, for $i = j$

$$\tilde{W}_{jj}(s, z) = \frac{1 - \tilde{\varphi}_{jj}(s)}{s} + z \tilde{\varphi}_{jj}(s) \tilde{W}_{jj}(s, z), \quad j \in S. \tag{5.11}$$

Hence,

$$\tilde{W}_{jj}(s, z) = \frac{1 - \tilde{\varphi}_{jj}(s)}{s\left[1 - z\tilde{\varphi}_{jj}(s)\right]}, \quad j \in S. \tag{5.12}$$

Formula (5.11) takes the following form:

$$\tilde{W}_{ij}(s, z) = \frac{1 - \tilde{\varphi}_{ij}(s)}{s} + z \tilde{\varphi}_{ij}(s) \frac{1 - \tilde{\varphi}_{jj}(s)}{s\left[1 - z\tilde{\varphi}_{jj}(s)\right]}, \quad i, j \in S. \tag{5.13}$$

We obtain the Laplace transform $\tilde{H}_{ij}(s)$ of the expectation $H_{ij}(t) = E\left[V_{ij}(t)\right]$ using the formula

$$\tilde{H}_{ij}(s) = \frac{\partial \tilde{W}_{ij}(s, z)}{\partial z}\Big|_{z=1}. \tag{5.14}$$

After calculation, we get

$$\tilde{H}_{ij}(s) = \frac{\tilde{\varphi}_{ij}(s)}{s\left(1 - \tilde{\varphi}_{jj}(s)\right)}. \tag{5.15}$$

The expectation $H_{ij}(t) = E\left[V_{ij}(t)\right]$ corresponds to the renewal function in the renewal theory. We calculate the Laplace transform of the second moment of the stochastic process $\{V_{ij}(t) : t \in \mathbb{R}_+\}$ applying the formula

$$\mathcal{L}\left[E\left[(V_{ij}(t))^2\right]\right] = \frac{\partial^2 \tilde{W}_{ij}(s, z)}{\partial z^2}\Big|_{z=1} + \frac{\partial \tilde{W}_{ij}(s, z)}{\partial z}\Big|_{z=1}. \tag{5.16}$$

Making a simple calculation, we have

$$\mathcal{L}\left[E\left[(V_{ij}(t))^2\right]\right] = \frac{2\tilde{\varphi}_{ij}(s)\tilde{\varphi}_{jj}(s)}{s\left(1 - \tilde{\varphi}_{jj}(s)\right)^2} + \frac{\tilde{\varphi}_{ij}(s)}{s(1 - \tilde{\varphi}_{jj}(s))}. \tag{5.17}$$

Applying properties of the Laplace transform, we obtain

$$E\left[(V_{ij}(t))^2\right] = 2\int_0^t H_{ij}(t - x) dH_{jj}(x) + H_{ij}(t). \tag{5.18}$$

Consequently, the variance is

$$D^2\left[V_{ij}(t)\right] = 2\int_0^t H_{ij}(t-x)\mathrm{d}H_{jj}(x) + H_{ij}(t) - H_{ij}^2(t). \tag{5.19}$$

5.2 Limiting distribution of the process

Adopting theorems of the renewal theory we can formulate a limit theorem concerning the stochastic process $\{V_{ij}(t) : t \in \mathbb{R}_+\}$ as $t \to \infty$. Let

$$E(\Theta_{ij}) = a_{ij}, \quad E(\Theta_{jj}) = a_{jj}, \quad V(\Theta_{ij}) = \sigma_{ij}^2, \quad V(\Theta_{jj}) = \sigma_{jj}^2.$$

Theorem 5.1 (Grabski [30]). If

- j is a positive recurrent state,
- j is strongly accessible from the state $i \in S$,
- the random variables that are independent copies of the random variable Θ_{jj} have the finite positive expected values and variances,

then

$$\lim_{t\to\infty} P\left(\frac{V_{ij}(t) - \frac{t - E(\Theta_{ij}) + E(\Theta_{jj})}{E(\Theta_{jj})}}{\sqrt{\frac{V(\Theta_{jj})t}{[E(\Theta_{jj})]^3}}} \leqslant x\right) = \frac{1}{\sqrt{2\pi}}\int_{-\infty}^x e^{u^2/2}\mathrm{d}u. \tag{5.20}$$

Proof: [30].

Example 5.1 (Reliability model from Section 3.5). We will explain and illustrate the concepts presented above using the example presented in Sections 3.5 and 3.5.4. Recall that a reliability model of an object is the semi-Markov process $\{X(t) : t \geqslant 0\}$ with the five elements state space $S = \{1, 2, 3, 4, 5\}$, where

1 denotes state of waiting for the tasks performing,
2 means performing of the task 1.

The kernel of the process is given by (3.116). Let $i = 1$ and $j = 2$. We want to calculate the necessary parameters of the process $\{V_{12}(t) : t \geqslant 0\}$ approximate distribution. Recall that a value of the process represents a number of the semi-Markov process visits in the state 2 in the interval $[0, t]$ if the initial state is 1. From Theorem 5.1 it follows that we have to calculate the expectations $E(\Theta_{12})$, $E(\Theta_{22})$ and the variance $V(\Theta_{22})$. To obtain these parameters, we apply (3.25), (3.26), (3.30), and (3.31). In this case, (3.25) takes the form

$$\begin{bmatrix} 1 & -q & 0 & 0 \\ -p_{21} & 1 & 0 & p_{35} \\ -1 & 0 & 1 & 0 \\ -1 & 0 & 0 & 1 \end{bmatrix} \begin{bmatrix} E(\Theta_{12}) \\ E(\Theta_{32}) \\ E(\Theta_{42}) \\ E(\Theta_{52}) \end{bmatrix} = \begin{bmatrix} E(T_1) \\ E(T_3) \\ E(T_4) \\ E(T5) \end{bmatrix}. \tag{5.21}$$

Using the procedure LinearSolve [a, c] in the MATHEMATICA program, for parameters presented in Section 3.5.4 we obtain

$$E(\Theta_{12}) = 9.60, \quad E(\Theta_{32}) = 17.76, \quad E(\Theta_{42}) = 22.94, \quad E(\Theta_{52}) = 19.60. \tag{5.22}$$

We get the expectation $E(\Theta_{22})$ using the equality (3.30), which in this case is

$$E(\Theta_{22}) = E(T_2) + p_{21}E(\Theta_{12}) + p_{24}E(\Theta_{42}). \tag{5.23}$$

Substituting suitable numbers from Section 3.5.4 we obtain the expected value $E(\Theta_{22}) = 26.37$. In a similar way, using (3.26) and (3.31), the second moment is $E(\Theta_{22}^2) = 830.73$. Thus, the variance is $V(\Theta_{22}) = 135.44$. From Theorem 5.1 it follows that the renewal process $\{V_{12}(t) : t \geqslant 0\}$ is approximately normally distributed with an expectation

$$E[V_{12}(t)] \approx \frac{t - E(\Theta_{12}) + E(\Theta_{22})}{E(\Theta_{22})} \tag{5.24}$$

and a standard deviation

$$\sigma[V_{12}(t)] \approx \sqrt{\frac{V(\Theta_{22})t}{[E(\Theta_{22})]^3}}. \tag{5.25}$$

Substituting the numerical values, we get

$$E[V_{12}(t)] \approx \frac{t + 16.77}{26.37}, \quad \sigma[V_{12}(t)] \approx 0.0859\sqrt{t}, \tag{5.26}$$

which represents the mean number and the standard deviation of performing task 1 during the time t if 1 is the initial state. For example, if $t = 1000$ [h], then $E[V_{12}(1000)] \approx 38.56$ [h] and $\sigma[V_{12}(1000)] = 2.72$ [h]. A probability that the number of performing task 1 during this time belongs to the interval [36, 44] is

$$P(36 \leqslant V_{12}(1000) \leqslant 44) \approx \Phi(2) + \Phi(0.941) - 1 = 0.8039.$$

The symbol $\Phi(\cdot)$ used above denotes CDF of a normal standard distribution.

5.3 Additive functionals of the alternating process

In the cumulative working time of a machine, wearing out an engine in period $[0, t]$ is an example of quantities that can be described by a so-called stochastic cumulative processes. We investigate only cumulative processes connected to semi-Markov processes. First we take under consideration a cumulative process defined by an alternating process. An example of that kind of process is a cumulative time of work of an machine during a period $[0, t]$.

Consider a sequence of mutually independent random variables $\zeta_1, \gamma_1, \zeta_2, \gamma_2 \ldots$. Assume that random variables $\zeta_1, \zeta_2, \ldots$ are copies of the random variable ζ with CDF $F_\zeta(t)$, $t \geqslant 0$ and random variables $\gamma_1, \gamma_2, \ldots$ are copies of the random variable

γ having CDF $F_\gamma(t)$. For convenience, the random variable ζ will represent the length of a working period and γ will denote the length of a service period of an object.

Let $\{(\xi_n, \vartheta_n) : n \in \mathbb{N}_0\}$ be a sequence of two-dimensional random variables defined by the following equalities:

$$\xi_n = \begin{cases} 1 & \text{for } n = 0, 2, 4, \ldots \\ 0 & \text{for } n = 1, 3, 5, \ldots \end{cases} \qquad \vartheta_n = \begin{cases} \zeta_n & \text{for } n = 1, 3, 5, \ldots \\ \gamma_n & \text{for } n = 2, 4, 6, \ldots \end{cases} \tag{5.27}$$

The sequence $\{(\xi_n, \vartheta_n) : n \in \mathbb{N}_0\}$ is a Markov renewal process with a kernel

$$Q(t) = \begin{bmatrix} 0 & F_\gamma(t) \\ F_\zeta(t) & 0 \end{bmatrix}. \tag{5.28}$$

That Markov renewal process generates the SM process $\{X(t) : t \geqslant 0\}$ with a state space $S = \{0, 1\}$ and an initial distribution

$$p = [0, 1]. \tag{5.29}$$

This process is called a *simple alternating process*. The stochastic process $\{K(t) : t \geqslant 0\}$ given by the formula

$$K(t) = \int_0^t X(u)\mathrm{d}u \tag{5.30}$$

means a global (cumulative) time of the object work in interval $[0, t]$. The process $\{K(t) : t \geqslant 0\}$ is an example of the *additive functional of the alternating process*. A trajectory of the process is shown in Figure 5.1.

Notice that for each trajectory $x(\cdot)$ of the alternating process, a value of the process $\{K(t) : t \geqslant 0\}$ trajectory at the instant t_0 is equal to the area of the shaded region $D = \{(t, x) : 0 \leqslant t < t_0, \quad 0 \leqslant x \leqslant x(t)\}$ (see Figure 5.2).

This remark allows us to notice that

$$K(t) = \begin{cases} t & \text{for } t \in [0, \tau_1) \\ \zeta_1 + \zeta_2 + \cdots + \zeta_k & \text{for } t \in [\tau_{2k-1}, \tau_{2k}) \\ t - (\gamma_1 + \gamma_2 + \cdots + \gamma_k) & \text{for } t \in [\tau_{2k}, \tau_{2k+1}) \end{cases} \tag{5.31}$$

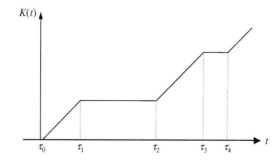

Figure 5.1 Trajectory of the process $\{K(t) : t \geqslant 0\}$.

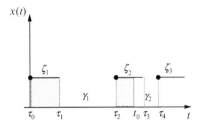

Figure 5.2 Trajectory of the process $\{X(t) : t \geqslant 0\}$.

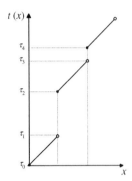

Figure 5.3 Trajectory of the process $\{T(x) : x \geqslant 0\}$.

where

$$\tau_{2k-1} = \zeta_1 + \gamma_1 + \zeta_2 + \gamma_2 + \cdots + \zeta_k$$
$$\tau_{2k} = \zeta_1 + \gamma_1 + \zeta_2 + \gamma_2 + \cdots + \zeta_k + \gamma_k \qquad (5.32)$$
$$\tau_{2k+1} = \zeta_1 + \gamma_1 + \zeta_2 + \gamma_2 + \cdots + \zeta_k + \gamma_k + \zeta_{k+1}$$
$$k = 1, 2, \ldots. \qquad (5.33)$$

If the random variable ζ_1 has a different distribution than the other random variables ζ_k, $k = 2, 3, \ldots$, then a stochastic process $\{X(t) : t \geqslant 0\}$ given by formula

$$X(t) = \xi_n \quad \text{dla } t \in [\tau_n, \tau_{n+1}), \quad \text{where } \tau_n = \vartheta_1 + \ldots + \vartheta_n$$

is called *the general alternating process*. This stochastic process enables us to define a random process $\{T(x) : x \geqslant 0\}$ in the following way:

$$T(x) = \inf\{t : K(t) > x\}. \qquad (5.34)$$

For fixed x, the random variable $T(x)$ denotes the moment that a random variable $K(t)$ achieves the level x. A trajectory of the random process $\{T(x) : x \geqslant 0\}$, which is shown in Figure 5.3, corresponds to the trajectory of the cumulative process that is shown in Figure 5.1. The trajectory of that process is the right continuous function.

An analysis of the process $\{T(x) : x \geqslant 0\}$ trajectory enables us to notice an important equality

$$T(x) = x + \gamma_0 + \gamma_1 + \gamma_2 + \cdots + \gamma_{N_\zeta(x)}, \tag{5.35}$$

where

$$N_\zeta(x) = \begin{cases} 0 & \text{dla } x \in [0, \zeta_1) \\ n & \text{dla } x \in [\zeta_1 + \zeta_2 + \cdots + \zeta_n, \ \zeta_1 + \zeta_2 + \cdots + \zeta_n + \zeta_{n+1}) \end{cases}$$

while $\gamma_0 = 0$ with probability 1. Hence, the cumulative distribution function of the random variable γ_0 is

$$F_{\gamma_0}(u) = P(\gamma_0 \leqslant u) = \begin{cases} 0 & \text{for } x \in (-\infty, 0) \\ 1 & \text{for } x \in [0, \infty) \end{cases}$$

A random process $\{N_\zeta(x) : x \geqslant 0\}$ is a renewal process with right continuous trajectories generated by a sequence of independent random variables $\zeta_0, \zeta_1, \ldots$. The equality (5.35) makes it easier to find distributions of the above-considered random processes.

Theorem 5.2. The CDF of a random process $\{T(x) : x \geqslant 0\}$ is

$$P(T(x) \leqslant t) = \sum_{n=0}^{\infty} F_\gamma^{(n)}(t - x) \left[F_\zeta^{(n)}(x) - F_\zeta^{(n+1)}(x) \right], \tag{5.36}$$

while the CDF of a stochastic process $\{K(t) : t \geqslant 0\}$ is given by

$$P(K(t) \leqslant x) = 1 - \sum_{n=0}^{\infty} F_\gamma^{(n)}(t - x) \left[F_\zeta^{(n)}(x) - F_\zeta^{(n+1)}(x) \right]. \tag{5.37}$$

We will present a theorem that allows us to evaluate the expected value and the variance of the random process $\{T(x) : x \geqslant 0\}$.

Theorem 5.3 (Kopociński [58]). If the probability distributions of the random variables ζ and γ have the finite positive expected values and variances, then there exists the finite positive expectation and variance of the random process $\{T(x) : x \geqslant 0\}$ and they are given by

$$E[T(x)] = x + E(\gamma)E[N_\zeta(x)], \tag{5.38}$$

$$V[T(x)] = [E(\gamma)]^2 V[N_\zeta(x)] + V(\gamma)E[N_\zeta(x)]. \tag{5.39}$$

$\square$

Proof: [32].

An evaluation of the renewal function is known [28].

$$E[N_\zeta(x)] = \frac{x}{E(\zeta)} + \frac{V(\zeta)}{2[E(\zeta)]^2} - \frac{1}{2} + o(1), \quad x \to \infty. \tag{5.40}$$

Substituting this equality to (5.38), we get the following proposition.

Proposition 5.1. For the aperiodic renewal process $\{N_\zeta(x) : x \geqslant 0\}$ the following equalities are fulfilled:

$$E[T(x)] = \frac{E(\gamma) + E(\zeta)}{E(\gamma)}x + E(\gamma)\frac{V(\zeta) - [E(\zeta)]^2}{2[E(\zeta)]^2} + o(1), \quad x \to \infty. \quad (5.41)$$

Theorem 5.4. If the probability distributions of the random variables ζ and γ have the finite positive expected values and variances, then $\{K(t) : t \geqslant 0\}$ has the asymptotically normal distribution

$$\lim_{t \to \infty} P\left(\frac{K(t) - m(t)}{\sigma(t)} \leqslant x\right) = \frac{1}{\sqrt{2\pi}}\int_{-\infty}^x e^{-u^2/2}du \quad (5.42)$$

where

$$m(t) = \frac{E(\zeta)}{E(\zeta) + E(\gamma)}t, \quad (5.43)$$

$$\sigma(t) = \sqrt{\frac{V(\zeta)[E(\gamma)]^2 + [V(\gamma)][E(\zeta)]^2}{[E(\zeta) + E(\gamma)]^3}}t. \quad (5.44)$$

$\square$

Proof: [28, 30].

Therefore, the stochastic process $\{K(t) : t \geqslant 0\}$ is asymptotically normal distributed. It means that the process has approximately normal distribution $N(m(t), \sigma(t))$ for large t.

As we know, a stochastic process $\{K(t) : t \geqslant 0\}$ is linked with a random process $\{T(x) : x \geqslant 0\}$ given by

$$T(x) = \inf\{t : K(t) > x\}. \quad (5.45)$$

At a particular number x, the random variable $T(x)$ denotes an instant of exceeding a value x by the random variable $K(t)$.

Theorem 5.5. Under assumptions of the previous theorem,

$$\lim_{t \to \infty} P\left(\frac{T(x) - m(x)}{\sigma(x)} \leqslant y\right) = \frac{1}{\sqrt{2\pi}}\int_{-\infty}^y e^{-u^2/2}du \quad (5.46)$$

where

$$m(x) = \frac{E(\zeta) + E(\gamma)}{E(\zeta)}x, \quad (5.47)$$

$$\sigma(x) = \sqrt{\frac{V(\zeta)[E(\gamma)]^2 + [V(\gamma)][E(\zeta)]^2}{[E(\zeta)]^3}}x. \quad (5.48)$$

Proof: [28, 30].

Example 5.2. The time between overhaul (TBO) on the type of AE300 diesel engines used in airplanes Diamond DA 40 and DA 42 is equal to 1500 h. At the same time, the life of engine components such as the EECU (Electronic Engine Control Unit) and gear that have to be replaced during operation have been equated with full engine TBO. This means that after $c = 1500$ h of work, a plane engine is regarded to be unable to use. We suppose that a sojourn time of one fly is a random variable ζ

with an expectation $E(\zeta) = 2.26$ [h] and a variance $V(\zeta) = 1.21$ $[\text{h}^2]$. A time of each plane stoppage is a positive random variable γ with the expectation $E(\gamma) = 5.20$ [h] and the variance $V(\gamma) = 3.24$ $[\text{h}^2]$. Under that assumption, the alternating process $\{X(t) : t \geqslant 0\}$ defined by the kernel (5.28) and the initial distribution (5.29) is a reliability model of the plane engine operation process. A random variable

$$T(c) = \inf\{t : K(t) > c\},$$

denotes an instant of exceeding a level c by the summary sojourn time $K(t) = \int_0^t X(u)du$ of the alternating process. From the above-presented theorem it follows that the random variable $T(c)$ has an approximately normal distribution $N(m(c), \sigma(c))$, where

$$m(c) = \frac{E(\zeta) + E(\gamma)}{E(\zeta)} c$$

and

$$\sigma(c) = \sqrt{\frac{V(\zeta)\,[E(\gamma)]^2 + [V(\gamma)]\,[E(\zeta)]^2}{[E(\zeta)]^3}} c. \qquad (5.49)$$

The random variable $T(c)$ can be treated as the lifetime of the plane engine. An approximation of the reliability function $R(t)$, $t \geqslant 0$ takes the form

$$R(t) = P(T(c) > t) \approx 1 - \Phi\left(\frac{t - m(c)}{\sigma(c)}\right) \qquad (5.50)$$

where $\Phi(\cdot)$ is CDF of the standard normal distribution. For the above-assumed parameters we obtain the expectation $m(1500) = 4951.33$ [h] ≈ 206.3 $[\text{days}]$ and the standard deviation $\sigma(1500) = 80.01$ [h] ≈ 3.33 $[\text{days}]$, for a lifetime of the plane engine.

The approximate reliability function for the above-presented parameters is given by the rule

$$R(t) \approx 1 - \Phi\left(\frac{t - 4951.33}{80.01}\right).$$

$$\triangle$$

5.4 Additive functionals of the Semi-Markov process

The above-presented concepts of the additive functionals of an alternating process can be extended to the kind of functionals of a semi-Markov process. Let $\{X(t) : t \geqslant 0\}$ be a SM process determined by the continuous type kernel (in sense of Definition 2.2). Let

$$I(X(u) = j | X(0) = i) = \begin{cases} 1 & \text{for } X(u) = j & \text{if } X(0) = i \\ 0 & \text{for } X(u) = k \neq j & \text{if } X(0) = i. \end{cases}$$

The stochastic process $\{K_{ij}(t) : t \geqslant 0\}$, defined as

$$K_{ij}(t) = \int_0^t I(X(u) = j|X(0) = i)\mathrm{d}u, \tag{5.51}$$

denotes a global (cumulative) sojourn time of the state j in interval $[0, t]$ if an initial state is i. Note, that

$$E\left[\int_0^t I(X(u) = j|X(0) = i)\mathrm{d}u\right] = \int_0^t E[I(X(u) = j|X(0) = i)]\,\mathrm{d}u$$

$$= \int_0^t P(X(u) = j|X(0) = i)\mathrm{d}u = \int_0^t P_{ij}(u)\mathrm{d}u$$

From definitions it follows that the process $\{K_{ij}(t) : t \geqslant 0\}$, defined by the SM process, is connected with the process $\{K(t) : t \geqslant 0\}$ generated by the alternating process. Suppose that ζ_n, $n = 2,\ldots$ represent the independent, identical distributed random variables with CDF

$$G_j(t) = P(\zeta_n \leqslant t) = P(\zeta \leqslant t) = P(T_j \leqslant t),$$

which denote the consecutive waiting times of a state j, while random variables γ_n, $n = 1, 2, \ldots$ are lengths of time intervals from the moment of nth exit of a state j to the instant of the next entrance to the same state. Then the definition of the process $\{K_{ij}(t) : t \geqslant 0\}$ which starts from i is almost identical to the definition of the process $\{K(t) : t \geqslant 0\}$. The significant difference is that the random variables ζ_n, γ_n, $n \in \mathbb{N}$ can be dependent. However, we can extract the class of SM processes for which these random variables are independent.

Theorem 5.6. If $\{X(t) : t \geqslant 0\}$ is a SM process defined by the kernel $Q(t) = \left[Q_{ij}(t) : i, j \in S\right]$ such that

$$Q_{ij}(t) = p_{ij}G_i(t), \quad \text{for } i, j \in S, \tag{5.52}$$

then random variables ζ_n, γ_n, $n \in \mathbb{N}$ are independent and

$$\Theta_{ii}^{(n)} = \zeta_n + \gamma_n. \tag{5.53}$$

$$\triangle$$

Proof: [32].

Korolyuk and Turbin in Ref. [60] have shown that each semi-Markov process $\{X(t) : t \geqslant 0\}$ with the kernel $Q(t) = \left[Q_{ij}(t) : i, j \in S\right]$, $Q_{ij}(t) = p_{ij}F_{ij}(t)$ can be replaced by the semi-Markov process $\{\hat{X}(t) : t \geqslant 0\}$ with the state space $E = S \times S$ and the kernel $\hat{Q}(t) = \left[\hat{Q}_{kr}(t) : k, r \in E\right]$, where $k = (i, j)$, $r = (l, m)$ and

$$\hat{Q}_{kr}(t) = \hat{Q}_{(i,j)(l,m)}(t) = \begin{cases} \delta_{jl}Q_{lm}(\infty)\dfrac{Q_{ij}(t)}{Q_{ij}(\infty)}, & Q_{ij}(t) > 0, \\ 0, & Q_{ij}(\infty) = 0. \end{cases} \tag{5.54}$$

Note that

$$\hat{Q}_{kr}(t) = \hat{Q}_{(i,j)(l,m)}(t) = P(\hat{\xi}_{n+1} = (l,m), \ \hat{\vartheta}_{n+1} \leqslant t | \hat{\xi}_n = (i,j)). \tag{5.55}$$

If $Q_{lm}(\infty) > 0, Q_{ij}(\infty) > 0$ for $j = l$ then, from (5.54) and (5.55), we get

$$P(\hat{\vartheta}_{n+1} \leqslant t | \hat{\xi}_{n+1} = (l,m), \ \hat{\xi}_n = (i,j)) = \hat{Q}_{kr}(t) = \hat{Q}_{(i,j)(l,m)}(t)$$

$$= \frac{Q_{ij}(t)}{Q_{ij}(\infty)} = F_{ij}(t)$$

$$= P(\hat{\vartheta}_{n+1} \leqslant t | \hat{\xi}_n = (i,j)) = \hat{G}_k(t). \tag{5.56}$$

Therefore, the kernel of the semi-Markov process $\{\hat{X}(t) : t \geqslant 0\}$ is

$$\hat{Q}(t) = \left[\hat{Q}_{kr}(t) : k, r \in E \right], \tag{5.57}$$

where $k = (i,j), r = (l,m), \hat{Q}_{kr}(t) = \hat{p}_{kr}\hat{G}_k(t)$, and

$$\hat{p}_{kr} = \hat{Q}_{(i,j)(l,m)}(\infty) = \begin{cases} \delta_{jl}Q_{lm}(\infty), & Q_{ij}(\infty) > 0, \\ 0, & Q_{ij}(\infty) = 0. \end{cases} \tag{5.58}$$

A disadvantage of this procedure lies in the fact that a state space of the semi-Markov process $\{\hat{X}(t) : t \geqslant 0\}$ is extended. Instead of the SM process with a state space S, we have to analyze the semi-Markov process with a set of states $E = S \times S$.

There is a primitive, approximate method of substitution SM process with the kernel $Q(t) = \left[p_{ij}F_{ij}(t) : i,j \right]$ by the SM process $\{\tilde{X}(t) : t \geqslant 0\}$ with the kernel $\tilde{Q}(t) = \left[p_{ij}\tilde{G}_i(t) : i,j \in S \right]$, where

$$\tilde{G}_i(t) = G_i(t) = \sum_{j \in S} p_{ij}F_{ij}(t).$$

The processes $\{\tilde{X}(t) : t \geqslant 0\}$ and $\{X(t) : t \geqslant 0\}$ have the identical transition probabilities of an embedded Markov chain, the identical distributions of all waiting times, and the same limit distributions.

We present the limit theorem concerning distribution of the global (cumulative) sojourn time of the state j in interval $[0, t]$ if an initial state is i.

Theorem 5.7 (Taga [79] and Grabski [30]). Let $\{X(t) : t \geqslant 0\}$ be a semi-Markov process determined by continuous type kernel $Q(t) = \left[Q_{ij}(t) : i,j \in S \right]$ such that

$$Q_{ij}(t) = p_{ij}G_i(t), \quad \text{where } p_{ii} = 0 \quad \text{for } i \in S. \tag{5.59}$$

Moreover, if the random variables T_j, Θ_{jj} have the positive finite expected values and variances, then

$$\lim_{t \to \infty} P\left(\frac{K_{ij}(t) - m_j(t)}{\sigma_j(t)} \leqslant x \right) = \frac{1}{\sqrt{2\pi}} \int_{-\infty}^{x} e^{-u^2/2}du \tag{5.60}$$

where

$$m_j(t) = P_j t = \frac{E(T_j)}{E(\Theta_{jj})} t \tag{5.61}$$

$$\sigma_j(t) = \sqrt{\frac{V(T_j)\left[E(\Theta_{jj}) - E(T_j)\right]^2 + \left[V(\Theta_{jj}) - V(T_j)\right]\left[E(T_j)\right]^2}{\left[E(\Theta_{jj})\right]^3} t}. \tag{5.62}$$

Proof: [30, 79].

From the theorem it follows that the total time spending in the state j by the process SM until a moment t has the approximate normal distribution $N(m_j(t), \sigma_j(t))$. We can get the expected value and the variance of the random variable Θ_{jj} using (3.30) and (3.31).

The stochastic process $\{K_{ij}(t) : t \geqslant 0\}$ is linked to the random process $\{T_{ij}(x) : x \geqslant 0\}$ given by

$$T_{ij}(x) = \inf\{t : K_{ij}(t) > x\}. \tag{5.63}$$

For a fixed x, the random variable $T_{ij}(x)$ represents the moment of exceeding a level x by a summary sojourn time of the state j of a SMP until an instant t if the initial state is i.

Theorem 5.8. Under the assumptions of Theorem 5.7, we have

$$\lim_{t \to \infty} P\left(\frac{T_{ij}(x) - m_j(x)}{\sigma_j(x)} \leqslant y\right) = \frac{1}{\sqrt{2\pi}} \int_{-\infty}^{y} e^{-u^2/2} du \tag{5.64}$$

where

$$m_j(x) = \frac{1}{P_j} x = \frac{E(\Theta_{jj})}{E(T_j)} x, \tag{5.65}$$

$$\sigma_j(t) = \sqrt{\frac{V(T_j)\left[E(\Theta_{jj}) - E(T_j)\right]^2 + \left[V(\Theta_{jj}) - V(T_j)\right]\left[E(T_j)\right]^2}{\left[E(T_j)\right]^3} x}. \tag{5.66}$$

$\square$

Proof: [30].

We adopt the concepts and theorems for general semi-Markov processes presented by Limnios and Oprisan in [72] for the discrete space SM processes. Suppose that $\{X(t) : t \geqslant 0\}$ is a discrete space SM regular process with a continuous kernel. Assume that $h : S \to \mathbb{R}_+$ is any Borel function taking its values on a subset $U = h(S) \subset \mathbb{R}_+ = [0, \infty)$.

Definition 5.1. The stochastic process $\{L(t) : t \geqslant 0\}$, defined as

$$L(t) = \int_0^t h\left[X(u)\right] du, \tag{5.67}$$

is said to be the integral functional of the SM process or the cumulative process of the SM process $\{X(t) : t \geqslant 0\}$.

If $\{(\xi_n, \vartheta_n) : n \in \mathbb{N}_0\}$ is the MRP defining the SM process $\{X(t) : t \geqslant 0\}$, then the process $\{L(t) : t \geqslant 0\}$ is given by

$$L(t) = h(\xi_0)\vartheta_1 + \cdots + h(\xi_{n-1})\vartheta_n + h(\xi_n)(t - \tau_n) \quad \text{for } t \in [\tau_n, \tau_{n+1}). \quad (5.68)$$

This formula allows us to generate the trajectories of the process. Using the definition of the counting process, we obtain an equivalent form of formula (5.68):

$$L(t) = \sum_{k=1}^{N(t)} h(\xi_{k-1})\vartheta_k + (t - \tau_{N(t)})h(\xi_{N(t)}). \quad (5.69)$$

Consider the joint distribution of the processes $\{X(t) : t \geqslant 0\}$ and $\{L(t) : t \geqslant 0\}$. Let

$$U_{iA}(t, x) = P(X(t) \in A, L(t) \leqslant x | X(0) = i), \quad i \in S. \quad (5.70)$$

Theorem 5.9 (Limnios and Oprisan [72]). The functions $U_{i,A}(t, x), i \in S$ satisfy the system of integral equations

$$U_{iA}(t, x) = I_{A \times [0, x]}(i, h(i)t)[1 - G_i(t)]$$
$$+ \sum_{j \in S} \int_0^t U_{jA}(t - v, x - h(i)v) \mathrm{d}Q_{ij}(v), \quad i \in S \quad (5.71)$$

Proof: We obtain the theorem and its proof as a conclusion from the considerations presented in Ref. [72].

Letting $A = S$, we get

$$U_{iS}(t, x) = P\{L(t) \leqslant x | X(0) = i\}, \quad i \in S. \quad (5.72)$$

Proposition 5.2. The conditional distribution of the cumulative process satisfies the system of equations

$$U_{iS}(t, x) = I_{[0, x]}(h(i)t)[1 - G_i(t)] + \sum_{j \in S} \int_0^t U_{jS}(t - v, x - h(i)v) \mathrm{d}Q_{ij}(v). \quad (5.73)$$

If $x \to \infty$, we get an important conclusion.

Proposition 5.3. Let

$$P_{iA}(t) = U_{iA}(t, \infty) = P(X(t) \in A | X(0) = i), \quad i \in S. \quad (5.74)$$

The interval transition probabilities $P_{iA}(t)$, $i \in S$ satisfy the system of equations

$$P_{iA}(t) = I_A(i)[1 - G_i(t)] + \sum_{j \in S} \int_0^t P_{jA}(t - v) \mathrm{d}Q_{ij}(v), \quad i \in S. \quad (5.75)$$

Notice that for $A = \{j\}$ we obtain the system of equations for the interval transition probabilities $P_{ij}(t)$, which is presented in Chapter 3. Now $i \in A \subset S$. Then

$$\Upsilon_{iA}(t, x) = P(X(u) \in A, \quad \forall u \in [0, t], L(t) \leqslant x | X(0) = i). \quad (5.76)$$

Similarly, the following theorem can be proved.

Theorem 5.10. The functions $\Upsilon_{iA}(t, x)$, $i \in A \subset S$ satisfy the system of equations

$$\Upsilon_{iA}(t, x) = I_{A \times [0,x]}(i, h(i)t) [1 - G_i(t)]$$

$$+ \sum_{j \in A} \int_0^t \Upsilon_{jA}(t - v, x - h(i)v) \mathrm{d}Q_{ij}(v), \quad i \in A. \qquad (5.77)$$

$\triangle$

Note that for $x \to \infty$.

$$\Upsilon_{iA}(t, \infty) = \lim_{x \to \infty} \Upsilon_{iA}(t, x) = P(X(u) \in A, \; \forall u \in [0, t] \,|\, X(0) = i), \quad i \in A.$$

$$(5.78)$$

If the subset A represents "up states" of the system, then

$$R_i(t) = \Upsilon_{iA}(t, \infty), \quad i \in A \qquad (5.79)$$

is the reliability function of it under a condition that an initial state is $i \in A$.

Proposition 5.4. The conditional reliability functions fulfill the system of equations

$$R_i(t) = 1 - G_i(t) + \sum_{j \in A} \int_0^t R_j(t - u) \mathrm{d}Q_{ij}(u), \quad i \in A. \qquad (5.80)$$

If T represents the lifetime of the system, then

$$R_i(t) = P(T > t | X(0) = i), \quad i \in A \qquad (5.81)$$

and

$$R_i(t, s) = P(T > t + s | T > t, \; X(0) = i) = \frac{R_i(t + s)}{R_i(t)}, \quad i \in A. \qquad (5.82)$$

There are many papers and books where we can find advanced theorems concerning limit properties of additive functionals of semi-Markov processes. Among these books, note monographs [60, 72, 90] and papers [1, 2, 79].

SM models of renewable cold standby system

6

Abstract

The semi-Markov reliability model of two different units of renewable cold standby system and the SM model of a hospital electrical power system consisting of mains, an emergency power system, and the automatic transfer switch with the generator starter are discussed in this chapter. The renewable cold standby system with series N components exponential subsystems is also presented here. The embedded semi-Markov process concept is applied for description of the system evolution. In our models, time to failure of the system is represented by a random variable denoting the first passage time from the given state to the subset of states. The appropriate theorems of the semi-Markov processes theory allow us to evaluate the reliability function and some reliability characteristics. In case of difficulties in calculating an exact reliability function of the system by means of the Laplace transform, we propose applying the theorem of semi-Markov processes perturbation theory that enables us to get an approximate reliability function of the system. Some illustrative examples in this chapter help to explain presented concepts.

Keywords: Semi-Markov model, Renewable cold standby system, Semi-Markov perturbed process, Reliability function, Mean time to failure, Electrical power system

6.1 Two different units of cold standby system with switch

6.1.1 Introduction

The models presented here are the extensions of the models that have been considered by Barlow and Proshan [7], Brodi and Pogosian [13], Korolyuk and Turbin [60], and Grabski [30, 32]. We construct the so-called embedded semi-Markov process by defining the renewal kernel of that one. This method was presented in Ref. [13]. The time to failure of the system is described by a random variable that means the first passage time from the given state to the subset of states. Applying Theorem 3.1 we get L-S transforms of these random variables by solving the appropriate system of linear equations. Using Theorem 3.2 we calculate the mean time to failure of the system. Very often, calculating an exact reliability function of the system by using Laplace transform is a complicated matter, but there is a possibility to apply the theorem of the theory of the semi-Markov processes perturbation to obtain an approximate reliability function of the system. We use the Pavlov and Ushakov concept of the perturbed SM process, which is presented in Chapter 4.

Semi-Markov Processes: Applications in System Reliability and Maintenance. http://dx.doi.org/10.1016/B978-0-12-800518-7.00006-5

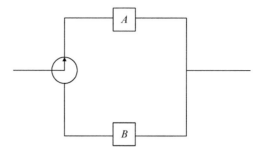

Figure 6.1 Diagram of the system.

6.1.2 Description and assumptions

We assume that the system consists of one operating unit A, the stand-by unit B, and a switch (Figure 6.1). We assume that a lifetime of a basic operating unit is represented by a random variable ζ_A, with distribution given by a probability density function (PDF) $f_A(x)$, $x \geqslant 0$. When the operating unit fails, the spare B is immediately put in motion by the switch. The failed unit is renewed by a single repair facility. A renewal time of a unit A is a random variable γ_A having distribution given by a cumulative distribution function (CDF) $H_A(x) = P(\gamma_A \leqslant x)$, $x \geqslant 0$. The lifetime of unit B is a random variable ζ_B, with PDF $f_B(x)$, $x \geqslant 0$. When unit B fails, unit A immediately starts to work by the switch (if it is "up") and unit B is repaired. A renewal time of unit B is a random variable γ_B having distribution given by the CDF $H_B(x) = P(\gamma_B \leqslant x)$, $x \geqslant 0$.

Let U be a random variable having a binary distribution

$$b(k) = P(U = k) = a^k (1 - a)^{1-k}, \quad k = 0, 1, 0 < a < 1,$$

where $U = 0$, if a switch is failed at the moment of the operating unit failure, and $U = 1$, if the switch works at that moment.

Failure of the system takes place when the operating unit fails and the component that has failed sooner is still not ready to work, or when both the operating subsystem and the switch have failed.

Moreover, we assume that all random variables mentioned above are independent.

6.1.3 Construction of Semi-Markov reliability model

To describe the reliability evolution of the system, we have to define the states and the renewal kernel. We introduce the following states:

    0—failure of the system,
    1—failure of unit A, unit B is working,
    2—failure of unit B, unit A is working,
    3—both operating and spare units are "up" and unit A is working.

We assume that 3 is an initial state.

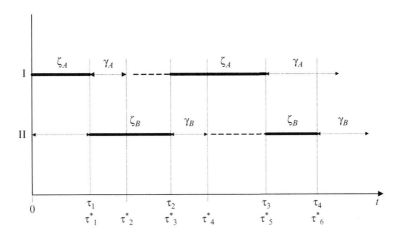

Figure 6.2 Reliability evolution of the standby system.

Figure 6.2 shows functioning of the system. Let $0 = \tau_0^*$, $\tau_1^*, \tau_2^*, \ldots$ denote the instants of the states changes, and $\{Y(t) : t \geqslant 0\}$ be a random process with the state space $S = \{0, 1, 2, 3\}$, which keeps constant values on the half-intervals $[\tau_n^*, \tau_{n+1}^*)$, $n = 0, 1, \ldots$ and it is right-hand continuous. This process is not semi-Markov, because a memoryless property is not satisfied for all instants of the state changes of it.

Let us construct a new random process in the following way. Let $0 = \tau_0$ and $\tau_1, \tau_2, \ldots$ denote instants of the unit failures or instants of the whole system failure. Reliability evolution of this standby system is shown in Figure 6.2.

The random process $\{X(t) : t \geqslant 0\}$, defined by equation

$$X(0) = 3, \quad X(t) = Y(\tau_n) \quad \text{for } t \in [\tau_n, \tau_{n+1}),\tag{6.1}$$

is the semi-Markov process. This process is called an *embedded semi-Markov process in the stochastic process* $\{Y(t) : t \geqslant 0\}$.

To determine a semi-Markov process as a model we have to define its initial distribution and all elements of its kernel. Recall that the semi-Markov kernel is the matrix of transition probabilities of the Markov renewal process

$$\mathbf{Q}(t) = \big[Q_{ij}(t) : i, j \in S\big],\tag{6.2}$$

where

$$Q_{ij}(t) = P(\tau_{n+1} - \tau_n \leqslant t, \, X(\tau_{n+1}) = j | X(\tau_n) = i), \quad t \geqslant 0.\tag{6.3}$$

Let us recall (Chapter 3) that the sequence $\{X(\tau_n) : n = 0, 1, \ldots\}$ is a homogeneous Markov chain with transition probabilities

$$p_{ij} = P(X(\tau_{n+1}) = j | X(\tau_n) = i) = \lim_{t \to \infty} Q_{ij}(t),\tag{6.4}$$

the function

$$G_i(t) = P(T_i \leqslant t) = P(\tau_{n+1} - \tau_n \leqslant t | X(\tau_n) = i) = \sum_{j \in S} Q_{ij}(t) \qquad (6.5)$$

is the CDF distribution of a waiting time T_i denoting the time spent in state i when the successor state is unknown, the function

$$F_{ij}(t) = P(\tau_{n+1} - \tau_n \leqslant t | X(\tau_n) = i, X(\tau_{n+1}) = j) = \frac{Q_{ij}(t)}{p_{ij}} \qquad (6.6)$$

is the CDF of a random variable T_{ij} that is called a holding time of a state i, if the next state will be j. We also know that

$$Q_{ij}(t) = p_{ij} F_{ij}(t). \qquad (6.7)$$

In this case, the semi-Markov kernel takes the form

$$\mathbf{Q}(t) = \begin{bmatrix} Q_{00}(t) & 0 & 0 & 0 & 0 \\ Q_{10}(t) & 0 & Q_{12}(t) & 0 \\ Q_{20}(t) & Q_{21}(t) & 0 & 0 \\ Q_{30}(t) & Q_{31}(t) & 0 & 0 \end{bmatrix}. \qquad (6.8)$$

From (6.3) and the assumptions we can calculate all elements of the semi-Markov kernel $\mathbf{Q}(t)$, $t \geqslant 0$. We start from the element $Q_{00}(t)$.

$$Q_{00}(t) = K(t), \qquad (6.9)$$

where $K(t)$ denotes an arbitrary CDF with a support $\mathbb{R}_+ = [0, \infty)$. An element $Q_{10}(t)$ of the second kernel row is

$$Q_{10}(t) = P(U = 0, \zeta_B \leqslant t) + P(U = 1, \zeta_B \leqslant t, \gamma_A > \zeta_B)$$

$$= (1-a)F_B(t) + a \iint_{C_{10}} f_B(x) \mathrm{d}x \mathrm{d}H_A(y),$$

where

$$C_{10} = \{(x, y) : x \leqslant t, y > x\}.$$

Hence,

$$Q_{10}(t) = (1-a)F_B(t) + a \int_0^t f_B(x) [1 - H_A(x)] \, \mathrm{d}x$$

$$= F_B(t) - a \int_0^t f_B(x) H_A(x) \mathrm{d}x. \qquad (6.10)$$

We obtain the transition probability from the state 1 to the state 2 during a time less or equal to t similarly:

$$Q_{12}(t) = P(U = 1, \zeta_B \leqslant t, \gamma_A < \zeta_B) = a \int_0^t f_B(x) H_A(x) \mathrm{d}x. \qquad (6.11)$$

The same way we get

$$Q_{20}(t) = F_A(t) - a \int_0^t f_A(x) H_B(x) dx,$$
$$Q_{21}(t) = P(U = 1, \zeta_A \leqslant t, \gamma_B < \zeta_A) = a \int_0^t f_A(x) H_B(x) dx,$$
$$Q_{31}(t) = P(U = 1, \zeta_A \leqslant t) = a F_A(t),$$
$$Q_{30}(t) = P(U = 0, \zeta_A \leqslant t) = (1 - a) F_A(t). \tag{6.12}$$

All elements of the kernel $\mathbf{Q}(t)$ have been defined, hence the semi-Markov process $\{X(t) : t \geqslant 0\}$ describing the reliability of the renewable cold standby system is constructed.

For all states we need to calculate the transition probabilities of the embedded Markov chain and also distributions of the waiting and holding times. The transition probabilities matrix of the embedded Markov chain $\{X(\tau_n) : n = 0, 1, \ldots\}$ is

$$\mathbf{P} = \begin{bmatrix} 1 & 0 & 0 & 0 \\ p_{10} & 0 & p_{12} & 0 \\ p_{20} & p_{21} & 0 & 0 \\ p_{30} & p_{31} & 0 & 0 \end{bmatrix} \tag{6.13}$$

where

$$p_{10} = 1 - a \int_0^\infty f_B(x) H_A(x) dx, \quad p_{12} = a \int_0^\infty f_B(x) H_A(x) dx, \tag{6.14}$$

$$p_{20} = 1 - a \int_0^\infty f_A(x) H_B(x) dx, \quad p_{21} = a \int_0^\infty f_A(x) H_B(x) dx,$$

$$p_{30} = 1 - a, \quad p_{31} = a.$$

Using formula (6.5), we obtain CDF of the waiting times for the states $i \in S$.

$$G_1(t) = Q_{10}(t) + Q_{12}(t) = F_B(t),$$

$$G_2(t) = Q_{20}(t) + Q_{21}(t) = F_A(t), \tag{6.15}$$

$$G_3(t) = Q_{30}(t) + Q_{31}(t) = F_A(t).$$

Applying the equality (6.6), we calculate CDF of the holding times for pairs of states $(i, j) \in S \times S$.

$$F_{10}(t) = \frac{F_B(t) - a \int_0^t f_B(x) H_A(x) dx}{1 - a \int_0^\infty f_B(x) H_A(x) dx}, \quad F_{12}(t) = \frac{\int_0^t f_B(x) H_A(x) dx}{\int_0^\infty f_B(x) H_A(x) dx}, \tag{6.16}$$

$$F_{20}(t) = \frac{F_A(t) - a \int_0^t f_A(x) H_B(x) dx}{1 - a \int_0^\infty f_A(x) H_B(x) dx}, \quad F_{21}(t) = \frac{\int_0^t f_A(x) H_B(x) dx}{\int_0^\infty f_A(x) H_B(x) dx}, \tag{6.17}$$

$$F_{31}(t) = F_A(t), \quad F_{30}(t) = F_A(t).$$

6.1.4   Reliability characteristics

The function

$$\Phi_{iD}(t) = P(\Theta_D \leqslant t | X(0) = i), \quad t \geqslant 0 \tag{6.18}$$

is the CDF of a random variable Θ_{iD} denoting the first passage time from the state $i \in D'$ to the subset D or the exit time of $\{X(t) : t \geqslant 0\}$ from the subset D' with an initial state i. From Theorem 4.1 it follows that L-S transforms of the unknown CDF of the random variables $\Theta_{iD}, i \in D'$ satisfy the system of linear equations (4.18), which is equivalent to the matrix equation

$$\left(\boldsymbol{I} - \tilde{\boldsymbol{q}}_{D'}(s)\right) \tilde{\boldsymbol{\varphi}}_{D'}(s) = \tilde{\boldsymbol{b}}(s), \tag{6.19}$$

where

$$\boldsymbol{I} = \left[\delta_{ij} : i,j \in D'\right] \tag{6.20}$$

is the unit matrix,

$$\tilde{\boldsymbol{q}}_{D'}(s) = \left[\tilde{q}_{ij}(s) : i,j \in D'\right] \tag{6.21}$$

is the square submatrix of the matrix $\tilde{\boldsymbol{q}}(s)$, while

$$\tilde{\boldsymbol{\varphi}}_{D'}(s) = \left[\tilde{\varphi}_{iD}(s) : i \in D'\right]^{\mathrm{T}}, \quad \tilde{\boldsymbol{b}}(s) = \left[\sum_{j \in D} \tilde{q}_{ij}(s) : i \in A'\right]^{\mathrm{T}} \tag{6.22}$$

are one-column matrices of the corresponding L-S transforms.

From Theorem 3.2 it follows that there exist expectations $E(\Theta_{iD})$, $i \in D'$ and they are unique solutions of the linear systems of equations that have the matrix form

$$(\boldsymbol{I} - \boldsymbol{P}_{D'})\overline{\boldsymbol{\Theta}}_{D'} = \overline{\boldsymbol{T}}_{D'}, \tag{6.23}$$

where

$$\boldsymbol{P}_{D'} = \left[p_{ij} : i,j \in D'\right], \quad \overline{\boldsymbol{\Theta}}_{D'} = \left[E(\Theta_{iD}) : i \in D'\right]^{\mathrm{T}}, \quad \overline{\boldsymbol{T}}_{D'} = \left[E(T_i) : i \in D'\right]^{\mathrm{T}} \tag{6.24}$$

and $\boldsymbol{I}$ is the unit matrix.

In our model $D = \{0\}$ and $D' = \{1, 2, 3\}$, the random variable Θ_{iD}, which denotes the first passage time from the state $i = 3$ to the subset D, represents the time to failure of the system in our model. The function

$$R(t) = P(\Theta_{30} > t) = 1 - \Phi_{30}(t), \quad t \geqslant 0 \tag{6.25}$$

is the reliability function of the considered cold standby system with repair.

In this case, (6.19) takes the form

$$\begin{bmatrix} 1 & -\tilde{q}_{12}(s) & 0 \\ -\tilde{q}_{21}(s) & 1 & 0 \\ -\tilde{q}_{31}(s) & 0 & 1 \end{bmatrix} \begin{bmatrix} \tilde{\varphi}_{10}(s) \\ \tilde{\varphi}_{20}(s) \\ \tilde{\varphi}_{30}(s) \end{bmatrix} = \begin{bmatrix} \tilde{q}_{10}(s) \\ \tilde{q}_{20}(s) \\ \tilde{q}_{30}(s) \end{bmatrix}. \tag{6.26}$$

From the solution of this equation, we get

$$\tilde{\varphi}_{10}(s) = \frac{\tilde{q}_{10}(s) + \tilde{q}_{12}(s)\tilde{q}_{10}(s)}{1 - \tilde{q}_{12}(s)\tilde{q}_{21}(s)}, \tag{6.27}$$

$$\tilde{\varphi}_{20}(s) = \frac{\tilde{q}_{20}(s) + \tilde{q}_{21}(s)\tilde{q}_{10}(s)}{1 - \tilde{q}_{12}(s)\tilde{q}_{21}(s)}, \tag{6.28}$$

$$\tilde{\varphi}_{30}(s) = \tilde{q}_{30}(s) + \frac{\tilde{q}_{12}(s)\tilde{q}_{31}(s)\tilde{q}_{20}(s) + \tilde{q}_{10}(s)\tilde{q}_{31}(s)}{1 - \tilde{q}_{12}(s)\tilde{q}_{21}(s)}. \tag{6.29}$$

Consequently, we obtain the Laplace transform of the reliability function

$$\tilde{R}(s) = \frac{1 - \tilde{\varphi}_{30}(s)}{s}. \tag{6.30}$$

We will calculate a mean time to failure by solving (6.23). In this case, this equation takes the form

$$\begin{bmatrix} 1 & -p_{12} & 0 \\ -p_{21} & 1 & 0 \\ -p_{31} & 0 & 1 \end{bmatrix} \begin{bmatrix} E(\Theta_{10}) \\ E(\Theta_{20}) \\ E(\Theta_{30}) \end{bmatrix} = \begin{bmatrix} E(T_1) \\ E(T_2) \\ E(T_3) \end{bmatrix}. \tag{6.31}$$

As a result, we get

$$E(\Theta_{10}) = \frac{E(T_1) + p_{12}E(T_1)}{1 - p_{12}p_{21}}, \tag{6.32}$$

$$E(\Theta_{20}) = \frac{E(T_2) + p_{21}E(T_1)}{1 - p_{12}p_{21}}, \tag{6.33}$$

$$E(\Theta_{30}) = \frac{E(T_3)(1 - p_{12}p_{21}) + p_{12}p_{31}E(T_2) + p_{31}E(T_1)}{1 - p_{12}p_{21}}. \tag{6.34}$$

Using (6.14) and (6.15) we obtain the mean time to failure of the system

$$E(\Theta_{30}) = E(\zeta_A) + a\frac{p_{12}E(\zeta_A) + E(\zeta_B)}{1 - p_{12}p_{21}}, \tag{6.35}$$

where

$$p_{12} = a \int_0^\infty f_B(x)H_A(x)dx, \quad p_{21} = a \int_0^\infty f_A(x)H_B(x)dx. \tag{6.36}$$

For the identical distribution of time to failure of components A and $B : f_A(x) = f_B(x) = f(x)$, and the identical distribution of the both components renewal time $H_A(x) = H_B(x) = H(x)$, from (6.35) we get

$$E(\Theta_{30}) = E(\zeta) + a\frac{(1 + p_{12})E(\zeta)}{1 - p_{12}^2} = E(\zeta) + a\frac{E(\zeta)}{1 - p_{12}}, \tag{6.37}$$

where

$$p_{12} = a \int_0^\infty f(x)H(x)dx. \tag{6.38}$$

This result was presented in Ref. [32, 34].

6.1.5   An approximate reliability function

In general, calculating an exact reliability function of the system by means of Laplace transform is a complicated matter. Finding an approximate reliability function of that system is possible by using some results from the theory of semi-Markov processes perturbations. We will apply the Pavlov and Ushakov [83] concept of the perturbed semi-Markov process presented in Section 4.3.

We can assume that the considered SM process $\{X(t) : t \geqslant 0\}$ with the state space $S = \{0, 1, 2, 3\}$ is the perturbed process with respect to the SM process $\{X^0(t) : t \geqslant 0\}$ with the state space $D' = \{1, 2, 3\}$ and the kernel

$$\mathbf{Q}^0(t) = \begin{bmatrix} 0 & Q_{12}^0(t) & 0 \\ Q_{21}^0(t) & 0 & 0 \\ Q_{31}^0(t) & 0 & 0 \end{bmatrix}, \tag{6.39}$$

where

$$Q_{12}^0(t) = p_{12}^0 F_{12}(t), \quad Q_{21}^0(t) = p_{21}^0 F_{21}(t), \quad Q_{31}^0(t) = p_{31}^0 F_{31}(t).$$

From (5.18) and (5.19) we obtain

$$p_{12}^0 = 1, \quad p_{21}^0 = 1, \quad p_{31}^0 = 1. \tag{6.40}$$

Therefore, taking under consideration (6.16), (6.17), we get

$$Q_{12}^0(t) = \frac{\int_0^t f_B(x)H_A(x)\mathrm{d}x}{\int_0^\infty f_B(x)H_A(x)\mathrm{d}x}, \quad Q_{21}^0(t) = \frac{\int_0^t f_A(x)H_B(x)\mathrm{d}x}{\int_0^\infty f_A(x)H_B(x)\mathrm{d}x}, \qquad Q_{31}^0(t) = F_A(t). \tag{6.41}$$

From (5.18) and (6.14), we have

$$\varepsilon_1 = p_{10} = 1 - a \int_0^\infty f_B(x)H_A(x)\mathrm{d}x, \varepsilon_2 = p_{20}$$

$$= 1 - a \int_0^\infty f_A(x)H_B(x)\mathrm{d}x, \varepsilon_3 = p_{30} = 1 - a. \tag{6.42}$$

The transition matrix of the embedded Markov chain of the SM process $\{X^0(t) : t \geqslant 0\}$ is

$$\mathbf{P}^0 = \begin{bmatrix} 0 & 1 & 0 \\ 1 & 0 & 0 \\ 1 & 0 & 0 \end{bmatrix}. \tag{6.43}$$

From the system of equations

$$\begin{bmatrix} \pi_1^0 & \pi_2^0 & \pi_3^0 \end{bmatrix} \begin{bmatrix} 0 & 1 & 0 \\ 1 & 0 & 0 \\ 1 & 0 & 0 \end{bmatrix} = \begin{bmatrix} \pi_1^0 & \pi_2^0 & \pi_3^0 \end{bmatrix}, \tag{6.44}$$

$$\pi_1^0 + \pi_2^0 + \pi_3^0 = 1, \tag{6.45}$$

we get $\left[\pi_1^0 = 0.5 \quad \pi_2^0 = 0.5 \quad \pi_3^0 = 0\right]$. It follows from Theorem 4.2 that for small ε

$$R(t) = P(\Theta_{iD} > t) = P(\varepsilon\Theta_{iD} > \varepsilon t) \approx \exp\left[-\frac{\varepsilon}{m^0}t\right], \quad t \geqslant 0, \tag{6.46}$$

where

$$\varepsilon = 0.5(\varepsilon_1 + \varepsilon_1) = 1 - 0.5a\left(\int_0^\infty f_B(x)H_A(x)dx + \int_0^\infty f_A(x)H_B(x)dx\right), \tag{6.47}$$

$$m^0 = 0.5(m_1^0 + m_2^0) = 0.5\frac{\int_0^\infty xf_B(x)H_A(x)dx}{\int_0^\infty f_B(x)H_A(x)dx} + 0.5\frac{\int_0^\infty xf_A(x)H_B(x)dx}{\int_0^\infty f_A(x)H_B(x)dx}. \tag{6.48}$$

The shape of parameter ε shows that we can apply this formula only if the numbers $P(\gamma_B \geqslant \zeta_A)$, $P(\gamma_A \geqslant \zeta_B)$, denoting probabilities of the components failure during the repair periods of earlier-failed components, are small.

6.1.6 Illustrative numerical examples

Example 1

We will consider a special case of the model assuming exponential times to failure and exponential repair times of components. We suppose that random variables $\zeta_A, \zeta_B, \gamma_A, \gamma_B$ have exponential distributions with parameters $\lambda_A, \lambda_B, \mu_A, \mu_B$ correspondingly. Substituting some proper characteristics of the exponential distributions in (6.10), (6.11), and (6.12), we obtain

$$Q_{10}(t) = (1 - a)(1 - e^{-\lambda_B t}) + a\frac{\lambda_B}{\lambda_B + \mu_A}(1 - e^{-(\lambda_B + \mu_A)t}), \tag{6.49}$$

$$Q_{12}(t) = a(1 - e^{-\lambda_B t}) - a\frac{\lambda_B}{\lambda_B + \mu_A}(1 - e^{-(\lambda_B + \mu_A)t}), \tag{6.50}$$

$$Q_{20}(t) = (1 - a)(1 - e^{-\lambda_A t}) + a\frac{\lambda_A}{\lambda_A + \mu_B}(1 - e^{-(\lambda_A + \mu_B)t}), \tag{6.51}$$

$$Q_{21}(t) = a(1 - e^{-\lambda_A t}) - a\frac{\lambda_A}{\lambda_A + \mu_B}(1 - e^{-(\lambda_A + \mu_B)t}), \tag{6.52}$$

$$Q_{31}(t) = a(1 - e^{-\lambda_A t}), \tag{6.53}$$

$$Q_{30}(t) = (1 - a)(1 - e^{-\lambda_A t}), \tag{6.54}$$

for $t \geqslant 0$. The Laplace-Stieltjes transforms of the above functions are

$$\tilde{q}_{10}(s) = (1 - a)\frac{\lambda_B}{\lambda_B + s} + a\frac{\lambda_B}{\lambda_B + \mu_A + s}, \tag{6.55}$$

$$\tilde{q}_{12}(s) = a\frac{\lambda_B}{\lambda_B + s} - a\frac{\lambda_B}{\lambda_B + \mu_A + s}, \tag{6.56}$$

$$\tilde{q}_{20}(s) = (1 - a)\frac{\lambda_A}{\lambda_A + s} + a\frac{\lambda_A}{\lambda_A + \mu_B + s}, \tag{6.57}$$

$$\tilde{q}_{21}(s) = a\frac{\lambda_A}{\lambda_A + s} - a\frac{\lambda_B}{\lambda_A + \mu_B + s}, \tag{6.58}$$

$$\tilde{q}_{31}(s) = a\frac{\lambda_A}{\lambda_A + s}, \tag{6.59}$$

$$\tilde{q}_{30}(s) = (1 - a)\frac{\lambda_A}{\lambda_A + s}. \tag{6.60}$$

From (6.19) and (6.30), we obtain the Laplace transform of the reliability function for the parameters

$$\lambda_A = 0.001, \quad \lambda_B = 0.004, \quad \mu_A = 0.4, \quad \mu_B = 0.2, \quad a = 0.98, \tag{6.61}$$

using the computer program written in MATHEMATICA. We get the reliability function by inverting this Laplace transform. The reliability function of the considered system is

$$R(t) = -2.95624 \times 10^{-15} - 0.0000237071\, e^{-0.404009t} + 9.19548 \times 10^{-7}\, e^{-0.200962t} +$$
$$-0.0027939\, e^{-0.00498584t} - 1.6822 \times 10^{-16}\, e^{-0.001t} + 1.00282\, e^{-0.0000432014t}$$

The reliability function is shown in Figure 6.3.

Notice that in this case

$$E(\zeta_A) = \frac{1}{\lambda_A}, \quad E(\zeta_B) = \frac{1}{\lambda_B}, \quad E(\gamma_A) = \frac{1}{\mu_A}, \quad E(\gamma_B) = \frac{1}{\mu_B}$$

and

$$p_{12} = a\frac{\mu_B}{\lambda_B + \mu_A}, \quad p_{21} = a\frac{\mu_A}{\lambda_A + \mu_B}.$$

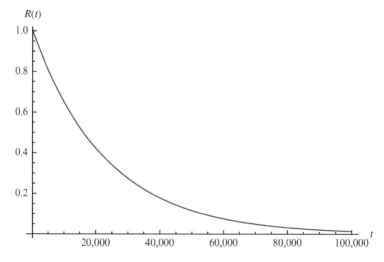

Figure 6.3 Reliability function of the system.

Therefore, from (6.35) and (6.36) we get the mean time to failure:

$$E(\Theta_{30}) = \frac{1}{\lambda_A} + a\frac{a\frac{\mu_B}{\lambda_B+\mu_A}\frac{1}{\lambda_A} + \frac{1}{\lambda_B}}{1 - a^2\frac{\mu_B}{\lambda_B+\mu_A}\frac{\mu_A}{\lambda_A+\mu_B}}. \tag{6.62}$$

For the parameters (6.61) we obtain

$$E(\Theta_{30}) = 23,212.1. \tag{6.63}$$

Example 2

We assume that random variables ζ_A, ζ_B, denoting the lifetimes of units A and B, have an Erlang distribution defined by PDFs

$$f_A(x) = \alpha_A^2 x e^{-\alpha_A x}, \quad f_B(x) = \alpha_B^2 x e^{-\alpha x}, \quad x \geqslant 0. \tag{6.64}$$

The repair times of the failed units that are represented by the random variables γ_A, γ_B have one-point distributions with PDF

$$H_A(x) = \begin{cases} 0 & \text{for } t \leqslant T_A \\ 1 & \text{for } t > T_A \end{cases}, \quad H_B(x) = \begin{cases} 0 & \text{for } t \leqslant T_B \\ 1 & \text{for } t > T_B \end{cases}. \tag{6.65}$$

This means that both components' (units') restoring times are deterministic. Now, the functions that are elements of the semi-Markov kernel (6.8) are given by the following equalities:

$$Q_{10}(t) = \begin{cases} 1 - (1 + \alpha_B t)e^{-\alpha_B t} & \text{for } t \leqslant T_A \\ 1 - (1-a)(1 + \alpha_B t)e^{-\alpha_B t} - a(1 + \alpha_B T_A)e^{-\alpha_B T_A} & \text{for } t > T_A \end{cases}, \tag{6.66}$$

$$Q_{12}(t) = \begin{cases} 0 & \text{for } t \leqslant T_A \\ a((1 + \alpha_B T_A)e^{-\alpha_B T_A} - (1 + \alpha_B t)e^{-\alpha_B t}) & \text{for } t > T_A \end{cases}, \tag{6.67}$$

$$Q_{20}(t) = \begin{cases} 1 - (1 + \alpha_A t)e^{-\alpha_A t} & \text{for } t \leqslant T_B \\ 1 - (1-a)(1 + \alpha_A t)e^{-\alpha_A t} - a(1 + \alpha_A T_B)e^{-\alpha_A T_B} & \text{for } t > T_B \end{cases}, \tag{6.68}$$

$$Q_{21}(t) = \begin{cases} 0 & \text{for } t \leqslant T_B \\ a((1 + \alpha_A T_B)e^{-\alpha_A T_B} - (1 + \alpha_A t)e^{-\alpha_A t}) & \text{for } t > T_B \end{cases}, \tag{6.69}$$

$$Q_{30}(t) = (1 - a)\left[1 - (1 + \alpha_A t)e^{-\alpha_A t}\right], \quad t \geqslant 0, \tag{6.70}$$

$$Q_{31}(t) = a\left[1 - (1 + \alpha_A t)e^{-\alpha_A t}\right], \quad t \geqslant 0. \tag{6.71}$$

From (6.14) we obtain the transition probabilities of the embedded Markov chain.

$$p_{10} = 1 - a(1 + \alpha_B T_A)e^{-\alpha_B T_A}, \quad p_{12} = a(1 + \alpha_B T_A)e^{-\alpha_B T_A}, \tag{6.72}$$

$$p_{20} = 1 - a(1 + \alpha_A T_B)e^{-\alpha_A T_B}, \quad p_{21} = a(1 + \alpha_A T_B)e^{-\alpha_A T_B}, \tag{6.73}$$

$$p_{30} = 1 - a, \quad p_{31} = a. \tag{6.74}$$

We obtain the mean time to failure of the system using formula (6.35). In this case,

$$E(\zeta_A) = \frac{2}{\alpha_A}, \quad E(\zeta_B) = \frac{2}{\alpha_B}.$$

Hence,

$$E(\Theta_{30}) = \frac{2}{\alpha_A} + \frac{\frac{2a^2}{\alpha_A}(1 + \alpha_B T_A)e^{-\alpha_B T_A} + \frac{2a}{\alpha_B}}{1 - a^2(1 + \alpha_B T_A)e^{-\alpha_B T_A}(1 + \alpha_A T_B)e^{-\alpha_A T_B}}. \tag{6.75}$$

To compare the mean time to failure with a result from the previous example we assume that all expectations are the same in both examples. This means that

$$\frac{2}{\alpha_A} = \frac{1}{\lambda_A}, \quad \frac{2}{\alpha_B} = \frac{1}{\lambda_B}, \quad T_A = \frac{1}{\mu_A}, \quad T_B = \frac{1}{\mu_B}.$$

For parameters (6.61) from Example 1 we get the values of parameters from Example 2:

$$\alpha_A = 0.002, \quad \alpha_B = 0.008, \quad T_A = 2.5, \quad T_B = 5 \quad a = 0.98.$$

In this case, the mean time to failure is

$$E(\Theta_{30}) = 31,253.4.$$

Recall that for the exponential distributions of the components' lifetimes and restoring times for corresponding parameters we obtained

$$E(\Theta_{30}) = 23,212.1.$$

From (6.46)–(6.48) we get an approximate reliability function

$$R(t) \approx \exp\left[-\frac{\varepsilon}{m^0}t\right], \quad t \geqslant 0, \tag{6.76}$$

where

$$\varepsilon = 0.5(\varepsilon_1 + \varepsilon_1) = 1 - 0.5a\left((1 + \alpha_B T_A)e^{-\alpha_B T_A} + (1 + \alpha_A T_B)e^{-\alpha_A T_B}\right), \tag{6.77}$$

$$m^0 = 0.5(m_1^0 + m_2^0) = 0.5\frac{2 + 2\alpha_B T_A + \alpha_B^2 T_A^2}{\alpha_B(1 + \alpha_B T_A)} + 0.5\frac{2 + 2\alpha_A T_B + \alpha_A^2 T_B^2}{\alpha_A(1 + \alpha_A T_B)}. \tag{6.78}$$

6.1.7 Conclusions

- The expectation $E(\Theta_{30})$, denoting the mean time to failure of the considered cold standby system, depends on both components' probability distribution of the lifetimes and elements renewal times:

$$E(\Theta_{30}) = E(\zeta_A) + a\frac{p_{12}E(\zeta_A) + E(\zeta_B)}{1 - p_{12}p_{21}},$$

where

$$p_{12} = a \int_0^\infty f_B(x) H_A(x) dx, \quad p_{21} = a \int_0^\infty f_A(x) H_B(x) dx.$$

- If distributions of times to failure and renewal times of components A and B are identical: $f_A(x) = f_B(x) = f(x)$, $H_A(x) = H_B(x) = H(x)$, we obtain the known result [32, 34]:

$$E(\Theta_{30}) = E(\zeta) + a \frac{E(\zeta)}{1-c},$$

where

$$c = p_{12} = a \int_0^\infty f(x) H(x) dx.$$

In this case, the cold standby determines the increase of the mean time to failure $1 + \frac{a}{1-c}$ times.

- Examples 1 and 2 show that for the deterministic renewal times of components, the mean time to failure of the cold standby system is essentially greater than in the case of the exponentially distributed restoring times.
- The approximate reliability function of the system is exponential

$$R(t) \approx \exp\left[-\frac{\varepsilon}{m^0} t\right], \quad t \geqslant 0,$$

where

$$\varepsilon = 1 - 0.5a \left(\int_0^\infty f_B(x) H_A(x) dx + \int_0^\infty f_A(x) H_B(x) dx \right),$$

$$m^0 = 0.5 \frac{\int_0^\infty x f_B(x) H_A(x) dx}{\int_0^\infty f_B(x) H_A(x) dx} + 0.5 \frac{\int_0^\infty x f_A(x) H_B(x) dx}{\int_0^\infty f_A(x) H_B(x) dx}.$$

6.2 Technical example

As a technical example of the cold standby system, we consider an electrical power system of a hospital consisting of the mains and the emergency power systems. Mains power of a hospital can be lost due to interruption of electrical lines during earthworks, or in the case of the overhead line, of its destruction as a result of strong wind, heavy icing, malfunctions at a substation, planned blackouts, or in extreme cases such as a gridwide failure. Loss of mains voltage may last from a few seconds to several hours or in extreme cases several days. The effects of such a situation may differ for different classes of devices. It leads to the automatic shutdown of equipment, and there is a high risk of equipment damage. It refers to lifesaving equipment, anesthesia apparatus, dialyzers in dialysis, or ultrasound devices. In the case of a medical device, usually any loss of power causes disruption as the device, after restoring the power supply, begins the process of self-testing and self-calibration and is not performing its basic functions. For other devices like elevators, refrigerators, or ventilators, a restoration of power is not a major problem. In hospitals, most emergency power systems have been and are still based on generators. An electric generator supplies electricity to selected electrical circuits and electric appliances in the hospital in an emergency (no power from the power grid). A hospital's power generator system includes a set of devices consisting of an internal combustion or Diesel engine, a generator, and an automatic detection system if there is no voltage in the power grid and the self-actuating generator. A separate issue is the switching time. Usually, it ranges from a few to several seconds. The generator starts automatically. With regular generators, an automatic transfer switch is used to connect emergency power. One side is connected to both the normal power feed and the emergency power feed, while the other side is connected to the load designated as emergency.

6.2.1 Assumptions

To evaluate the reliability characteristics of the hospital electrical power system, we will apply the significant modification of the models presented in this chapter. In this case an operating unit A denotes mains power system (subsystem), the stand-by unit B means emergency power system (subsystem), and a switch is the automatic transfer switch with the generator starter. The time to failure of the mains power subsystems is an exponential distributed random variable ζ_A with a PDF

$$f_A(x) = \alpha_A \, e^{-\alpha_A x}, \quad x \geqslant 0, \; \alpha_A > 0. \tag{6.79}$$

The failure rate α_A is given by

$$\alpha_A = \frac{1}{m_A}, \tag{6.80}$$

where m_A denotes a mean time to failure of the mains power subsystem.

When the mains A fails, the emergency power subsystem B is immediately put in motion by the switch (switching time is omitted). The failed system is repaired. A repair time of the basic power system A is a random variable γ_A having distribution given by the PDF

$$h_A(x) = \mu_A^2 x e^{-\mu_A x}, \quad x \geqslant 0. \tag{6.81}$$

After repairing the mains power, subsystem A is put in motion at once by the switch.

A time to failure of the emergency power system B is a random variable ζ_B, with an exponential PDF

$$f_B(x) = \alpha_B \, e^{-\alpha_B x}, \quad x \geqslant 0. \tag{6.82}$$

If the emergency power subsystem B fails during the repair period of the mains power subsystem A, then it follows that damage to the whole electricity power system will occur.

The failure of the system takes place when the mains power subsystem fails and the emergency subsystem fails before repairing the basic subsystem A or when the subsystem A fails and the switch fails.

Let U be a random variable having a binary distribution

$$b(k) = P(U = k) = a^k (1 - a)^{1-k}, \quad k = 0, 1, 0 < a < 1,$$

where $U = 0$ if a switch is "down" at the moment of the subsystem A failure or renewal and $U = 1$ otherwise.

A restoring time of the whole power system is the random variable γ having distribution given by the PDF

$$h(x) = \mu^2 x e^{-\mu x}, \quad x \geq 0. \tag{6.83}$$

Moreover, we assume that all random variables mentioned above are independent.

6.2.2 Model construction

To describe the reliability evolution of the system, we have to define the states and the renewal kernel. We introduce the following states:

0—failure of the whole system due to a failure of a switch,
1—failure of the whole system due to the failure of the subsystem B during repair period of the subsystem A,
2—failure of the mains power subsystem A, the emergency power subsystem B is working,
3—both the mains power subsystem and emergency subsystem are "up" and system A is working.

We assume that 3 is the initial state.

We construct a random process in the following way. Let $0 = \tau_0$ and $\tau_1, \tau_2, \ldots$ denote instants of the power subsystems failures or instants of the subsystem A or whole system repair. Let $\{X(t); \ t \geq 0\}$ be a stochastic process with the state space $S = \{0, 1, 2, 3\}$, keeping its constant values on the half-intervals $[\tau_n, \tau_{n+1})$, $n = 0, 1, \ldots$ with the right-continuous sample paths. This stochastic process is semi-Markov. Its kernel takes the form

$$\mathbf{Q}(t) = \begin{bmatrix} 0 & 0 & 0 & Q_{03}(t) \\ 0 & 0 & 0 & Q_{13}(t) \\ Q_{20}(t) & Q_{21}(t) & 0 & Q_{23}(t) \\ Q_{30}(t) & 0 & Q_{32}(t) & 0 \end{bmatrix}. \tag{6.84}$$

From (6.3) and from the assumptions, we can calculate all elements of the semi-Markov kernel $\mathbf{Q}(t)$, $t \geq 0$. The elements $Q_{03}(t)$ and $Q_{13}(t)$ are CDF of the system renewal time.

$$Q_{03}(t) = Q_{13}(t) = H(t) = \int_0^t h(x)dx. \tag{6.85}$$

From the system description and assumptions we get the following equalities:

$$Q_{20}(t) = P(U = 0, \gamma_A \leq t, \zeta_B > \gamma_A) = (1 - a) \int_0^t h_A(x) [1 - F_B(x)] dx,$$

$$Q_{21}(t) = P(\zeta_B \leq t, \zeta_B < \gamma_A) = \int_0^t f_B(x) [1 - H_A(x)] dx, \tag{6.86}$$

$$Q_{23}(t) = P(U = 1, \gamma_A \leq t, \zeta_B > \gamma_A) = a \int_0^t h_A(x) [1 - F_B(x)] dx.$$

In the same way, we obtain

$$Q_{30}(t) = P(U = 0, \zeta_A \leq t) = (1 - a)F_A(t), \tag{6.87}$$

$$Q_{32}(t) = P(U = 1, \zeta_A \leq t) = aF_A(t).$$

All elements of $\mathbf{Q}(t)$ have been defined; hence, the semi-Markov model describing the hospital power system evolution, in a reliability sense, is constructed. It is necessary to calculate the transition probabilities of the embedded Markov chain. The transition probabilities matrix of the embedded Markov chain $\{X(\tau_n) : n = 0, 1, \ldots\}$ is

$$P = \begin{bmatrix} 0 & 0 & 0 & 1 \\ 0 & 0 & 0 & 1 \\ p_{20} & p_{21} & 0 & p_{23} \\ 1-a & 0 & a & 0 \end{bmatrix}, \tag{6.88}$$

where

$$p_{20} = (1-a) \int_0^\infty h_A(x)\,[1 - F_B(x)]\,\mathrm{d}x, p_{21} = \int_0^\infty f_B(x)\,[1 - H_A(x)]\,\mathrm{d}x,$$

$$p_{23} = a \int_0^\infty h_A(x)\,[1 - F_B(x)]\,\mathrm{d}x.$$

For supposing distributions for $t \geqslant 0$, we get

$$Q_{03}(t) = Q_{13}(t) = H(t) = 1 - (1 + \mu t)\mathrm{e}^{-\mu t}, \tag{6.89}$$

$$Q_{20}(t) = \frac{(1-a)\mu_A^2}{(\alpha_B + \mu_A)^2}\left[1 - (1 + (\alpha_B + \mu_A)t)\,\mathrm{e}^{-(\alpha_B + \mu_A)t}\right],$$

$$Q_{21}(t) = \frac{\alpha_B}{(\alpha_B + \mu_A)^2}\left[\alpha_A + 2\mu_B - (\alpha_B + \alpha_B\mu_A t + \mu_A(2 + \mu_A t))\,\mathrm{e}^{-(\alpha_B + \mu_A)t}\right],$$

$$Q_{23}(t) = \frac{a\mu_A^2}{(\alpha_B + \mu_A)^2}\left[1 - (1 + (\alpha_B + \mu_A)t)\,\mathrm{e}^{-(\alpha_B + \mu_A)t}\right],$$

$$Q_{30}(t) = (1-a)(1 - \mathrm{e}^{-\alpha_A t}),$$

$$Q_{32}(t) = a(1 - \mathrm{e}^{-\alpha_A t}).$$

The Laplace-Stieltjes transform of these functions are

$$\tilde{q}_{03}(s) = \frac{\mu^2}{(s + \mu)^2}, \quad \tilde{q}_{13}(s) = \frac{\mu^2}{(s + \mu)^2}, \tag{6.90}$$

$$\tilde{q}_{20}(s) = \frac{(1-a)\mu_A^2}{(s + \alpha_B + \mu_A)^2}, \quad \tilde{q}_{21}(s) = \frac{\alpha_B(s + \alpha_B + 2\mu_A)}{(s + \alpha_B + \mu_A)^2},$$

$$\tilde{q}_{23}(s) = \frac{a\mu_A^2}{(s + \alpha_B + \mu_A)^2},$$

$$\tilde{q}_{30}(s) = \frac{(1-a)\alpha_A}{s + \alpha_A}, \quad \tilde{q}_{32}(s) = \frac{a\alpha_A}{s + \alpha_A}.$$

The CDF of the waiting times are

$$G_0(t) = G_1(t) = Q_{03}(t) = Q_{13}(t) = 1 - (1 + \mu t)e^{-\mu t}, \tag{6.91}$$

$$G_2(t) = Q_{20}(t) + Q_{21}(t) + Q_{23}(t) = 1 - (1 + \mu_A t)e^{-(\alpha_B + \mu_A)t},$$

$$G_3(t) = 1 - e^{-\alpha_A t}.$$

The expected values of waiting times are

$$E(T_0) = \frac{2}{\mu}, \quad E(T_1) = \frac{2}{\mu}, \quad E(T_2) = \frac{\alpha_B + 2\mu_A}{(\alpha_B + \mu_A)^2}, \quad E(T_3) = \frac{1}{\alpha_A}. \tag{6.92}$$

6.2.3 Reliability characteristic

A main goal of this subsection is an evaluation of a reliability function and a mean time to failure of the hospital power system. To get the reliability function, we have to solve the matrix equation (6.19). In this case, a set of "down" states is $D = \{1, 2\}$ and the set of "up" states is $\{2, 3\}$. The equation takes the form of

$$\begin{bmatrix} 1 & -\tilde{q}_{23}(s) \\ -\tilde{q}_{32}(s) & 1 \end{bmatrix} \begin{bmatrix} \tilde{\varphi}_{2D}(s) \\ \tilde{\varphi}_{3D}(s) \end{bmatrix} = \begin{bmatrix} \tilde{q}_{20}(s) + \tilde{q}_{21}(s) \\ \tilde{q}_{30}(s) \end{bmatrix}. \tag{6.93}$$

The solution of the equivalent linear equations system is

$$\tilde{\varphi}_{2D}(s) = \frac{\tilde{q}_{20}(s) + \tilde{q}_{21}(s) + \tilde{q}_{23}(s)\tilde{q}_{30}(s)}{1 - \tilde{q}_{23}(s)\tilde{q}_{32}(s)}, \tag{6.94}$$

$$\tilde{\varphi}_{3D}(s) = \frac{\tilde{q}_{30}(s) + \tilde{q}_{32}(s)\tilde{q}_{20}(s) + \tilde{q}_{32}(s)\tilde{q}_{21}(s)}{1 - \tilde{q}_{23}(s)\tilde{q}_{32}(s)}. \tag{6.95}$$

Consequently, we obtain the Laplace transform of the reliability function

$$\tilde{R}(s) = \frac{1 - \tilde{\varphi}_{3D}(s)}{s}. \tag{6.96}$$

The parameters of the model were estimated on the basis of few data and the opinion of the technical staff at one of the hospital in Poland. Therefore, the values of these parameters should be treated as approximate. The values of the model parameters are

$$\mu = 5.4, \quad a = 0.99, \quad \alpha_A = 0.0054, \quad \alpha_B = 0.164, \quad \mu_A = 4.2, \quad \mu_B = 3.1. \tag{6.97}$$

For these parameters, substituting transforms (6.90) to (6.95) we obtain the L-S transform $\tilde{\varphi}_{3D}(s)$. Using the formula (6.96) we get the Laplace transform of the reliability function $\tilde{R}(s)$. We get the reliability function $R(t)$ of the system using the procedure InverseLaplaceTransform [r, s, t] in the MATHEMATICA program. Finally, we obtain

$$R(t) = 1.000156e^{-0.000497t}$$

$$- 2[0.000078\cos(0.146252t) + 0.000825\sin(0.146252t)]e^{-4.36645t}$$

$$\approx e^{-0.000497t}. \tag{6.98}$$

From the well-known equality

$$E(T) = \int_0^\infty R(t)\mathrm{d}t \tag{6.99}$$

we get the mean time to failure of the system

$$E(T) = 2013.77\,[\mathrm{days}] \approx 5.5\,[\mathrm{years}]. \tag{6.100}$$

We calculate the second moment using the following formula:

$$E(T^2) = 2\int_0^\infty tR(t)\mathrm{d}t. \tag{6.101}$$

We get the standard deviation using the well-known equality

$$D(T_2) = \sqrt{E(T^2) - [E(T)]^2}. \tag{6.102}$$

And finally, for presented parameters, we get

$$D(T_2) = 2013.45\,[\mathrm{days}]. \tag{6.103}$$

We can also calculate a mean time to failure by solving equation (6.23). In this case, this equation takes the form

$$\begin{bmatrix} 1 & -p_{23} \\ -p_{21} & 1 \end{bmatrix} \begin{bmatrix} E(\Theta_{20}) \\ E(\Theta_{30}) \end{bmatrix} = \begin{bmatrix} E(T_2) \\ E(T_3) \end{bmatrix}. \tag{6.104}$$

By solving it, we get

$$E(\Theta_{2D}) = \frac{E(T_2) + p_{23}E(T_3)}{1 - p_{23}p_{32}}, \tag{6.105}$$

$$E(\Theta_{3D}) = \frac{E(T_3) + p_{32}E(T_1)}{1 - p_{23}p_{32}}. \tag{6.106}$$

The number $E(\Theta_{3D}) = E(T)$ represents the mean time to failure of the considered system.

6.3 Cold standby system with series exponential subsystems

6.3.1 Description and assumptions

We suppose that the system consists of one operating series subsystem, an identical stand-by subsystem, and a switch. Each subsystem consists of N components. We assume that time to failure of those elements are represented by nonnegative mutually independent random variables $\zeta_k, k = 1, \ldots, N$ with exponential PDF

$$f_k(x) = \lambda_k e^{-\lambda_k}, \quad x \geqslant 0, \ k = 1, \ldots, N.$$

When the operating subsystem fails, the spare is put in motion by the switch immediately. The failed subsystem is restoring or changing the failed component.

Because of the memoryless property of the exponential distribution of components lifetime, the renewal of a failed element means the renewal of the whole subsystem. We assume that there is a single repair facility. A renewal time is a random variable having distribution depending on a failed component. We suppose that the lengths of restoring periods of components are represented by identical copies of nonnegative random variables $\gamma_k, k = 1, \ldots, N$, which have the CDF

$$H_k(x) = P(\gamma_k \leqslant x), \quad x \geqslant 0.$$

The failure of the system occurs when the operating subsystem fails and the subsystem that has failed sooner is not ready to work or when the operating subsystem fails and the switch also fails. Let U be a random variable having a binary distribution

$$b(k) = P(U = k) = a^k(1 - a)^{1-k}, \quad k = 0, 1, 0 < a < 1,$$

where $U = 0$, if a switch is failed at the moment of the operating unit failure, and $U = 1$, if the switch works at that moment. We also assume that the whole failed system is replaced by the new identical one. The replacing time is a nonnegative random variable η with CDF

$$K(x) = P(\eta \leqslant x), \quad x \geqslant 0.$$

Moreover, we assume that all random variables mentioned above are independent.

6.3.2   Construction of Semi-Markov reliability model

To describe the reliability evolution of the system, we have to define the states and the renewal kernel. We introduce the following states:

0—failure of the system,
k—the work of the spare subsystem and repair (renewal) of the failed subsystem after a failure of kth, $k = 1, \ldots, N$ component and
$N + 1$—both an operating unit and a spare are "up."

Let $0 = \tau_0^*, \tau_1^*, \tau_2^*, \ldots$ denote the instants of the states changes, and $\{Y(t) : t \geqslant 0\}$ be a random process with the state space $S = \{0, 1, \ldots N, N + 1\}$, which keeps constant values on the half-intervals $[\tau_n^*, \tau_{n+1}^*)$, $n = 0, 1, \ldots$ and its trajectories are the right-continuous functions. This process is not semi-Markov, as the memoryless property is not satisfied for any instants of the state changes of it. As in the previous case, we construct a new stochastic process. Let $0 = \tau_0$ and $\tau_1, \tau_2, \ldots$ denote instants of the subsystems failure or instants of the subsystems renewal. The stochastic process $\{X(t) : t \geqslant 0\}$ is defined as follows:

$$X(0) = 0, \quad X(t) = Y(\tau_n) \quad \text{for } t \in [\tau_n, \tau_{n+1}) \tag{6.107}$$

is the semi-Markov one. Now we define initial distribution and all elements of the semi-Markov process kernel. In this case, the semi-Markov kernel has the form

$$\boldsymbol{Q}(t) = \begin{bmatrix} 0 & 0 & \cdots & 0 & Q_{0N+1}(t) \\ Q_{1\,0}(t) & Q_{1\,1}(t) & \cdots & Q_{1\,N}(t) & 0 \\ \vdots & \vdots & \vdots & \vdots & \vdots \\ Q_{N\,0}(t) & Q_{N\,1}(t) & \cdots & Q_{N\,N}(t) & 0 \\ Q_{N+1\,0}(t) & Q_{N+1\,1}(t) & \cdots & Q_{N+1N}(t) & 0 \end{bmatrix}. \tag{6.108}$$

We know that the semi-Markov process $\{X(t) : t \geqslant 0\}$ is determined if all elements of the matrix $\boldsymbol{Q}(t)$ are defined. From assumption it follows that random variables $\zeta_k, k = 1, \ldots, N$ are exponentially distributed with parameters $\lambda_k, k = 1, \ldots, N$, correspondingly. It means that CDFs are

$$F_k(x) = 1 - e^{-\lambda_k x}, \quad x \geqslant 0.$$

From (6.2.2), we calculate elements of the above matrix.

For $j = 1, \ldots, N$, we get

$$\begin{aligned} Q_{N+1j}(t) &= P(X(\tau_{n+1}) = j, \tau_{n+1} - \tau_n \leqslant t | X(\tau_n) = N + 1) \\ &= P(U = 1, \zeta_j \leqslant t, \zeta_i > \zeta_j \text{ for } i \neq j) \\ &= a \int_0^t \prod_{i \neq j}^N [1 - F_i(x)] f_j(x) dx. \end{aligned} \tag{6.109}$$

Substituting the suitable exponential CDFs and PDF, we get

$$Q_{N+1j}(t) = a \int_0^t \prod_{i \neq j}^N \left[e^{-\lambda_i x}\right] \lambda_j e^{-\lambda_j x} dx = a \frac{\lambda_j}{\Lambda}(1 - e^{-\Lambda t}), \quad t \geqslant 0 \tag{6.110}$$

for $j = 1, \ldots, N$, where

$$\Lambda = \lambda_1 + \cdots + \lambda_N.$$

For $j = 0$, we obtain

$$Q_{N+10}(t) = (1 - a)(1 - e^{-\Lambda t}), \quad t \geqslant 0. \tag{6.111}$$

For $i, j = 1, \ldots, N$, we have

$$Q_{ij}(t) = a\lambda_j \int_0^t H_i(x) e^{-\Lambda x} dx. \tag{6.112}$$

For $j = 0$,

$$Q_{i0}(t) = 1 - e^{-\Lambda t} - a\Lambda \int_0^t H_i(x) e^{-\Lambda x} dx. \tag{6.113}$$

From the assumption, it follows that

$$Q_{0N+1}(t) = K(t).$$

All elements of the SM process kernel are determined, so the SM model is constructed. Reliability characteristics and parameters can be calculated in a manner similar to the previous cases.

SM models of multistage operation

7

Abstract

Semi-Markov models of multistage operation are presented in this chapter. It is assumed that stages of the operation are realized in turn. An execution time of each stage is assumed to be a positive random variable. Each step of the operation may be perturbed or failed. The perturbations extend the operation time and the probability of failure as well. In this chapter, two models are presented: the first one concerns a multistage operation with possible perturbations of task execution at different stages, the second one concerns a multistage operation without this kind of perturbations. The second model is applied as a model of a multi-modal transport operation. To explain the problems considered, we present three illustrative numerical examples.

Keywords: Semi-Markov model, Multistage operation, Transport operation, Reliability function, Mean time to failure

7.1 Introduction

Lots of operations consist of stages that are realized sequentially. An execution time of the operation stage is assumed to be a positive random variable. Each stage of the operation can include possible perturbations or failures. The perturbations extend the duration of the operation and they cause an increase of failure probability as well. Semi-Markov processes theory enables constructing some useful models of the multistage operations. The SM models constructed and investigated in this chapter are some modification and extension of the results presented in Refs. [12, 33, 34].

7.2 Description and assumptions

Suppose that the operation consists of n stages that follow in turn. We assume that a duration of an ith stage of operation, $(i = 1, \ldots, n)$ is a nonnegative random variable $\xi_i, i = 1, \ldots, n$ with distribution given by *CDF*

$$F_{\xi_i}(t) = P(\xi_i \leqslant t). \tag{7.1}$$

Time to failure of the operation on the ith stage is the nonnegative random variable $\eta_i, i = 1, \ldots, n$ with the exponential distribution

$$P(\eta_i \leqslant t) = 1 - e^{-\lambda_i t}, \quad i = 1, \ldots, n. \tag{7.2}$$

Semi-Markov Processes: Applications in System Reliability and Maintenance. http://dx.doi.org/10.1016/B978-0-12-800518-7.00007-7

The operation may be perturbed on each step. We assume that on each stage no more than one event causing perturbation of the operation may occur. Time to an event causing of an operation perturbation on ith stage is a nonnegative random variable $\zeta_i, i = 1, \ldots, n$ with the exponential distribution

$$F_{\zeta_i} = P(\zeta_i \leqslant t) = 1 - e^{-\alpha_i t}, i = 1, \ldots, n. \tag{7.3}$$

The perturbation decreases the probability of the operation success. We suppose that time to failure on ith stage after the danger event is a nonnegative random variable $v_i, i = 1, 2, \ldots, n$ exponentially distributed with a parameter $\beta_i > 0$, where $\beta_i > \lambda_i$,

$$P(v_i \leqslant t) = 1 - e^{-\beta_i t}, \quad i = 1, \ldots, n. \tag{7.4}$$

The duration of the perturbed ith stage of operation is a nonnegative random variable $\kappa_i = c_i [\xi_i - \zeta_i]_+, i = 1, \ldots, n, c_i \geqslant 1$, where c_i is a factor of the time elongation of the ith perturbed stage of the operation. We assume that the operation is cyclic. We also assume that random variables $\xi_i, \zeta_i, v_i, \eta_i, i = 1, \ldots, n$ and their copies are mutually independent.

7.3 Construction of Semi-Markov model

To construct the reliability model of operation, we have to start from the definition of states. We introduce the following states:

$$\begin{array}{ll} i & \text{—}i\text{th not perturbed stage of the operation, } i = 1, \ldots n, \\ i + n & \text{—}i\text{th perturbed stage of the operation,} \qquad i = 1, \ldots n, \\ 2n + 1 & \text{—total failure of the operation.} \end{array} \tag{7.5}$$

Under the assumptions above, a stochastic process describing the overall operation in reliability aspect is a semi-Markov process $\{X(t) : t \geqslant 0\}$ with a space of states $S = \{1, 2, \ldots, 2n, 2n + 1\}$ and a transition graph shown in Figure 7.1.

To obtain a semi-Markov model, we have to define all nonnegative elements of the semi-Markov kernel

$$Q(t) = [Q_{ij}(t) : i, j \in S],$$

$$Q_{ij}(t) = P(X(\tau_{n+1}) = j, \tau_{n+1} - \tau_n \leqslant t | X(\tau_n) = i).$$

First, we have to define the transition probabilities from the state i to the state j at the time not greater than t for $i = 1, \ldots, n - 1$:

$$Q_{ii+1}(t) = P(\xi_i \leqslant t, \eta_i > \xi_i, \zeta_i > \xi_i)$$

$$= \iiint_D \alpha_i e^{-\alpha_i y} \lambda_i e^{-\lambda_i z} dF_{\xi_i}(x) dy dz,$$

where

$$\begin{aligned} D = \{ &(x, y, z) : x \geqslant 0, y \geqslant 0, z \geqslant 0, \\ &x \leqslant t, z > x, y > x \}. \end{aligned}$$

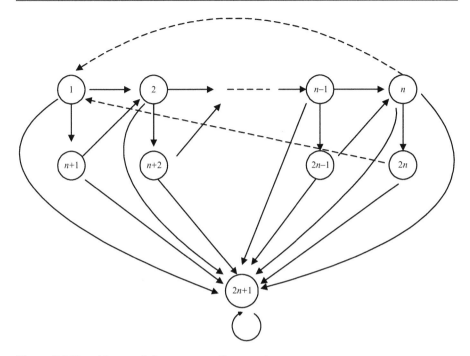

Figure 7.1 Transition graph for n-stage cyclic operation.

Therefore we have

$$Q_{ii+1}(t) = \int_0^t dF_{\xi_i}(x) \int_x^\infty \alpha_i e^{-\alpha_i y} dy \int_x^\infty \lambda_i e^{-\lambda_i z} dz = \int_0^t e^{-(\lambda_i+\alpha_i)x} dF_{\xi_i}(x).$$

$$(7.6)$$

The same way we get

$$Q_{n1}(t) = P(\xi_n \leqslant t, \eta_n > \xi_n, \zeta_n > \xi_n) = \int_0^t e^{-(\lambda_n+\alpha_n)x} dF_{\xi_n}(x). \qquad (7.7)$$

For $i = 1, \ldots, n - 1$, we obtain

$$Q_{ii+n}(t) = P(\zeta_i \leqslant t, \eta_i > \zeta_i, \xi_i > \zeta_i)$$

$$= \int_0^t \alpha_i e^{-(\lambda_i+\alpha_i)x} \left[1 - F_{\xi_i}(x)\right] dx. \qquad (7.8)$$

For $i = 1, \ldots, n$, we get

$$Q_{i2n+1}(t) = P(\eta_i \leqslant t, \eta_i < \zeta_i, \eta_i < \xi_i)$$

$$= \int_0^t \lambda_i e^{-(\lambda_i+\alpha_i)u} \left[1 - F_{\xi_i}(x)\right] dx. \qquad (7.9)$$

If on ith stage some perturbation has happened, then the transition probability to the next state for time less than or equal to t is

$$Q_{n+ii+1}(t) = P(\kappa_i = c_i(\xi_i - \zeta_i) \leqslant t, \nu_i > \kappa_i | \xi_i > \zeta_i).$$

Thus we get

$$Q_{n+ii+1}(t) = \frac{\int_0^t e^{-\beta_i x} dx \left\{ \int_0^\infty \alpha_i e^{-\alpha_i y} \left[F_{\xi_i}(y + \frac{x}{c}) - F_{\xi_i}(y) \right] dy \right\}}{\int_0^\infty \alpha_i e^{-\alpha_i y} \left[1 - F_{\xi_i}(y) \right] dy} \tag{7.10}$$

for $i = 1, \ldots, n - 1$. Moreover,

$$Q_{2n1}(t) = \frac{\int_0^t e^{-\beta_n x} dx \left\{ \int_0^\infty \alpha_n e^{-\alpha_n y} \left[F_{\xi_n}(y + \frac{x}{c}) - F_{\xi_n}(y) \right] dy \right\}}{\int_0^\infty \alpha_n e^{-\alpha_n y} \left[1 - F_{\xi_n}(y) \right] dy}. \tag{7.11}$$

For $i = 1, \ldots, n$, we have

$$Q_{n+i2n+1}(t) = P(\nu_i \leqslant t, \nu_i < \kappa_i = c_i(\xi_i - \zeta_i) | \xi_i > \zeta_i). \tag{7.12}$$

In a similar way, we get

$$Q_{n+i2n+1}(t) = \frac{\int_0^t \beta_i e^{-\beta_i z} \int_z^\infty \left\{ dx \left[\int_0^\infty \alpha_i e^{-\alpha_i y} \left[F_{\xi_i}(y + \frac{x}{c}) - F_{\xi_i}(y) \right] dy \right] \right\} dz}{\int_0^\infty \alpha_i e^{-\alpha_i y} \left[1 - F_{\xi_i}(y) \right] dy}. \tag{7.13}$$

Therefore, the semi-Markov reliability model of the operation has been constructed.

7.4   Illustrative numerical examples

We will investigate a particular case of that model, assuming $n = 3$. In this case the semi-Markov kernel is

$$\mathbf{Q}(t) = \begin{bmatrix} 0 & Q_{12}(t) & 0 & Q_{14}(t) & 0 & 0 & Q_{17}(t) \\ 0 & 0 & Q_{23}(t) & 0 & Q_{25}(t) & 0 & Q_{27}(t) \\ Q_{31}(t) & 0 & 0 & 0 & 0 & Q_{36}(t) & Q_{37}(t) \\ 0 & Q_{42}(t) & 0 & 0 & 0 & 0 & Q_{47}(t) \\ 0 & 0 & Q_{53}(t) & 0 & 0 & 0 & Q_{57}(t) \\ Q_{61}(t) & 0 & 0 & 0 & 0 & 0 & Q_{67}(t) \\ 0 & 0 & 0 & 0 & 0 & 0 & Q_{77}(t) \end{bmatrix}. \tag{7.14}$$

We will present two extreme cases. In the first one we assume that a duration of the ith stage of the operation $(i = 1, \ldots, 3)$, is represented by the random variable ξ_i, which is exponentially distributed, and in the second case we suppose that this random variable has one-point distribution. This means that the duration of the ith stage of the operation is deterministic.

7.4.1 Example 1

We suppose that

$$F_{\xi_i}(t) = 1 - e^{-\gamma_i t}, \quad t \geq 0, i = 1, 2, 3.$$

Using (7.6)–(7.13), we get

$$Q_{ii+1}(t) = \frac{\gamma_i}{\lambda_i + \alpha_i + \gamma_i} 1 - e^{-(\lambda_i + \alpha_i + \gamma_i)t}, \quad i = 1, 2, \tag{7.15}$$

$$Q_{31}(t) = \frac{\gamma_3}{\lambda_3 + \alpha_3 + \gamma_3} 1 - e^{-(\lambda_3 + \alpha_3 + \gamma_3)t}, \tag{7.16}$$

$$Q_{ii+3} = \frac{\alpha_i}{\lambda_i + \alpha_i + \gamma_i} 1 - e^{-(\lambda_i + \alpha_i + \gamma_i)t}, \quad i = 1, 2, \tag{7.17}$$

$$Q_{i7}(t) = Q_{ii+3} = \frac{\lambda_i}{\lambda_i + \alpha_i + \gamma_i} 1 - e^{-(\lambda_i + \alpha_i + \gamma_i)t}, \quad i = 1, 2, 3, \tag{7.18}$$

$$Q_{3+ii+1}(t) = \frac{\gamma_i}{c_i \beta_i + \gamma_i} \left(1 - e^{-\left(\beta_i + \frac{\gamma_i}{c_i}\right)t}\right), \quad i = 1, 2, \tag{7.19}$$

$$Q_{61}(t) = \frac{\gamma_3}{c_3 \beta_3 + \gamma_3} \left(1 - e^{-\left(\beta_3 + \frac{\gamma_3}{c_3}\right)t}\right), \tag{7.20}$$

$$Q_{3+i7}(t) = \frac{c_i \beta_i}{c_i \beta_i + \gamma_i} \left(1 - e^{-\left(\beta_i + \frac{\gamma_i}{c_i}\right)t}\right), \quad i = 1, 2, 3. \tag{7.21}$$

The Laplace-Stieltjes transforms of these functions are

$$\tilde{q}_{ii+1}(s) = \frac{\gamma_i}{s + \lambda_i + \alpha_i + \gamma_i}, \quad i = 1, 2, \tag{7.22}$$

$$\tilde{q}_{31}(s) = \frac{\gamma_3}{s + \lambda_3 + \alpha_3 + \gamma_3}, \tag{7.23}$$

$$\tilde{q}_{ii+3} = \frac{\alpha_i}{s + \lambda_i + \alpha_i + \gamma_i}, \tag{7.24}$$

$$\tilde{q}_{i7}(s) = \frac{\lambda_i}{s + \lambda_i + \alpha_i + \gamma_i}, \quad i = 1, 2, 3, \tag{7.25}$$

$$\tilde{q}_{3+ii+1}(s) = \frac{\gamma_i}{s + c_i \beta_i + \gamma_i} \quad i = 1, 2, \tag{7.26}$$

$$\tilde{q}_{61}(s) = \frac{\gamma_3}{s + c_3 \beta_3 + \gamma_3}, \tag{7.27}$$

$$\tilde{q}_{3+i7}(s) = \frac{c_i \beta_i}{s + c_i \beta_i + \gamma_i}, \quad i = 1, 2, 3. \tag{7.28}$$

The Laplace-Stieltjes transform of the matrix (7.14) is

$$
q(s) = \begin{bmatrix}
0 & \tilde{q}_{12}(s) & 0 & \tilde{q}_{14}(s) & 0 & 0 & \tilde{q}_{17}(s) \\
0 & 0 & \tilde{q}_{23}(s) & 0 & \tilde{q}_{25}(s) & 0 & \tilde{q}_{27}(s) \\
\tilde{q}_{31}(s) & 0 & 0 & 0 & 0 & \tilde{q}_{36}(s) & \tilde{q}_{37}(s) \\
0 & \tilde{q}_{42}(s) & 0 & 0 & 0 & 0 & \tilde{q}_{47}(s) \\
0 & 0 & \tilde{q}_{53}(s) & 0 & 0 & 0 & \tilde{q}_{57}(s) \\
\tilde{q}_{61}(s) & 0 & 0 & 0 & 0 & 0 & \tilde{q}_{67}(s) \\
0 & 0 & 0 & 0 & 0 & 0 & \tilde{q}_{77}(s)
\end{bmatrix}.
\tag{7.29}
$$

Reliability characteristics and parameters

The constructed SM model allows us to obtain some reliability characteristics of the operation. The states $1, 2, 3$ represent "up" reliability states, the states $4, 5, 6$ are the perturbed reliability states (partly "up"), the state 7 is a "down" reliability state. The random variable $\Theta_{iA}, A = \{7\}, i \in A' = \{1, \ldots, 6\}$ denoting the first passage time from a state $i \in A'$ to the state 7 means the time to the total failure of the operation if the initial state is i. The Laplace-Stieltjes transform for the CDF of the random variables $\Theta_{iA}, i \in A'$ we obtain from a matrix equation from Theorem 4.1. In this case, we have $A' = \{1, 2, 3, 4, 5, 6\}, A = \{7\}$, and

$$
\begin{bmatrix}
1 & -\tilde{q}_{12}(s) & 0 & -\tilde{q}_{14}(s) & 0 & 0 \\
0 & 1 & -\tilde{q}_{23}(s) & 0 & -\tilde{q}_{25}(s) & 0 \\
-\tilde{q}_{31}(s) & 0 & 1 & 0 & 0 & -\tilde{q}_{36}(s) \\
0 & -\tilde{q}_{42}(s) & 0 & 1 & 0 & 0 \\
0 & 0 & -\tilde{q}_{53}(s) & 0 & 1 & 0 \\
-\tilde{q}_{61}(s) & 0 & 0 & 0 & 0 & 1
\end{bmatrix}
\begin{bmatrix}
\tilde{\varphi}_{17}(s) \\
\tilde{\varphi}_{27}(s) \\
\tilde{\varphi}_{37}(s) \\
\tilde{\varphi}_{47}(s) \\
\tilde{\varphi}_{57}(s) \\
\tilde{\varphi}_{67}(s)
\end{bmatrix}
=
\begin{bmatrix}
\tilde{q}_{17}(s) \\
\tilde{q}_{27}(s) \\
\tilde{q}_{37}(s) \\
\tilde{q}_{47}(s) \\
\tilde{q}_{57}(s) \\
\tilde{q}_{67}(s)
\end{bmatrix}.
$$

From the solution of this equation we get the Laplace-Stieltjes transform of the density functions $\tilde{\varphi}_{i7}(s)$ of the random variable $\Theta_{i7}, i = 1, \ldots, 6$. The Laplace transform of the reliability function is given by the formula

$$
\tilde{R}(s) = \frac{1 - \tilde{\varphi}_{17}(s)}{s}
\tag{7.30}
$$

under the assumption that the initial state of the operation is 1. The reliability function we get by using the procedure InverseLaplaceTransform $\left[\tilde{R}(s), s, t\right]$ in the MATHEMATICA program. For a numerical example we accept the following values of parameters:

$$c_1 = 1.2 \qquad c_2 = 1.4 \qquad c_3 = 1.2$$
$$\lambda_1 = 0.0002 \quad \lambda_2 = 0.0001 \quad \lambda_1 = 0.0001$$
$$\alpha_1 = 0.001 \quad \alpha_2 = 0.006 \quad \alpha_3 = 0.004 \tag{7.31}$$
$$\beta_1 = 0.0008 \quad \beta_2 = 0.0006 \quad \beta_3 = 0.0004$$
$$\gamma_1 = 0.125 \quad \gamma_2 = 0.06 \qquad \gamma_3 = 0.2.$$

The reliability function calculated by the computer is

$$
\begin{aligned}
R(t) = {} & 3.37703 \times 10^{-14} - 0.0000264945\, e^{-0.204472t} + 0.0000196922 e^{-0.126962t} \\
& + 1.0005\, e^{-0.00018378t} - 0.000716071\, e^{-0.0668177t} \\
& + 0.000112677\, e^{-0.192672t} \cos(0.0862698t) \\
& - 0.00013604\, e^{-0.192672t} \sin(0.0862698t) \\
& + 0.000112677\, e^{-0.192672t} \cos(0.0862698t)\cos(0.17254t) \\
& + 0.00013604\, e^{-0.192672t} \cos(0.17254t)\sin(0.0862698t) \\
& - 0.00013604\, e^{-0.192672} \cos(0.0862698t)\sin(0.17254t) \\
& - 0.000112677\, e^{-0.192672t} \sin(0.0862698t)\sin(0.17254t). \tag{7.32}
\end{aligned}
$$

As we see, the formula is too long for a quick analysis. This reliability function is shown in Figure 7.2. We can calculate basic reliability parameters of the system more easily. To obtain mean time to failure of the operation we have to solve the matrix

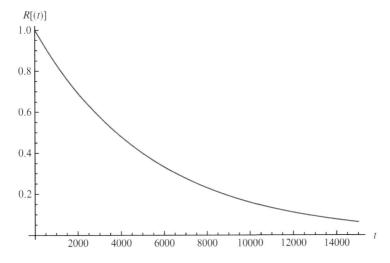

Figure 7.2 Reliability function of the operation.

equation (3.25). In this case, it is of the form

$$
\begin{bmatrix}
1 & -p_{12} & 0 & -p_{14} & 0 & 0 \\
0 & 1 & -p_{23} & 0 & -p_{25} & 0 \\
-p_{31} & 0 & 1 & 0 & 0 & -p_{36} \\
0 & -p_{42} & 0 & 1 & 0 & 0 \\
0 & 0 & -p_{53} & 0 & 1 & 0 \\
-p_{61} & 0 & 0 & 0 & 0 & 1
\end{bmatrix}
\begin{bmatrix}
E(\Theta_{17}) \\
E(\Theta_{27}) \\
E(\Theta_{37}) \\
E(\Theta_{47}) \\
E(\Theta_{57}) \\
E(\Theta_{67})
\end{bmatrix}
=
\begin{bmatrix}
E(T_1) \\
E(T_2) \\
E(T_3) \\
E(T_4) \\
E(T_5) \\
E(T_6)
\end{bmatrix},
\tag{7.33}
$$

where

$$
p_{12} = \frac{\gamma_1}{\lambda_1 + \alpha_1 + \gamma_1}, \quad p_{14} = \frac{\alpha_1}{\lambda_1 + \alpha_1 + \gamma_1}, \quad p_{23} = \frac{\gamma_2}{\lambda_2 + \alpha_2 + \gamma_2},
$$
$$
p_{25} = \frac{\alpha_2}{\lambda_2 + \alpha_2 + \gamma_2}, \quad p_{31} = \frac{\gamma_3}{\lambda_3 + \alpha_3 + \gamma_3}, \quad p_{36} = \frac{\alpha_3}{\lambda_3 + \alpha_3 + \gamma_3},
\tag{7.34}
$$
$$
p_{42} = \frac{\gamma_1}{c_1\beta_1 + \gamma_1}, \quad p_{53} = \frac{\gamma_2}{c_2\beta_2 + \gamma_2}, \quad p_{61} = \frac{\gamma_3}{c_3\beta_3 + \gamma_3},
$$

$$
E(T_1) = \frac{1}{\lambda_1 + \alpha_1 + \gamma_1}, \quad E(T_2) = \frac{1}{\lambda_2 + \alpha_2 + \gamma_2}, \quad E(T_3) = \frac{1}{\lambda_3 + \alpha_3 + \gamma_3},
$$
$$
E(T_4) = \frac{c_1}{c_1\beta_1 + \gamma_1}, \quad E(T_5) = \frac{c_2}{c_2\beta_2 + \gamma_2}, \quad E(T_6) = \frac{c_3}{c_3\beta_3 + \gamma_3}.
\tag{7.35}
$$

The numerical solution of (7.33) is

$$
\begin{bmatrix}
E(\Theta_{17}) \\
E(\Theta_{27}) \\
E(\Theta_{37}) \\
E(\Theta_{47}) \\
E(\Theta_{57}) \\
E(\Theta_{67})
\end{bmatrix}
=
\begin{bmatrix}
5559.63 \\
5560.77 \\
5558.93 \\
5527.92 \\
5505.19 \\
5552.3
\end{bmatrix}.
$$

Under the assumption that the initial state of the operation is 1, the mean time to the operation failure is

$$
E(T) = \bar{T} = 5559.63.
$$

The second moments of the random variables $\Theta_{i7}, i \in \{1, \ldots, 6\}$ can be obtained by solving the matrix equation (3.26):

$$
\begin{bmatrix}
1 & -p_{12} & 0 & -p_{14} & 0 & 0 \\
0 & 1 & -p_{23} & 0 & -p_{25} & 0 \\
-p_{31} & 0 & 1 & 0 & 0 & -p_{36} \\
0 & -p_{42} & 0 & 1 & 0 & 0 \\
0 & 0 & -p_{53} & 0 & 1 & 0 \\
-p_{61} & 0 & 0 & 0 & 0 & 1
\end{bmatrix}
\begin{bmatrix}
E(\Theta_{17}^2) \\
E(\Theta_{27}^2) \\
E(\Theta_{37}^2) \\
E(\Theta_{47}^2) \\
E(\Theta_{57}^2) \\
E(\Theta_{67}^2)
\end{bmatrix}
=
\begin{bmatrix}
b_{17} \\
b_{27} \\
b_{37} \\
b_{47} \\
b_{57} \\
b_{67}
\end{bmatrix},
\tag{7.36}
$$

where

$$b_{i7} = E(T_i^2) + 2 \sum_{k \in A'} p_{ik} E(T_{ik}) E(\Theta_{k7}),$$

$$E(T_1^2) = \frac{2}{(\lambda_1 + \alpha_1 + \gamma_1)^2}, \quad E(T_2^2) = \frac{2}{(\lambda_2 + \alpha_2 + \gamma_2)^2}, \quad E(T_3^2) = \frac{2}{(\lambda_3 + \alpha_3 + \gamma_3)^2},$$

$$E(T_4^2) = \frac{c_1^2}{(c_1 \beta_1 + \gamma_1)^2}, \quad E(T_5^2) = \frac{c_2^2}{(c_2 \beta_2 + \gamma_2)^2}, \quad E(T_6^2) = \frac{c_3^2}{(c_3 \beta_3 + \gamma_3)^2},$$

$$(7.37)$$

$$E(T_{12}) = E(T_{14}) = \frac{1}{\lambda_1 + \alpha_1 + \gamma_1}, \quad E(T_{23}) = E(T_{25}) = \frac{1}{\lambda_2 + \alpha_2 + \gamma_2},$$

$$E(T_{31}) = E(T_{36}) = \frac{1}{\lambda_3 + \alpha_3 + \gamma_3}, \quad E(T_{42}) = \frac{c_1}{c_1 \beta_1 + \gamma_1}, \quad (7.38)$$

$$E(T_{53}) = \frac{c_2}{c_2 \beta_2 + \gamma_2}, \quad\quad E(T_{61}) = \frac{c_3}{c_3 \beta_3 + \gamma_3}.$$

The numerical solution of this equation is

$$
\begin{bmatrix}
E(\Theta_{17}^2) \\
E(\Theta_{27}^2) \\
E(\Theta_{37}^2) \\
E(\Theta_{47}^2) \\
E(\Theta_{57}^2) \\
E(\Theta_{67}^2)
\end{bmatrix}
=
\begin{bmatrix}
6.1781 \times 10^7 \\
6.17937 \times 10^7 \\
6.17733 \times 10^7 \\
6.14281 \times 10^7 \\
6.11738 \times 10^7 \\
6.16995 \times 10^7
\end{bmatrix}.
$$

Hence, the standard deviation of the operation time to failure given by

$$D(T) = \sqrt{D(\Theta_{17}^2) - [E(\Theta_{17})]^2}$$

is

$$D(T) = 5556.22$$

in this case.

7.4.2 Example 2

Now we assume that

$$F_i(t) = \begin{cases} 0 & \text{for } t \leqslant d_i \\ 1 & \text{for } t > d_i \end{cases}, \quad i = 1, 2, 3. \quad (7.39)$$

This means that the duration of stage i is determined and is equal to d_i for $i = 1, 2, 3$.

In this case, the elements of $Q(t)$ are

$$Q_{ii+1}(t) = \begin{cases} 0 & \text{for } t \leqslant d_i, \\ e^{-(\lambda_i+\alpha_i)d_i} & \text{for } t > d_i, \end{cases} \quad i = 1, 2,$$

$$Q_{31}(t) = \begin{cases} 0 & \text{for } t \leqslant d_i, \\ e^{-(\lambda_3+\alpha_3)d_3} & \text{for } t > d_3, \end{cases}$$

$$Q_{ii+3}(t) = \begin{cases} \dfrac{\alpha_i}{\lambda_i+\alpha_i}\left(1-e^{(-\lambda_i+\alpha_i)t}\right) & \text{for } t \leqslant d_i, \\ \dfrac{\alpha_i}{\lambda_i+\alpha_i}\left(1-e^{-(\lambda_i+\alpha_i)d_i}\right) & \text{for } t > d_i, \end{cases} \quad i = 1, 2, 3$$

$$Q_{i7}(t) = \begin{cases} \dfrac{\lambda_i}{\lambda_i+\alpha_i}\left(1-e^{-(\lambda_i+\alpha_i)t}\right) & \text{for } t \leqslant d_i, \\ \dfrac{\lambda_i}{\lambda_i+\alpha_i}\left(1-e^{-(\lambda_i+\alpha_i)d_i}\right) & \text{for } t > d_i, \end{cases} \quad i = 1, 2, 3,$$

$$Q_{3+ii+1}(t) = \begin{cases} \dfrac{\alpha_i e^{-\alpha_i d_i}}{(c_i\beta_i-\alpha_i)(1-e^{-\alpha_i d_i})}\left(1-e^{-(\beta_i-\alpha_i/c_i)t}\right) \\ \quad \text{for } 0 \leqslant t \leqslant c_i d_i, \\ \dfrac{\alpha_i e^{-\alpha_i d_i}}{(c_i\beta_i-\alpha_i)(1-e^{-\alpha_i d_i})}\left(1-e^{-(\beta_i-\alpha_i/c_i)c_i d_i}\right) \\ \quad \text{for } t > c_i d_i, \end{cases} \quad i = 1, 2,$$

$$Q_{61}(t) = \begin{cases} \dfrac{\alpha_3 e^{-\alpha_3 d_3}}{(c_3\beta_3-\alpha_3)(1-e^{-\alpha_i d_i})}\left(1-e^{-(\beta_i-\alpha_i/c_3)t}\right) \\ \quad \text{for } 0 \leqslant t \leqslant c_i d_i, \\ \dfrac{\alpha_3 e^{-\alpha_3 d_3}}{(c_3\beta_3-\alpha_3)(1-e^{-\alpha_3 d_3})}\left(1-e^{-(\beta_i-\alpha_3/c_3)c_3 d_3}\right) \\ \quad \text{for } t > c_3 d_3, \end{cases} \quad i = 1, 2,$$

$$Q_{3+i7}(t) = \begin{cases} \dfrac{1}{(1-e^{-\alpha_i d_i})}\left[\left(1-e^{-\beta_i t}\right) - \dfrac{c_i\beta_i e^{-\alpha_i d_i}}{(c_i\beta_i-\alpha_i)}\left(1-e^{-(\beta_i-\alpha_i/c_i)t}\right)\right] \\ \quad \text{for } 0 \leqslant t \leqslant c_i d_i, \\ \dfrac{1}{(1-e^{-\alpha_i d_i})}\left[\left(1-e^{-\beta_i c_i d_i}\right) - \dfrac{c_i\beta_i e^{-\alpha_i d_i}}{(c_i\beta_i-\alpha_i)}\left(1-e^{-(\beta_i-\alpha_i/c_i)c_i d_i}\right)\right] \\ \quad \text{for } t > c_i d_i. \end{cases}$$

The transition probabilities of the embedded Markov chain have the following form:

$$p_{12} = e^{-(\lambda_1+\alpha_1)d_1}, \quad p_{14} = \frac{\alpha_1}{\lambda_1+\alpha_i}\left(1-e^{-(\lambda_1+\alpha_1)d_1}\right),$$

$$p_{23} = e^{-(\lambda_2+\alpha_2)d_2}, \quad p_{25} = \frac{\alpha_2}{\lambda_2+\alpha_2}\left(1-e^{-(\lambda_2+\alpha_2)d_2}\right),$$

$$p_{31} = e^{-(\lambda_3+\alpha_3)d_3}, \quad p_{36} = \frac{\alpha_3}{\lambda_3+\alpha_i}\left(1-e^{-(\lambda_3+\alpha_3)d_3}\right),$$

$$p_{42} = \frac{\alpha_1 e^{-\alpha_i d_1}}{(c_1\beta_1 - \alpha_i)(1 - e^{-\alpha_1 d_1})}\left(1 - e^{-(\beta_1 - \alpha_1/c_i)c_1 d_1}\right),$$

$$p_{53} = \frac{\alpha_2 e^{-\alpha_i d_2}}{(c_i\beta_2 - \alpha_2)(1 - e^{-\alpha_2 d_2})}\left(1 - e^{-(\beta_2 - \alpha_2/c_2)c_2 d_2}\right),$$

$$p_{61} = \frac{\alpha_3 e^{-\alpha_3 d_3}}{(c_i\beta_3 - \alpha_3)(1 - e^{-\alpha_3 d_3})}\left(1 - e^{-(\beta_3 - \alpha_3/c_3)c_3 d_3}\right).$$

For the same parameters (7.31) and

$$d_1 = \frac{1}{\gamma_1}, \quad d_2 = \frac{1}{\gamma_2}, \quad d_3 = \frac{1}{\gamma_3},$$

$$d_4 = \frac{1}{\gamma_1}, \quad d_5 = \frac{1}{\gamma_2}, \quad d_6 = \frac{1}{\gamma_3},$$

the numerical solution of (7.33) is

$$\begin{bmatrix} E(\Theta_{17}) \\ E(\Theta_{27}) \\ E(\Theta_{37}) \\ E(\Theta_{47}) \\ E(\Theta_{57}) \\ E(\Theta_{67}) \end{bmatrix} = \begin{bmatrix} 6139.39 \\ 6141.35 \\ 6138.18 \\ 6123.5 \\ 6107.17 \\ 6135.08 \end{bmatrix}.$$

Under an assumption that the initial state of the operation is 1 the mean time to the operation failure is

$$E(T) = \bar{T} = 6139.39.$$

Recall that in Example 1 this parameter was equal to $E(T) = \bar{T} = 5559.63$. Therefore, from presented examples it follows that we can formulate a hypothesis that for the determined duration of the stages a mean time to failure of the operation is essentially greater than for exponentially distributed duration of the stages with identical expectations.

The numerical solution of (7.36) is

$$\begin{bmatrix} E(\Theta_{17}^2) \\ E(\Theta_{27}^2) \\ E(\Theta_{37}^2) \\ E(\Theta_{47}^2) \\ E(\Theta_{57}^2) \\ E(\Theta_{67}^2) \end{bmatrix} = \begin{bmatrix} 7.51967 \times 10^7 \\ 7.52205 \times 10^7 \\ 7.5182 \times 10^7 \\ 7.49908 \times 10^7 \\ 7.47939 \times 10^7 \\ 7.51431 \times 10^7 \end{bmatrix}.$$

Therefore, in this case the standard deviation of the operation time to failure is

$$D(T) = D(\Theta_{17}) = 6124.1.$$

7.5   Model of multimodal transport operation

7.5.1   Introduction

The tasks of transport are realized by the transport operation systems. Some of them realized by any one of the carriers (by truck, by train, by ship, or by plane) we would contractually agree to call "simple transport." The task of transport realized by a combined means (carriers) of delivery we would contractually call "complex transport," which is a combination of the above- defined basic tasks of transport.

Multimodal transport is the transport of objects through at least two different carriers of any combination of simple tasks of transport carriers (by truck, by train, by ship, or by plane). The carrier might change any provided container meaning that it might be repackaged to another type of container to suit the requirements of any given carrier, as it might be required or practiced for logistic reasons.

Intermodal transport is the transport of objects through at least two different carriers of any combination of transport basic media (by truck, by train, by ship, or by plane). The carrier might not change the provided container, meaning that it might not be repackaged to another type of container, and has to be shipped in the originator's container. To suit the requirements of any air transport carrier (air transport should be used), it must be required to be packaged in a special approved for air transport specific container, usually leased or loaned from the air carrier.

Bimodal transport is the transport using a bimodal (truck, train) special truck trailer hook up container built to be transported by truck trailer (tractor) or on a railroads on a railroad platform car. The cargo is never repackaged between destinations.

The term reliability of the transport operation at the given moment t, means the probability of ability of the transport tasks realization at the instant t by the complex transport system. A reliability model of the intermodal system was investigated by M. Zając and it was presented in his Ph.D. thesis [100].

7.5.2   SM model

We will construct a SM model of the multimodal transport operation under the assumption that there is no perturbation during execution of the elementary tasks. In this case, we assume that the state space of the process is

$$S = \{1, 2, \ldots, n, n+1, n+2\}, \tag{7.40}$$

where $i = 1, \ldots, n$ denotes ith stage of the operation, $n + 1$ represents a state of the operation final part, and $n + 2$ means a failure of the operation.

We assume, as in the previous model, that a duration of an ith stage of operation, $(i = 1, \ldots, n, n+1)$ is a nonnegative random variable $\xi_i, i = 1, \ldots, n+1$ having a distribution given by CDF

$$F_{\xi_i}(t) = P(\xi_i \leqslant t). \tag{7.41}$$

We also assume that the time to failure of the operation on the ith stage is the nonnegative random variable $\eta_i, i = 1, \ldots, n$ with the exponential distribution

$$P(\eta_i \leqslant t) = 1 - e^{-\lambda_i t}, \quad i = 1, \ldots, n. \tag{7.42}$$

We also suppose that the operation interrupted due to failure can be carried out from the beginning. Time to resume the operation is a nonnegative random variable $\zeta_i, i = 1, \ldots, n$ with CDF

$$F_{\zeta_i}(t) = P(\zeta_i \leqslant t). \tag{7.43}$$

In this case, we suppose that the state space of the process is

$$S = \{1, 2, \ldots, n, n+1, n+2\},$$

where $i, i = 1, \ldots n$ denotes ith stage of the operation, $n+1$ represents a state of the operation final part, and $n+2$ means a failure of the operation. The possible state changes of the process are represented by a flow graph, which is shown in Figure 7.3.

The corresponding SM kernel is of the form

$$Q(t) = \begin{bmatrix} 0 & Q_{12}(t) & 0 & \ldots & 0 & 0 & 0 & Q_{1n+2}(t) \\ 0 & 0 & Q_{23}(t) & 0 & \ldots & 0 & 0 & Q_{2n+2}(t) \\ & & & \cdots\cdots\cdots & & & & \\ 0 & 0 & 0 & \ldots & 0 & 0 & Q_{nn+1}(t) & Q_{nn+2}(t) \\ Q_{n+11}(t) & 0 & 0 & \ldots & 0 & 0 & 0 & Q_{n+1n+2}(t) \\ Q_{n+21}(t) & 0 & 0 & \ldots & 0 & 0 & 0 & Q_{n+2n+2} \end{bmatrix}, \tag{7.44}$$

where

$$Q_{ii+1}(t) = P(\xi_i \leqslant t, \eta_i > \xi_i) = \int_0^t e^{-\lambda_i x} dF_{\xi_i}(x), \quad i = 1, \ldots, n, \tag{7.45}$$

$$Q_{in+2}(t) = P(\eta_i \leqslant t, \eta_i < \xi_i) = \int_0^t \lambda_i e^{-\lambda_i x} \left[1 - F_{\xi_i}(x) \right] dx, \quad i = 1, \ldots, n+1, \tag{7.46}$$

$$Q_{n+11}(t) = P(\xi_n \leqslant t) = F_{\xi_{n+1}}(t), \tag{7.47}$$

$$Q_{n+21}(t) = P(\zeta \leqslant t) = F_\zeta(t). \tag{7.48}$$

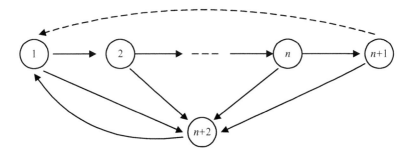

Figure 7.3 The graph of the state changes.

7.5.3 Example 3

A container with cargo is transported from Warsaw to Stockholm. From Warsaw to Gdynia the container is transported by lorry, from Gdynia to Karlscorona by ferry and from Karlscorona to Stockholm by truck. The transport operation final part is unloading the container. To describe the transport operation we apply the model presented above assuming $n = 3$. Random variables denoting duration of the operation stages mean:

     ξ_1—duration of the container transport from Warsaw to Gdynia and loading on the ferry,
     ξ_2—duration of the container transport from Gdynia to Karlscorona and unloading,
     ξ_3—duration of the container transport from Karlscorona to Stockholm,
     ξ_4—duration the operation final part (unloading).

We assume that values of expectations and standard deviations of these random variables are

$$E(\xi_1) = 6.5, \quad E(\xi_2) = 11.2, \quad E(\xi_3) = 8.6, \quad E(\xi_4) = 1.5\,[\text{h}]\,.$$

We also suppose that the failure rates on the transport stages are

$$\lambda_1 = 000028, \quad \lambda_2 = 0.0000076, \quad \lambda_3 = 0.0000014, \quad \lambda_4 = 0.000022\left[\frac{1}{\text{h}}\right].$$

The first passage time from the state $i, i = 1, 2, 3, 4$ to state 5, which is denoted as Θ_{i5}, represents a time to failure of the operation if the initial state is i. We will compute the expected values, the second moments and the standard deviations of these random variables. Equation (3.25) allows us to compute the expected values of these random variables. The equation with unknown expectations $E(\Theta_{i5}), i = 1, 2, 3, 4$, denoting the mean time to the operation failure, is

$$\begin{bmatrix} 1 & -p_{12} & 0 & 0 \\ 0 & 1 & -p_{23} & 0 \\ 0 & 0 & 1 & -p_{34} \\ -p_{41} & 0 & 0 & 1 \end{bmatrix} \begin{bmatrix} E(\Theta_{15}) \\ E(\Theta_{25}) \\ E(\Theta_{35}) \\ E(\Theta_{45}) \end{bmatrix} = \begin{bmatrix} E(T_1) \\ E(T_2) \\ E(T_3) \\ E(T_4) \end{bmatrix}, \tag{7.49}$$

where

$$p_{ii+1} = \int_0^\infty e^{-\lambda_i x}dF_{\xi_i}(x), \quad i = 1, 2, 3, \tag{7.50}$$

$$p_{i5} = \int_0^\infty \lambda_i e^{-\lambda_i x}\left[1 - F_{\xi_i}(x)\right]dx, \quad i = 1, \ldots, 4, \tag{7.51}$$

$$p_{41} = \int_0^\infty e^{-\lambda_4 x}dF_{\xi_4}(x). \tag{7.52}$$

We also assume that

$$F_{\xi_i}(t) = \begin{cases} 0 \text{ for } t \leqslant d_i, \\ 1 \text{ for } t > d_i, \end{cases}, \quad i = 1, 2, 3. \tag{7.53}$$

This means that the duration of each stage is deterministic and it is equal to $\xi_i = d_i = E(\xi)$ for $i = 1, 2, 3$. The numerical solution of (7.49) is

$$\begin{bmatrix} E(\Theta_{15}) \\ E(\Theta_{25}) \\ E(\Theta_{35}) \\ E(\Theta_{45}) \end{bmatrix} = \begin{bmatrix} 23,426.5 \\ 23,424.3 \\ 23,433.1 \\ 23,427.3 \end{bmatrix}.$$

The initial state of the operation is 1. Therefore, the mean time to failure of the transport operation is

$$E(T) = \bar{T} = E(\Theta_{15}) = 23,426.5 \,[\mathrm{h}].$$

In a manner similar to Example 1, we compute the second moments. We get

$$\begin{bmatrix} E(\Theta_{15}^2) \\ E(\Theta_{25}^2) \\ E(\Theta_{35}^2) \\ E(\Theta_{45}^2) \end{bmatrix} = \begin{bmatrix} 1.09769 \times 10^9 \\ 1.09758 \times 10^9 \\ 1.09799 \times 10^9 \\ 1.09772 \times 10^9 \end{bmatrix}.$$

Because we obtain a standard deviation of time to failure of the transport operation,

$$D(T) = D(\Theta_{15}) = 23,428.3 \,[\mathrm{h}].$$

Notice that the expected value and standard deviation are almost equal. Taking under consideration this fact and theorems of the perturbation theory from Chapter 5, we can assume that time to the transport operation failure is approximately exponentially distributed with parameter

$$\lambda = \frac{1}{E(T)}.$$

Therefore, the approximate reliability function of the transport operation is

$$R(t) \approx e^{-0.0000426t}.$$

SM model of working intensity process

Abstract

A working intensity is one of the important operation process characteristics. Mostly, the working intensity is randomly changeable and it may be modeled to be a finite state space semi-Markov process. Particularly, the working intensity is modeled by a semi-Markov random walk. For this process, lots of interesting characteristics are obtained. A semi-Markov model of a ship engine load allowed us to construct a random research test of the toxicity of exhaust gases.

Keywords: Working intensity process, Semi-Markov model, Ship engine load process, Research test of toxicity of exhaust gases

8.1   Introduction

A working intensity (working rate) is one of the important characteristics of an operation process. The measures of the working rate may be quite different. A load of a ship engine and a car speed are examples of it. Piasecki [84] and Olearczuk [77] have described the working rate by a deterministic real function taking nonnegative values. A semi-Markov model of the working rate process was introduced by Grabski [29, 30]. This kind of model has been applied to the mathematical description of a ship engine load process [31] and a random car speed [32]. The semi-Markov model of the ship engine load process allowed us to construct the research tests of the toxicity of the exhaust gases. Two- and three-phase research tests of toxic compounds in the exhaust gas of an engine installed in an engine room is presented by Kniaziewicz and Piaseczny in Ref. [66]. In the case of the car engine, during the chassis engine test bench, the running test is realized. These tests are suitable to the typical traffic conditions in large cities. Various running tests are elaborated.

8.2   Semi-Markov model of the ship engine load process

This section presents a semi-Markov model of the engine load of the main drive of the corvette, a warship. The model is a modification and extension of the results presented by Grabski et al. in Ref. [31]. The model is applied for the SULCER engine of 16ASV25/30 type (four-stroke, 16-cylinder engine supercharged in V system, $P_n = 3200\,\text{kW}$, $M_n = 1000\,\text{rot/min}$) installed in the power transmission system (with the adjusting screw) of the corvette.

Semi-Markov Processes: Applications in System Reliability and Maintenance. http://dx.doi.org/10.1016/B978-0-12-800518-7.00008-9

8.2.1 Model

We assume eight states of the motor action load due to the setting of the control system corresponding to the states of the load moment and rotational speed:

1—the motor action "with light running" $A_1 = [0 - 250)$ kW
2—the engine run with the load from the interval $A_2 = [250 - 270)$ kW,
3—the engine run with the load from the interval $A_3 = [270 - 280)$ kW,
4—the engine run with the load from the interval $A_4 = [280 - 300)$ kW,
5—the engine run with the load from the interval $A_5 = [300 - 350)$ kW,
6—the engine run with the load from the interval $A_6 = [350 - 560)$ kW,
7—the engine run with the load from the interval $A_7 = [560 - 1270)$ kW,
8—the engine run with the load from the interval $A_8 = [1270 - 3500)$ kW

A set $S = \{1, 2, \ldots, 8\}$ denotes the state space of the stochastic load process of the ship engine. A semi-Markov process $\{X(t) : t \geqslant 0\}$ seems to be the natural stochastic process with discrete state space describing the changeable load of the ship's engine during its work. We assume that a kernel of the process is of the form

$$\boldsymbol{Q}(t) = \left[Q_{ij}(t) : i, j \in S \right],$$

$$Q_{ij}(t) = p_{ij} G_i(t) : i, j \in S,$$

where p_{ij} represents a transition probability of a state change from a state i to j and $G_i(t)$ means the CDF of a random variable T_i that denotes a waiting time in state i. Based on the results of research on the engine load while the motor is running, it follows that the possible state changes are represented by the transition probability matrix:

$$\boldsymbol{P} = \begin{bmatrix} 0 & p_{12} & p_{13} & p_{14} & 0 & 0 & 0 & 0 \\ p_{21} & 0 & p_{23} & p_{24} & p_{25} & 0 & 0 & 0 \\ p_{31} & p_{32} & 0 & p_{34} & p_{35} & 0 & 0 & 0 \\ p_{41} & p_{42} & p_{43} & 0 & p_{45} & 0 & 0 & 0 \\ p_{51} & p_{52} & p_{53} & p_{54} & 0 & p_{56} & p_{57} & 0 \\ 0 & 0 & 0 & p_{64} & p_{65} & 0 & p_{67} & 0 \\ 0 & 0 & 0 & 0 & p_{75} & p_{76} & 0 & p_{78} \\ 0 & 0 & 0 & 0 & 0 & 0 & 1 & 0 \end{bmatrix}. \tag{8.1}$$

The matrix of CDFs of waiting times is

$$\boldsymbol{G}(t) = [G_i(t) : i = 1, \ldots, 8]^{\mathrm{T}}. \tag{8.2}$$

The results of observations allow us to estimate transition probabilities, expected values, and second moments of random variables T_i denoting the appropriate waiting times. The value of the estimator defined by (2.64) enables us to calculate the estimate of transition probability p_{ij}:

$$\hat{p}_{ij} = \frac{n_{ij}}{n_i}, \quad i, j \in S, \tag{8.3}$$

where n_{ij} denotes the semi-Markov process direct number of transitions from the state i to j and n_i means the number of jumps to the state i among all observations coming from a sample path of the process. We compute the estimates of expected values and second moments of the random variables T_i, $i \in S$ using the well-known formulas:

$$\hat{t}_i = \frac{1}{n_i} \sum_{k=1}^{n_i} t_{i,k}, \quad i \in S, \tag{8.4}$$

$$\hat{t}_i^2 = \frac{1}{n_i} \sum_{k=1}^{n_i} t_{i,k}^2, \quad i \in S, \tag{8.5}$$

where $t_{i,k}, k = 1, 2, \ldots, n_i$ is a sequence of realizations of the random variable T_i. These estimates we treat as parameters of the model. Hence, we get the matrix of transition probabilities

$$P = \begin{bmatrix} 0 & 0.68 & 0.23 & 0.09 & 0 & 0 & 0 & 0 \\ 0.22 & 0 & 0.35 & 0.39 & 0.04 & 0 & 0 & 0 \\ 0.32 & 0.44 & 0 & 0.19 & 0.05 & 0 & 0 & 0 \\ 0.17 & 0.26 & 0.28 & 0 & 0.22 & 0.07 & 0 & 0 \\ 0.06 & 0.03 & 0.19 & 0.37 & 0 & 0.28 & 0.07 & 0 \\ 0 & 0 & 0 & 0.25 & 0.67 & 0 & 0.08 & 0 \\ 0 & 0 & 0 & 0 & 0.14 & 0.39 & 0 & 0.47 \\ 0 & 0 & 0 & 0 & 0 & 0 & 1 & 0 \end{bmatrix}, \tag{8.6}$$

and matrices of expected values and second moments of the random variables T_i, $\in S$,

$$[E(T_i)]^{\mathrm{T}} = \begin{bmatrix} 9.98 & 8.19 & 12.12 & 11.34 & 20.06 & 7.98 & 7.35 & 14.62 \end{bmatrix}, \tag{8.7}$$

$$\left[E(T_i^2)\right]^{\mathrm{T}} = \begin{bmatrix} 117.88 & 104.30 & 249.60 & 257.19 & 444.77 & 78.33 & 72.59 & 363.34 \end{bmatrix}. \tag{8.8}$$

Statistical analysis of the data comes to the conclusion that distribution of waiting times can be chosen from a family of Weibull distributions. This means that the CDF of the random variable T_i is

$$G_i(x) = 1 - e^{-(x/\beta_i)^{\alpha_i}}, \quad x \geqslant 0, \ i \in S. \tag{8.9}$$

We have estimated unknown parameters α_i, β_i, $i \in S$ using the moments method. The computed parameters are presented in Table 8.1.

Table 8.1 **Computed Parameters**

State i	1	2	3	4	5	6	7	8
α_i	2.5	1.4	1.2	1.0	1.8	2.2	2.0	1.2
β_i	11.25	8.99	12.88	11.34	20.56	9.01	8.52	15.54

8.2.2 Characteristics of engine load process

An important characteristic of the engine load process is the limiting distribution P_j, $j \in S$, of the load states. To find it we have to apply Theorem 3.6. From this theorem, it follows that

$$P_j = \frac{\pi_j E(T_j)}{\sum_{k \in S} \pi_k E(T_k)}, \tag{8.10}$$

where π_i, $i \in S$ form the stationary distribution of Markov chain with transition probability matrix $\boldsymbol{P} = \left[p_{ij} : i,j \in S \right]$. This means that they are a solution of the linear system of equations

$$\sum_{i \in S} \pi_i p_{ij} = \pi_j, \quad j \in S, \quad \sum_{j \in S} \pi_j = 1. \tag{8.11}$$

Using a computer program written in MATHEMATICA, we get

$$\begin{aligned} \pi_1 &= 0.1605, \quad \pi_2 = 0.2532, \quad \pi_3 = 0.2013, \quad \pi_4 = 0.2017, \\ \pi_5 &= 0.1015, \quad \pi_6 = 0.0508, \quad \pi_7 = 0.0211, \quad \pi_8 = 0.0099 \end{aligned} \tag{8.12}$$

and

$$\begin{aligned} P_1 &= 0.1438, \quad P_2 = 0.1860, \quad P_3 = 0.2189, \quad P_4 = 0.2052, \\ P_5 &= 0.1828, \quad P_6 = 0.0364, \quad P_7 = 0.0139, \quad P_8 = 0.0130. \end{aligned} \tag{8.13}$$

This distribution is sometimes called the "spectrum" of the engine load process. For a better illustration of the engine load distribution, we introduce a random variable L with a density function

$$h(x) = \sum_{k=1}^{8} \frac{P_k}{|A_k|} I_{A_k}(x), \tag{8.14}$$

where $|A_k|$ denotes the length of the interval A_k and $I_{A_k}(x)$ means its indicator.

8.2.3 Model of research test of toxicity of exhaust gases

Contemporary research into the emission of exhaust gases toxicity by internal-combustion engines characterizes frequent modifications of the examination tests. The main goal is to obtain a more appropriate evaluation of the gases emission. The obligatory research tests of the exhaust gases toxicity take into consideration the real conditions of the engines' work by using the average values of their load. The basic fault of those tests is their deterministic character, because the load of the engine is usually randomly changeable. On one hand deterministic tests ensure the simplicity in conducting an examination of toxicity, but on the other hand those tests do not imitate the randomness of the engines loads. We propose that the research test be based on the semi-Markov model of the ship engine loads.

To construct a research rest, we propose as an application a sample path of semi-Markov model of the engine load process, which we can obtain using Monte-Carlo simulation procedure. The inconvenience of this method is the long time of examination. Therefore, instead of simulating the process in which the waiting times T_i, $i = 1, \ldots, 8$, have Weibull distributions with scale parameters β_i, $i = 1, \ldots, 8$, we can simulate the SM process in which these parameters are

$$\tilde{\beta}_i = c\beta_i, \quad 0 < c < 1.$$

Notice that for corresponding random variables $\tilde{T}_i$ $i = 1, \ldots, 8$, we have

$$E(\tilde{T}_i) = cE(T_i) \quad \text{and} \quad E(\tilde{T}_i^2) = c^2 E(T_i^2),$$

because

$$E(T_i) = \beta_i \Gamma \left[1 + \frac{1}{\alpha} \right] \quad \text{and} \quad E(T_i^2) = \beta_i^2 \Gamma \left[1 + \frac{2}{\alpha} \right].$$

Also notice that

$$\tilde{P}_j = \frac{\pi_j E(\tilde{T}_j)}{\sum_{k \in S} \pi_k E(\tilde{T}_k)} = \frac{\pi_j E(cT_j)}{\sum_{k \in S} \pi_k E(cT_k)} = \frac{\pi_j E(T_j)}{\sum_{k \in S} \pi_k E(T_k)} = P_j.$$

This means that in both cases the "spectrum" of the engine load process is identical. To obtain a research test we have to simulate the SM process $\{\tilde{X}(t) : t \geqslant 0\}$.

The algorithm of the SM process $\{\tilde{X}(t) : t \geqslant 0\}$ simulation

1. Take $\tilde{X}(0) = 1$.
2. Generate value t_1 of the waiting time T_1 with Weibull distribution $\mathcal{W}(\alpha_1, c\beta_1)$:
   - generate u from the uniform distribution $\mathcal{U}(0, 1)$,
   - calculate $t_1 = c\beta_1 [-\ln(1 - u)]^{1/\alpha_1}$.
3. Generate state $i = \tilde{X}(\tau_1)$ from distribution p_{1k}, $k = 1 \ldots 8$:
   - generate u from the uniform distribution $\mathcal{U}(0, 1)$,
   - take $i = \min\{r : u > \sum_{k=1}^{r-1} p_{1k}\}$.
4. Generate value t_i of the waiting time T_i with Weibull distribution $\mathcal{W}(\alpha_i, c\beta_i)$:
   - generate u from the uniform distribution $\mathcal{U}(0, 1)$,
   - calculate $t_i = c\beta_i [-\ln(1 - u)]^{1/\alpha_i}$.
5. Generate state $j = \tilde{X}(\tau_2)$ from distribution p_{ik}, $k = 1 \ldots 8$:
   - generate u from the uniform distribution $\mathcal{U}(0, 1)$,
   - take $j = \min\{r : u > \sum_{k=1}^{r-1}\}p_{ik}$.
6. Substitute $i = j$ and repeat points 4 and 5 m times.

According to this algorithm the computer program written in MATHEMATICA allowed us to construct the research test of toxicity of exhaust gases of the ship engine. We assume a probability transition matrix P given by (8.6), Weibull distributions parameters presented in Table 8.1 and parameters $c = 0.2$, $m = 100$. Finally, we obtain the following research test. Tables 8.2 and 8.3 represent part of the tests.

We can modify this test by substituting the expectations $E(\tilde{T}_i)$, $i = 1, \ldots, 8$ instead of random variables $\tilde{T}_i$, $i = 1, \ldots, 8$ realizations. The expectations $E(\tilde{T}_i)$, $i = 1, \ldots, 8$ are

$$\left[E(\tilde{T}_i) \right]^{\mathrm{T}} = \left[\begin{array}{cccccccc} 1.99 & 1.64 & 2.42 & 2.27 & 4.01 & 1.60 & 1.47 & 2.92 \end{array} \right].$$

Table 8.2 First Research Test of Toxicity of Exhaust Gases

State	1	2	3	2	4	3	4	5	4	5
Time	2.16	1.47	2.18	1.63	2.85	2.28	1.89	3.93	1.85	4.46
State	6	7	8	7	5	3	4	5	4	3
Time	1.84	2.14	2.27	1.44	2.23	3.18	2.39	4.23	1.85	2.56
State	2	1	2	3	4	3	2	1	3	1
Time	1.88	3.02	1.38	1.63	2.85	2.28	1.79	3.43	1.85	5.46
State	2	3	4	5	4	3	4	5	4	3
Time	1.88	3.78	2.12	1.63	1.77	2.99	2.30	5.75	2.01	0.95
State	2	1	3	2	1	3	4	5	4	5
Time	1.75	1.90	3.12	1.63	2.05	2.95	2.39	3.83	1.96	5.36

Table 8.3 Second Research Test of Toxicity of Exhaust Gases

State	1	2	3	2	4	3	4	5	4	5
Time	1.99	1.64	2.42	1.64	2.27	2.42	2.27	4.01	2.27	4.01
State	6	7	8	7	5	3	4	5	4	3
Time	1.60	1.47	2.92	1.47	4.01	2.42	2.27	4.01	2.27	2.42
State	2	1	2	3	4	3	2	1	3	1
Time	1.64	1.99	1.64	2.42	2.27	2.42	1.64	1.99	2.42	1.99
State	2	3	4	5	4	3	4	5	4	3
Time	1.64	2.42	2.27	4.01	2.27	2.42	2.27	4.01	2.27	2.42
State	2	1	3	2	1	3	4	5	4	5
Time	1.64	1.99	2.42	1.64	1.99	2.42	2.27	4.01	2.27	4.01

8.3   SM model for continuous working intensity process

Frequently, the working rate is randomly changeable and it may be treated as a stochastic process having nonnegative, continuous, and bounded trajectories. Let $\{U(t) : t \geqslant 0\}$ denote this stochastic process and let $\{u(t) : t \geqslant 0\}$, $u(t) \in U$ be its trajectory. One method of simplifying analysis of the process is state space discretization. A family of sets $\mathcal{A} = \{A_k = [u_{k-1}, u_k), \ k = 1, 2, \ldots, r\}$ such that

$$A_i \cap A_j = \emptyset, \quad \text{for } i \neq j \text{ and} \quad U = \bigcup_{k=1}^{r} A_k$$

is said to be partition of a state space U. A stochastic process $\{X(t) : t \geqslant 0\}$ with trajectories defined by a formula

$$x(t) = s_k \quad \text{iff} \quad u(t) \in A_k, \ k = 1, 2, \ldots, r \tag{8.15}$$

is a discrete state process with a state space $S = \{s_1, x_2, \ldots, s_r\}$. From the Darboux property of the continuous functions it follows that the graph of the process $\{X(t) : t \geqslant 0\}$ state changes has the form shown in Figure 8.1.

The instants τ_n, $n = 1, 2, \ldots$, denote moments of the fixed levels crossing by the stochastic process $\{U(t) : t \geqslant 0\}$. A random sequence $\{X(\tau_n) : n = 0, 1, 2, \ldots\}$ has a Markov property. The process $\{X(t) : t \geqslant 0\}$ takes the constant values on intervals

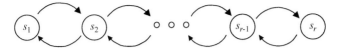

Figure 8.1 Graph of the process state changes.

$[\tau_0, \tau_1), [\tau_1, \tau_2), \ldots, [\tau_n, \tau_{n+1}) \ldots$ Lengths of those intervals are the positive random variables. These arguments allow us to assume that $\{X(t) : t \geqslant 0\}$ is a semi-Markov process with finite state space $S = \{s_1, s_2, \ldots, s_r\}$ and the flow graph presented in Figure 8.1. Let $J = \{1, \ldots r\}$ be a set of the states numbers. We can see that transition probabilities for all neighboring states are positive, that is, $p_{ii+1} > 0$ and $p_{ii-1} > 0$ for all $i - 1, i, i + 1 \in J$. That kind of process is called a *finite state semi-Markov random walk*.

8.3.1 Model

The model of a working rate stochastic process is a semi-Markov random walk. According to the graph shown in Figure 8.1, the kernel of the SM process is

$$\boldsymbol{Q}(t) = \begin{bmatrix} 0 & Q_{12}(t) & 0 & \cdots & 0 \\ Q_{21}(t) & 0 & Q_{23}(t) & \cdots & 0 \\ & & \cdots\cdots\cdots\cdots\cdots & & \\ 0 & \cdots & Q_{r-1r-2}(t) & 0 & Q_{r-1r}(t) \\ 0 & \cdots & & 0 & Q_{r,r-1}(t) & 0 \end{bmatrix}. \tag{8.16}$$

A matrix of transition probabilities of an embedded Markov chain $\{X(\tau_n) : n \in \mathbb{N}_0\}$ is

$$\boldsymbol{P} = \begin{bmatrix} 0 & 1 & 0 & \cdots & 0 \\ p_{21} & 0 & p_{23} & \cdots & 0 \\ & & \cdots\cdots\cdots\cdots & & \\ 0 & \cdots & p_{r-1r-2} & 0 & p_{r-1r} \\ 0 & \cdots & & 0 & 1 & 0 \end{bmatrix}, \tag{8.17}$$

where

$$p_{ij} = P\{X(\tau_{n+1}) = s_j | X(\tau_n) = s_i\} = \lim_{t \to \infty} Q_{ij}(t). \tag{8.18}$$

Let

$$p_{kk-1} = a_k \quad \text{and} \quad p_{kk+1} = b_k, \quad k = 2, \ldots, r - 1. \tag{8.19}$$

From the property of matrix $\boldsymbol{P}$, it follows that

$$a_k + b_k = 1.$$

We assume that kernel elements of this process are of the form

$$Q_{ij}(t) = p_{ij} G_i(t), \quad t \in [0, \infty), \tag{8.20}$$

where $G_i(\cdot)$ is CDF of a waiting time in a state s_i.

8.3.2 Estimation of the model parameters

The working intensity process trajectories (path functions) $\{u(t) : t \geqslant 0\}$ allow us to construct the trajectories $\{x(t) : t \geqslant 0\}$ of the discrete state space semi-Markov model. Having a continuous path function of the working rate process and the numbers $u_0, u_1, \ldots, u_r$, which define the partition of the space U, we can calculate the instants of the state changes τ_k, $k = 1, 2, \ldots, n$ of the process $\{X(t) : t \geqslant 0\}$. We can do it by solving the equations

$$u(t) = u_k, \quad k = 0, 1, 2, \ldots, r. \tag{8.21}$$

The increasing sequence of roots of those equations enables us to obtain a number n_{ij} of the transitions from the state s_i to s_j and the values $t_{i,k}$, of the waiting time T_i. A value of the maximum likelihood estimator of the transition probability p_{ij} is given by (8.3).

8.3.3 Analysis of the SM working intensity

An analysis of the random working rate process in this case consists of investigation of the probability characteristics and parameters of its semi-Markov model. One of the important characteristics is a limiting probability distribution of the process, which states

$$P_j = \lim_{t \to \infty} P\{X(t) = s_j\}. \tag{8.22}$$

Recall that

$$m_k = E(T_k) = \int_0^\infty [1 - G_k(t)] \, dt \tag{8.23}$$

is the expected value of the waiting time in state k. We know that

$$P_j = \frac{\pi_j m_j}{\sum_{k \in S} \pi_k m_k}, \tag{8.24}$$

and the stationary distribution $\pi = [\pi_1, \pi_2, \ldots, \pi_r]$ of an embedded Markov chain satisfies the linear system of equations

$$\pi P = \pi, \tag{8.25}$$

$$\sum_{j=1}^r \pi_j = 1,$$

which now takes the form

$$
\begin{aligned}
a_2 \pi_2 &= \pi_1 \\
\pi_1 + a_3 \pi_3 &= \pi_2 \\
b_2 \pi_2 + a_4 \pi_4 &= \pi_3 \\
&\vdots \\
b_{r-1} \pi_{r-1} &= \pi_r \\
\pi_1 + \pi_2 + \cdots + \pi_r &= 1.
\end{aligned}
\tag{8.26}
$$

Expressing the probabilities $\pi_2, \pi_3, \ldots, \pi_r$ by π_1, we have

$$\pi_j = \frac{b_1 b_2 \ldots b_{j-1}}{a_2 a_3 \ldots a_j} \pi_1 \quad \text{for } j = 2, 3, \ldots, r, \tag{8.27}$$

where

$$b_1 = 1, \quad a_r = 1.$$

From the equation $\sum_{j=1}^{r} \pi_j = 1$, we obtain

$$\pi_1 = \left(1 + \sum_{j=2}^{r} \prod_{k=2}^{j} \frac{b_{k-1}}{a_k} \right)^{-1}. \tag{8.28}$$

Using (8.24), we obtain a limit distribution of the working rate process $\{X(t) : t \geqslant 0\}$

$$P_1 = \frac{m_1}{m_1 + \sum_{j=2}^{r} \left[\prod_{k=2}^{j} \frac{b_{k-1}}{a_k} \right] m_j}, \tag{8.29}$$

$$P_j = \frac{\prod_{k=2}^{j} \frac{b_{k-1}}{a_k} m_j}{m_1 + \sum_{j=2}^{r} \left[\prod_{k=2}^{j} \frac{b_{k-1}}{a_k} \right] m_j}, \quad j = 2, \ldots, r.$$

A knowledge of the limit distribution is essential to the practice of the machine's exploitation. Notice that

$$P_j = \lim_{t \to \infty} P\{X(t) \in A_j\}, \quad j = 1, 2, \ldots, r.$$

If the measure of the ship engine working rate at the moment t is the load, then P_j denotes a probability that at any instant, distant from the initial moment, the load will have the values from the interval A_j.

Knowing the probability distribution $[P_1, P_2, \ldots, P_r]$ enables us to, among other things, anticipate fuel consumption and the emission of the exhaust gases toxicity.

If z_j denotes a mean value of an engine fuel consumption in a time unit under load from interval A_j, then the approximate expected value of the fuel consumption in the time interval $[0, t]$ is

$$E[Z(t)] \approx \sum_{j=1}^{n} P_j z_j t.$$

A length of intervals between entering the states s_i and s_j, signified as a Θ_{ij}, is the next characteristic of the process $\{X(t) : t \geqslant 0\}$. A value of this random variable denotes the time interval from the moment of beginning the work with a rate s_i to an instant of beginning the work with the rate s_j. As we know from Chapter 3, the random variable Θ_{ij} is called the first passage time from a state s_i to s_j. If $j = i$ we obtain the random variable Θ_{ii} that is said to be a return time to the state s_i. From (3.21) it follows that the Laplace-Stieltjes transforms of CDF $\Phi_{ij}(t) = P\{\Theta_{ij} \leqslant t\}$ are the elements of solutions of the matrix equation

$$\left(\boldsymbol{I} - \tilde{\boldsymbol{q}}_{A'}(s)\right) \tilde{\boldsymbol{\varphi}}_{A'}(s) = \tilde{\boldsymbol{b}}(s),$$ (8.30)

where

$$A' = J - \{j\} \quad \text{and} \quad \boldsymbol{I} = \left[\delta_{ik} : i, k \in A'\right]$$

is unit matrix,

$$\tilde{\boldsymbol{q}}_{A'}(s) = \left[\tilde{q}_{ik}(s) : i, k \in A'\right]$$ (8.31)

is a submatrix of

$$\tilde{q}(s) = \begin{bmatrix} 0 & \tilde{q}_{12}(t) & 0 & \cdots & 0 \\ \tilde{q}_{21}(s) & 0 & \tilde{q}_{23}(s) & \cdots & 0 \\ & & \cdots\cdots\cdots\cdots\cdots & & \\ 0 & \cdots & \tilde{q}_{r-1r-2}(s) & 0 & \tilde{q}_{r-1r}(s) \\ 0 & \cdots & 0 & \tilde{q}_{rr-1}(s) & 0 \end{bmatrix},$$ (8.32)

which is constructed by removing jth row, and jth column and

$$\tilde{\boldsymbol{\varphi}}_{A'}(s) = \begin{bmatrix} \tilde{\phi}_{1j}(s) \\ \vdots \\ \tilde{\phi}_{j-1j}(s) \\ \tilde{\phi}_{j+1j}(s) \\ \vdots \\ \tilde{\phi}_{rj}(s) \end{bmatrix}, \quad \tilde{\boldsymbol{b}}(s) = \begin{bmatrix} \tilde{q}_{1j}(s) \\ \vdots \\ \tilde{q}_{j-1j}(s) \\ \tilde{q}_{j+1j}(s) \\ \vdots \\ \tilde{q}_{rj}(s) \end{bmatrix}$$ (8.33)

are the one-column matrices of the appropriate Laplace-Stieltjes transforms. It is much easier to calculate the moments of the random variables $\Theta_{ij}, i, j \in J$. From Theorems 3.2 and 3.3 it follows that the expectations and second moments of these random variables for $i \neq j$ are the only solutions of the system of linear equations that have the following matrix form:

$$(\boldsymbol{I} - \boldsymbol{P}_{A'})\overline{\boldsymbol{\Theta}}_{A'} = \overline{\boldsymbol{T}}_{A'},$$ (8.34)

$$(\boldsymbol{I} - \boldsymbol{P}_{A'})\overline{\boldsymbol{\Theta}}_{A'}^{2} = \boldsymbol{B}_{A'},$$ (8.35)

where $\boldsymbol{I}$ is the appropriate unit matrix, $\boldsymbol{P}_{A'}$ is a submatrix that is obtained by deleting the ith row and jth column of the matrix $\boldsymbol{P}$.

$$\overline{\boldsymbol{\Theta}}_{A'} = \begin{bmatrix} E(\Theta_{1j}) \\ \vdots \\ E(\Theta_{j-1j}) \\ E(\Theta_{j+1j}) \\ \vdots \\ E(\Theta_{rj}) \end{bmatrix}, \quad \overline{\boldsymbol{T}}_{A'} = \begin{bmatrix} m_1 \\ \vdots \\ m_{j-1} \\ m_{j+1} \\ \vdots \\ m_r \end{bmatrix}, \quad \overline{\boldsymbol{\Theta}}_{A'}^{2} = \begin{bmatrix} E(\Theta_{1j}^{2}) \\ \vdots \\ E(\Theta_{j-1j}^{2}) \\ E(\Theta_{j+1j}^{2}) \\ \vdots \\ E(\Theta_{rj}^{2}) \end{bmatrix}, \quad \boldsymbol{B}_{A'} = \begin{bmatrix} b_{1j} \\ \vdots \\ b_{j-1j} \\ b_{j+1j} \\ \vdots \\ b_{rj} \end{bmatrix},$$

$$b_{ij} = E(T_i^2) + 2 \sum_{k \in E, k \neq j} p_{ik} E(T_i) E(\Theta_{kj}),$$

$$i = 1, 2, \ldots, j-1, j+1, \ldots, r, \quad j = 1, 2, \ldots, r.$$

For $i = j$, we have

$$E(\Theta_{jj}) = E(T_j) + \sum_{k \in E, k \neq j} p_{jk} E(\Theta_{rj}), \tag{8.36}$$

$$E(\Theta_{jj}^2) = E(T_j^2) + 2 \sum_{k \in J, k \neq j} p_{jk} \left[2 E(T_j) E(\Theta_{kj}) + E(\Theta_{kj}^2) \right]. \tag{8.37}$$

Solutions of the above equations enable us to compute parameters

$$E(\Theta_{ij}), \quad E(\Theta_{ij}^2), \quad V(\Theta_{ij}), \quad \sigma(\Theta_{ij}).$$

Variability of the working rate is characterized by the stochastic process $\{V_{ij}(t) : t \geqslant 0\}$, the value of which at the instant t denotes the number of inputs to the state j in the interval $[0, t]$ under assumption that the process $\{X(t) : t \geqslant 0\}$ starts from the state i. As we know from Chapter 6, the process $\{V_{ij}(t) : t \geqslant 0\}$ is a general renewal process, which is defined by a sequence

$$\Theta_{ij}^{(1)}, \Theta_{jj}^{(2)}, \ldots, \Theta_{jj}^{(n)}, \ldots$$

of the independent random variables in the following way:

$$V_{ij}(t) = \sup\{n : \Theta_{ij}^{(1)} + \Theta_{jj}^{(2)} + \cdots + \Theta_{jj}^{(n)} \leqslant t\}. \tag{8.38}$$

It has asymptotically normal distribution with the expected value

$$E\left[V_{ij}(t)\right] \approx \frac{t - E(\Theta_{ij}) + E(\Theta_{jj})}{E(\Theta_{jj})} \tag{8.39}$$

and the standard deviation

$$\sigma\left[V_{ij}(t)\right] \approx \sqrt{\frac{t V(\Theta_{jj})}{E\left(\Theta_{jj}\right]^3)}}. \tag{8.40}$$

Formula (5.20) enables us to calculate an approximate probability of the number of inputs to the state j in the interval $[0, t]$.

The theory of semi-Markov processes allows us to determine many other characteristics of the working rate process.

8.4   Model of car speed

Let $U(t)$ denote a car speed at the moment t. Suppose U is a set of the possible car speeds. We shall divide the set U into disjoint subsets

$$A_1 = [y_0, y_1), \quad A_2 = [y_1, y_2), \quad A_3 = [y_2, y_3), \quad A_4 = [y_3, y_4),$$
$$A_5 = [y_4, y_5), \quad A_6 = [y_5, y_6],$$

$$U = \bigcup_{k=1}^{6} A_k.$$

We assume

$$y_0 = 0, \quad y_1 = 10, \quad y_2 = 20, \quad y_3 = 60, \quad y_4 = 90, \quad y_5 = 120,$$
$$y_6 = 200 \text{ [km/h]}.$$

The stochastic process $\{U(t) : t \geqslant 0\}$ we replace by the semi-Markov process $\{X(t) : t \geqslant 0\}$ with the set of states J $J = \{1, \ldots, 6\}$, corresponding to the speed intervals A_k, $k \in J$.

The transition matrix of the embedded Markov chain is of the form

$$P = \begin{bmatrix} 0 & 1 & 0 & 0 & 0 & 0 \\ a_2 & 0 & b_2 & 0 & & \\ 0 & a_3 & 0 & b_3 & 0 & 0 \\ 0 & 0 & a_4 & 0 & b_4 & 0 \\ 0 & 0 & 0 & a_5 & 0 & b_5 \\ 0 & 0 & 0 & 0 & 1 & 0 \end{bmatrix}, \tag{8.41}$$

where $a_i = 1 - b_i$. We assume

$$a_2 = 0.12, \quad a_3 = 0.25, \quad a_4 = 0.34, \quad a_5 = 0.64.$$

We assume that the distributions of the random variables T_i denoting durations of the process states are unknown, while we only know the expected values $E(T_i)$ and the second moments $E(T_{ij}^2), i, j \in S$. Let

$$\overline{T} = [E(T_i) : i, \in J]^{\text{T}}, \quad \overline{T^2} = \left[E(T_i^2) : i, j \in S \right]^{\text{T}}.$$

We assume

$$T = \begin{bmatrix} 12 \\ 18 \\ 126 \\ 186 \\ 224 \\ 140 \end{bmatrix} \text{ [s]}, \quad \overline{T}^2 = \begin{bmatrix} 196 \\ 400 \\ 16,900 \\ 42,600 \\ 71,200 \\ 25,000 \end{bmatrix} \text{ [s}^2\text{]}.$$

We calculate the limiting probabilities of the working rate process $\{X(t) : t \geqslant 0\}$, applying (8.27) and (8.28). Finally, we obtain

j	1	2	3	4	5	6
P_j	0.0003	0.0044	0.1078	0.3509	0.4357	0.1009

(8.42)

Table 8.4 **Parameters of Random Variables** $\Theta_{i4}, i = 1, \ldots, 6$

State i	1	2	3	4	5	6
$E(\Theta_{i4})$ [s]	209.45	197.45	175.36	428.75	528.60	568.75
$E(\Theta_{i4}^2)$ [s²]	55,346.2	50,411.3	41,942.5	427,052.0	336,166.0	481,216.0
$D(\Theta_{i4})$ [s]	107.12	106.88	105.78	493.18	390.30	397.16

We can obtain further conclusions on the car speed by computing parameters of the random variables Θ_{ij}, $i, j \in S$. Applying 3.25, 3.26, 3.30 and 3.31 we can calculate the expectations $E(\Theta_{ij})$, second moments $E(\Theta_{ij}^2)$, and standard deviations $D(\Theta_{ij})$. For illustration, we calculate these parameters for Θ_{i4}, $i = 1, \ldots, 6$. Using the author's program in MATHEMATICA for assuming the above parameters we get following results: Numbers $E(\Theta_{i4})$, $i = 1, \ldots, 6$ denote the expectations of time from the moment when the vehicle speed begins to move at A_i to the instant when it reaches a speed of A_4. Especially $E(\Theta_{44}) = 493.18$ [s] means the expectation of a time interval between two successive moments of achieving speed from A_4.

Variability of the car speed in the working intensity model is characterized by the stochastic processes $\{V_{ij}(t) : t \geqslant 0\}$, $i, j \in S$, the value of which at the instant t denotes the number of inputs to the state j in the interval $[0, t]$ under the assumption that the process $\{X(t) : t \geqslant 0\}$ starts from the state i. We suppose that an initial state is $i = 1$ and $j = 4$. We know from Chapter 5, that the process $\{V_{14}(t) : t \geqslant 0\}$ is a general renewal process, which has an asymptotically normal distribution with the expected value

$$E[V_{14}(t)] \approx \frac{t - E(\Theta_{14}) + E(\Theta_{44})}{E(\Theta_{jj})} \tag{8.43}$$

and the standard deviation

$$\sigma[V_{14}(t)] \approx \sqrt{\frac{tV(\Theta_{44})}{[E(\Theta_{44})]^3}}. \tag{8.44}$$

Substituting the appropriate numbers from Table 8.4, we get

$$E[V_{14}(t)] \approx 0.002332t + 0.511487 , \quad \sigma[V_{14}(t)] \approx 0.003086\sqrt{t}.$$

The number $E[V_{14}(21,600)] = 50.37$ means that during 6 h a car achieves its speed from the interval A_4 in average 50.37 times with the standard deviation $D[V_{14}(21,600)] = 0.45354$.

Multitask operation process

Abstract

An object (device, machine, mean of transport) can work by executing different randomly occurring tasks. A SM model of the object multitask operation process is constructed in this chapter. The SM model construction relies on determining the process kernel. The concepts and theorems of semi-Markov processes theory allow us to compute the object reliability characteristics and parameters such as limit distribution of the state, the mean time to failure, the second moment of the time to failure, and approximate reliability function.

Keywords: Semi-Markov model, Multitask operation, Maintenance service, Emergency service, Reliability function, Mean time to failure, Limiting distribution

9.1 Introduction

A lot of technical objects are intended to perform different tasks. The means of transport could be one of them. Distinct tasks determine different work intensities. The working intensity influences the reliability. Semi-Markov processes theory enables us to construct and analyze the multitask operation process. To examine the model we use concepts and theorems from Chapters 3 and 4. The model constructed and investigated here is an extension of the results discussed in Ref. [32, 33].

9.2 Description and assumptions

An object (device, machine, mean of transport) can work by executing the different tasks, say $z_1, \ldots, z_r$, which are treated as the values of a random variable Z with discrete distribution $P(Z = z_k) = a_k$, $k = 1, \ldots, r$. We suppose that z_k task realization begins at the moments $\tau_{2n}, n \in \mathbb{N}_0$ with probability a_k.

We assume that the duration of the z_k task realization $(k = 1, \ldots, r)$ is a nonnegative random variable ξ_k, $k = 1, \ldots, r$ with a CDF

$$F_{\xi_k}(x) = P(\xi_k \leqslant x).$$

After realization of the task z_k, the technical maintenance service is executed by the time η_k, which is a nonnegative random variable with probability distribution determined by PDF $f_\eta(x)$.

We assume that the object may be damaged during operation. We suppose that time to the object failure that executes the task z_k is a nonnegative random variable ζ_k governed by PDF $f_{\zeta_k}(x)$. An emergency repair begins immediately after failure, which lasts for a random time κ with PDF $f_\kappa(x)$. As a result of emergency service, the object's

Semi-Markov Processes: Applications in System Reliability and Maintenance. http://dx.doi.org/10.1016/B978-0-12-800518-7.00009-0

technical state is restored. We assume that all random variables presented here are independent of each other and the operation process is described by the independent copies of the random variables. We also assume that the all random variables have finite and positive expected values and variances.

9.3 Model construction

We introduce the following states:

k - executing the task z_k, $k = 1, \ldots, r$,
$k + 1$ - maintenance service after realization of the task z_k,
$2r + 1$ - emergency service after the failure of the object (down state).

Let $\{X(t) : t \geqslant 0\}$ be a stochastic process with the piecewise constant and right-continuous sample paths and a state space $S = \{1, \ldots, 2r + 1\}$. The jumps of the process follow at the moments $\tau_0, \tau_1, \tau_2, \ldots$, the work of the object begins at the instants, while the maintenance service starts at the moments $\tau_0, \tau_2, \ldots$.

From this assumption, it follows that the state change is at the moment τ_n, the duration of this state. The state that is reached at the instant τ_{n+1} does not stochastically depend on states trial in moments $\tau_0, \ldots, \tau_{n-1}$ and the time of their duration. This means that $\{X(t) : t \geqslant 0\}$ is the semi-Markov process.

The initial distribution of the process is given by

$$p_k = P(X(0) = k) = \begin{cases} a_k & \text{for } k = 1, \ldots, r \\ 0 & \text{for } k = r + 1, \ldots, 2r + 1 \end{cases}, \tag{9.1}$$

and the SM process kernel is of the form

$$\boldsymbol{Q}(t) = \begin{bmatrix} 0 & \cdots & 0 & Q_{1r+1}(t) & 0 & \cdots & 0 & Q_{12r+1}(t) \\ 0 & \cdots & 0 & 0 & Q_{2r+2}(t) & \cdots & 0 & Q_{22r+1}(t) \\ \vdots & \vdots & \vdots & \vdots & \vdots & \vdots & \vdots & \vdots \\ 0 & \cdots & 0 & 0 & 0 & \cdots & Q_{r2r}(t) & Q_{r2r+1}(t) \\ Q_{r+11}(t) & \cdots & Q_{r+1r}(t) & 0 & 0 & \cdots & 0 & 0 \\ \vdots & \vdots & \vdots & \vdots & \vdots & \vdots & \vdots & \vdots \\ Q_{2r1}(t) & \cdots & Q_{2rr}(t) & 0 & 0 & \cdots & 0 & 0 \\ Q_{2r+11}(t) & \cdots & Q_{2r+1r}(t) & 0 & 0 & \cdots & 0 & 0 \end{bmatrix}. \tag{9.2}$$

The SM process will be determined if we calculate all positive elements of the matrix $\boldsymbol{Q}(t)$. We will present he way of determining the kernel elements using the function

$$Q_{kr+k}(t) = P(\tau_{n+1} - \tau_n \leqslant t, \, X(\tau_{n+1}) = r + k | X(\tau_n) = k), \quad k = 1, \ldots, r.$$

Note that a change of the state from k (denoting the object execution of the task k) to the state $r + k$ (meaning corresponding maintenance service) at the time no greater than t happens if $\xi_k \leqslant t$ and $\zeta_k > \xi_k$. Because

$$Q_{k\,r+k}(t) = P(\xi_k \leqslant t, \zeta_k > \xi_k) = \iint_{D_{kr+k}} dF_{\xi_k}(x) f_{\zeta_k}(y) dy,$$

where

$$D_{k\,r+k} = \{(x, y) : \ 0 \leqslant x \leqslant t, \ y > x\}.$$

Hence

$$Q_{k\,r+k}(t) = \int_0^t dF_{\xi_k}(x) \int_x^\infty f_{\zeta_k}(y) dy = \int_0^t \left[1 - F_{\zeta_k}(x)\right] dF_{\xi_k}(x), \quad k = 1, \dots, r.$$

$$\tag{9.3}$$

We obtain the rest of the kernel elements in a similar way:

$$Q_{k2r+1}(t) = P(\zeta_k \leqslant t, \ \xi_k > \zeta_k) = \int_0^t \left[1 - F_{\xi_k}(y)\right] f_{\zeta_k}(y) dy, \quad k = 1, \dots, r,$$

$$\tag{9.4}$$

$$Q_{r+kj}(t) = P(\eta_k \leqslant t, Z = z_j) = a_j F_{\eta_k}(t), \quad k = 1, \dots, r, j = 1, \dots, r, \tag{9.5}$$

$$Q_{2r+1j}(t) = P(\kappa \leqslant t, Z = z_j) = a_j F_\kappa(t), \quad j = 1, \dots, r. \tag{9.6}$$

All elements of the matrix $Q(t)$ have been defined, therefore the semi-Markov model is constructed.

9.4 Reliability characteristics

For simplicity we assume $r = 3$. A graph shown in Figure 9.1 presents possible state changes of the process.

The corresponding probability transition matrix of the embedded Markov chain $\{X(\tau_n) : n \in \mathbb{N}_0\}$ is of the form

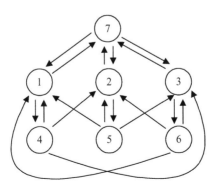

Figure 9.1 Possible state changes of the process.

$$P = \begin{bmatrix} 0 & 0 & 0 & p_{14} & 0 & 0 & p_{17} \\ 0 & 0 & 0 & 0 & p_{25} & 0 & p_{27} \\ 0 & 0 & 0 & 0 & 0 & p_{36} & p_{27} \\ p_{41} & p_{42} & p_{43} & 0 & 0 & 0 & 0 \\ p_{51} & p_{52} & p_{53} & 0 & 0 & 0 & 0 \\ p_{61} & p_{62} & p_{63} & 0 & 0 & 0 & 0 \\ p_{71} & p_{72} & p_{73} & 0 & 0 & 0 & 0 \end{bmatrix}, \tag{9.7}$$

where

$$p_{k\,r+k} = \int_0^\infty \left[1 - F_{\zeta_k}(x)\right] \mathrm{d}F_{\xi_k}(x), \quad k = 1, 2, 3,$$

$$p_{k7} = \int_0^\infty \left[1 - F_{\xi_k}(y)\right] f_{\zeta_k}(y)\mathrm{d}y, \quad k = 1, 2, 3, \tag{9.8}$$

$$p_{r+k\,j} = a_j, \quad k = 1, 2, 3, \, j = 1, 2, 3,$$

$$p_{7j} = a_j, \quad j = 1, 2, 3.$$

In this case, (1.21) for a stationary distribution of the embedded Markov chain is

$$\begin{bmatrix} -1 & 0 & 0 & p_{14} & 0 & 0 & p_{17} \\ 0 & -1 & 0 & 0 & p_{25} & 0 & p_{27} \\ 0 & 0 & -1 & 0 & 0 & p_{36} & p_{37} \\ 1 & 1 & 1 & 1 & 1 & 1 & 1 \\ p_{51} & p_{52} & p_{53} & 0 & -1 & 0 & 0 \\ p_{61} & p_{62} & p_{63} & 0 & 0 & -1 & 0 \\ p_{71} & p_{72} & p_{73} & 0 & 0 & 0 & -1 \end{bmatrix} \begin{bmatrix} \pi_1 \\ \pi_2 \\ \pi_3 \\ \pi_4 \\ \pi_5 \\ \pi_6 \\ \pi_7 \end{bmatrix} = \begin{bmatrix} 0 \\ 0 \\ 0 \\ 1 \\ 0 \\ 0 \\ 0 \end{bmatrix}. \tag{9.9}$$

We calculate limit distribution of the states probabilities

$$P_j = \lim_{t \to \infty} P(X(t) = j), \quad j \in S = \{1, \dots, 7\}$$

using Theorem 3.6. Then, the formula (3.73) takes the form of

$$P_j = \frac{\pi_j m_j}{\sum_{i=1}^7 \pi_i m_i}, \quad j \in S = \{1, \dots, 7\}, \tag{9.10}$$

where

$$m_k = E(T_k) = \int_0^\infty t\left[1 - F_{\zeta_k}(t)\right] \mathrm{d}F_{\xi_k}(t)$$
$$+ \int_0^\infty t\left[1 - F_{\xi_k}(t)\right] f_{\zeta_k}(t)\mathrm{d}t, \quad k = 1, 2, 3,$$

$$m_{3+k} = E(T_{3+k}) = \int_0^\infty t f_{\eta_k}(t)\mathrm{d}t, \quad k = 1, 2, 3, \tag{9.11}$$

$$m_7 = E(T_7) = \int_0^\infty t f_\kappa(t)\mathrm{d}t.$$

Recall that state 7 means the emergency service after the object failure. The functions $\Phi_{17}(t)$, $\Phi_{27}(t)$, $\Phi_{37}(t)$ are CDF of the random variables Θ_{17}, Θ_{27}, Θ_{37} denoting a first passage times from states $1, 2, 3$ to the "down" state 7. The function

$$F(t) = a_1 \Phi_{17}(t) + a_2 \Phi_{27}(t) + a_3 \Phi_{37}(t) \tag{9.12}$$

denotes CDF of lifetime T while

$$R(t) = 1 - F(t), \quad t \geqslant 0 \tag{9.13}$$

is a reliability function of the object. The Laplace-Stieltjes transforms of the functions $\Phi_{17}(t)$, $\Phi_{27}(t)$, $\Phi_{37}(t)$ we obtain by solving (4.20), which in this case takes the form of

$$
\begin{bmatrix}
1 & 0 & 0 & -\tilde{q}_{14}(s) & 0 & 0 \\
0 & 1 & 0 & 0 & -\tilde{q}_{25}(s) & 0 \\
0 & 0 & 1 & 0 & 0 & -\tilde{q}_{36}(s) \\
-\tilde{q}_{41}(s) & -\tilde{q}_{42}(s) & -\tilde{q}_{43}(s) & 1 & 0 & 0 \\
-\tilde{q}_{51}(s) & -\tilde{q}_{52}(s) & -\tilde{q}_{53}(s) & 1 & 0 & 0 \\
-\tilde{q}_{61}(s) & -\tilde{q}_{62}(s) & -\tilde{q}_{63}(s) & 1 & 0 & 0
\end{bmatrix}
\begin{bmatrix}
\tilde{\varphi}_{17}(s) \\
\tilde{\varphi}_{27}(s) \\
\tilde{\varphi}_{37}(s) \\
\tilde{\varphi}_{47}(s) \\
\tilde{\varphi}_{57}(s) \\
\tilde{\varphi}_{67}(s)
\end{bmatrix}
=
\begin{bmatrix}
\tilde{q}_{17}(s) \\
\tilde{q}_{27}(s) \\
\tilde{q}_{37}(s) \\
0 \\
0 \\
0
\end{bmatrix}.
$$

From the solution of this equation we get the Laplace-Stieltjes transform of the density functions $\tilde{\varphi}_{i7}(s)$ of the random variable Θ_{i7}, $i = 1, \ldots, 6$. The Laplace transform of the reliability function is given by the formula

$$\tilde{R}(s) = \frac{1 - (a_1 \tilde{\varphi}_{17}(s) + a_1 \tilde{\varphi}_{27}(s) + a_3 \tilde{\varphi}_{37}(s))}{s}. \tag{9.14}$$

We get the mean time to failure of the object by solving a matrix equation (4.25), which now has the form

$$
\begin{bmatrix}
1 & 0 & 0 & -p_{14} & 0 & 0 \\
0 & 1 & 0 & 0 & -p_{25} & 0 \\
0 & 0 & 1 & 0 & 0 & -p_{36} \\
-p_{41} & -p_{42} & -p_{43} & 1 & 0 & 0 \\
-p_{51} & -p_{52} & -p_{53} & 1 & 0 & 0 \\
-p_{61} & -p_{62} & -p_{63} & 1 & 0 & 0
\end{bmatrix}
\begin{bmatrix}
E(\Theta_{17}) \\
E(\Theta_{27}) \\
E(\Theta_{37}) \\
E(\Theta_{47}) \\
E(\Theta_{57}) \\
E(\Theta_{67})
\end{bmatrix}
=
\begin{bmatrix}
m_1 \\
m_2 \\
m_3 \\
m_4 \\
m_5 \\
m_6
\end{bmatrix}.
\tag{9.15}
$$

The mean time to the object failure is

$$E(T) = a_1, \quad E(\Theta_{17}) + a_2, \quad E(\Theta_{27}) + a_3, \quad E(\Theta_{37}). \tag{9.16}$$

We obtain the second moments of the random variables $\Theta_{i7}, i \in \{1, \ldots, 6\}$ by solving the matrix equation

$$\begin{bmatrix} 1 & 0 & 0 & -p_{14} & 0 & 0 \\ 0 & 1 & 0 & 0 & -p_{25} & 0 \\ 0 & 0 & 1 & 0 & 0 & -p_{36} \\ -p_{41} & -p_{42} & -p_{43} & 1 & 0 & 0 \\ -p_{51} & -p_{52} & -p_{53} & 1 & 0 & 0 \\ -p_{61} & -p_{62} & -p_{63} & 1 & 0 & 0 \end{bmatrix} \begin{bmatrix} E(\Theta_{17}^2) \\ E(\Theta_{27}^2) \\ E(\Theta_{37}^2) \\ E(\Theta_{47}^2) \\ E(\Theta_{57}^2) \\ E(\Theta_{67}^2) \end{bmatrix} = \begin{bmatrix} b_{17} \\ b_{27} \\ b_{37} \\ b_{47} \\ b_{57} \\ b_{67} \end{bmatrix}, \qquad (9.17)$$

where

$$b_{i7} = E(T_i^2) + 2 \sum_{k \in A'} p_{ik} E(T_{ik}) E(\Theta_{k7}),$$

and

$$E(T_{kr+k}) = \frac{1}{p_{kr+k}} \int_0^\infty t \left[1 - F_{\zeta_k}(t) \right] dF_{\xi_k}(t), \quad k = 1, 2, 3,$$

$$E(T_{kr+k}) = \frac{1}{p_{k2r+1}} \int_0^\infty t \left[1 - F_{\xi_k}(t) \right] f_{\zeta_k}(t) dt, \quad k = 1, 2, 3,$$

$$E(T_{r+kj}) = \int_0^\infty t f_{\eta_k}(t) dt, \qquad\qquad\qquad k = 1, 2, 3, \quad j = 1, 2, 3,$$

$$E(T_{2r+1j}) = \int_0^\infty t f_\kappa(t) dt, \qquad\qquad\qquad j = 1, 2, 3. \qquad (9.18)$$

The second moment of the time to the object failure is

$$E(T^2) = a_1 E(\Theta_{17}^2) + a_2 E(\Theta_{27}^2) + a_3 E(\Theta_{37}^2). \qquad (9.19)$$

9.5   Approximate reliability function

In this section we want to find an approximate reliability function using the Shpak concept of the perturbed SM process [89], which is presented in Chapter 4. In this case $A' = \{1, 2, \ldots, 6\}$, $A = \{7\}$, and $S = A \cup A'$. According to Definition 4.1 we can treat the investigated SM process determined by the kernel

$$\boldsymbol{Q}(t) = \begin{bmatrix} 0 & 0 & 0 & Q_{14}(t) & 0 & 0 & Q_{17}(t) \\ 0 & 0 & 0 & 0 & Q_{25}(t) & 0 & Q_{27}(t) \\ 0 & 0 & 0 & 0 & 0 & Q_{36}(t) & Q_{37}(t) \\ Q_{41}(t) & Q_{42}(t) & Q_{43}(t) & 0 & 0 & 0 & 0 \\ Q_{51}(t) & Q_{52}(t) & Q_{53}(t) & 0 & 0 & 0 & 0 \\ Q_{61}(t) & Q_{62}(t) & Q_{63}(t) & 0 & 0 & 0 & 0 \end{bmatrix}, \qquad (9.20)$$

where

$$Q_{k\,3+k}(t) = \int_0^t \left[1 - F_{\zeta_k}(x) \right] dF_{\xi_k}(x), \quad k = 1, 2, 3, \tag{9.21}$$

$$Q_{k7}(t) = \int_0^t \left[1 - F_{\xi_k}(y) \right] f_{\zeta_k}(y) dy, \quad k = 1, 2, 3, \tag{9.22}$$

$$Q_{3+kj}(t) = a_j F_{\eta_k}(t), \quad k = 1, 2, 3, \quad j = 1, 2, 3, \tag{9.23}$$

$$Q_{7j}(t) = a_j F_\kappa(t), \quad j = 1, 2, 3 \tag{9.24}$$

as a perturbed SM process with respect to SM process $\{X^0(t) : t \geqslant 0\}$ with the state space $A = \{1, \dots, 6\}$ and the kernel

$$\boldsymbol{Q}^0(t) = \begin{bmatrix} 0 & 0 & 0 & Q_{14}^0(t) & 0 & 0 \\ 0 & 0 & 0 & 0 & Q_{25}^0(t) & 0 \\ 0 & 0 & 0 & 0 & 0 & Q_{36}^0(t) \\ Q_{41}^0(t) & Q_{42}^0(t) & Q_{43}^0(t) & 0 & 0 & 0 \\ Q_{51}^0(t) & Q_{52}^0(t) & Q_{53}^0(t) & 0 & 0 & 0 \\ Q_{61}^0(t) & Q_{62}^0(t) & Q_{63}^0(t) & 0 & 0 & 0 \end{bmatrix}, \tag{9.25}$$

where

$$Q_{14}^0(t) = F_{\xi_1}(t), \quad Q_{25}^0(t) = F_{\xi_2}(t), \quad Q_{36}^0(t) = F_{\xi_3}(t),$$
$$Q_{41}^0(t) = a_1 F_{\eta_1}(t), \quad Q_{42}^0(t) = a_2 F_{\eta_1}(t), \quad Q_{43}^0(t) = a_3 F_{\eta_1}(t),$$
$$Q_{51}^0(t) = a_1 F_{\eta_2}(t), \quad Q_{52}^0(t) = a_2 F_{\eta_2}(t), \quad Q_{53}^0(t) = a_3 F_{\eta_2}(t), \tag{9.26}$$
$$Q_{61}^0(t) = a_1 F_{\eta_3}(t), \quad Q_{62}^0(t) = a_2 F_{\eta_3}(t), \quad Q_{63}^0(t) = a_3 F_{\eta_3}(t).$$

The transition probability matrix of the embedded Markov chain corresponding to this kernel is

$$\boldsymbol{P}^0 = \begin{bmatrix} 0 & 0 & 0 & 1 & 0 & 0 \\ 0 & 0 & 0 & 0 & 1 & 0 \\ 0 & 0 & 0 & 0 & 0 & 1 \\ a_1 & a_2 & a_3 & 0 & 0 & 0 \\ a_1 & a_2 & a_3 & 0 & 0 & 0 \\ a_1 & a_2 & a_3 & 0 & 0 & 0 \end{bmatrix}. \tag{9.27}$$

The stationary distribution of the Markov chain defined by this transition probability matrix is uniform:

$$\pi^0 = \left[\frac{1}{6}, \frac{1}{6}, \frac{1}{6}, \frac{1}{6}, \frac{1}{6}, \frac{1}{6} \right]. \tag{9.28}$$

We can obtain the approximate reliability function by applying Theorem 4.1. To use the theorem we have to calculate all parameters that we need. The CDF of the waiting times of the SM process $\{X^0(t) : t \geqslant 0\}$ are given by the formula

$$G_i^0(t) = \sum_{j \in A'} Q_{ij}^0(t). \tag{9.29}$$

Because

$$G_i^0(t) = F_{\xi_i}(t) \qquad \text{for } i = 1, 2, 3,$$
$$G_i^0(t) = F_{\eta_{i-3}}(t)) \quad \text{for } i = 4, 5, 6. \tag{9.30}$$

Hence, the expectations of the corresponding waiting times are

$$m_i^0 = E(T_i^0) = E(\xi_i) = \int_0^\infty t \, dF_{\xi_i}(t) \qquad \text{for } i = 1, 2, 3,$$
$$m_i^0 = E(T_i^0) = E(\eta_{i-3}) = \int_0^\infty t f_{\eta_{i-3}}(t) dt \quad \text{for } i = 4, 5, 6. \tag{9.31}$$

The parameter

$$\varepsilon_i = p_{i7} = \int_0^\infty \left[1 - F_{\xi_i}(y) \right] f_{\zeta_i}(y) dy, \quad i = 1, 2, 3, \tag{9.32}$$

denotes the transition probability of the embedded Markov chain $\{X(\tau_n) : n \in \mathbb{N}_0\}$ from the state $i \in A'$ to state 7. The parameter

$$\varepsilon = \sum_{i \in A'} \pi_i^0 \varepsilon_i \tag{9.33}$$

in this case takes the form

$$\varepsilon = \frac{1}{6} \sum_{i=1}^3 \int_0^\infty \left[1 - F_{\xi_i}(y) \right] f_{\zeta_i}(y) dy \tag{9.34}$$

while a parameter

$$m^0 = \sum_{i \in A'} \pi_i^0 m_i^0, \tag{9.35}$$

is

$$m^0 = \frac{1}{6} \left(E(\xi_1) + E(\eta_1) + E(\xi_2) + E(\eta_2) + E(\xi_3) + E(\eta_3) \right). \tag{9.36}$$

From Theorem 4.1 we have

$$\lim_{\varepsilon \to 0} P(\varepsilon \Theta_{i7} > t) = \exp\left[-\lambda t \right], \quad t \geqslant 0, \tag{9.37}$$

where

$$\lambda = \frac{6}{E(\xi_1) + E(\eta_1) + E(\xi_2) + E(\eta_2) + E(\xi_3) + E(\eta_3)}. \tag{9.38}$$

This means that the approximate reliability function of the object is

$$R(t) \approx e^{-\alpha t}, \quad \text{where } \alpha = \varepsilon \lambda. \tag{9.39}$$

Therefore,

$$R(t) \approx \exp\left[-\frac{\int_0^\infty [1-F_{\xi_1}(y)]f_{\zeta_1}(y)dy + \int_0^\infty [1-F_{\xi_2}(y)]f_{\zeta_2}(y)dy + \int_0^\infty [1-F_{\xi_3}(y)]f_{\zeta_3}(y)dy}{E(\xi_1)+E(\eta_1)+E(\xi_2)+E(\eta_2)+E(\xi_3)+E(\eta_3)} t \right].$$
$$\tag{9.40}$$

9.6 Numerical example

9.6.1 Input characteristics and parameters

For assuming characteristics of the model presented in the previous section, we determine their distributions and numerical values of parameters:

- $P(Z = z_i) = a_i, \quad i = 1, 2, 3,$

  $a_1 = 0.2, \quad a_2 = 0.4, \quad a_3 = 0.4;$

- $F_{\xi_i}(x) = 1 - e^{-(x/\beta_i)^{\alpha_i}}, \quad x \geqslant 0, \quad i = 1, 2, 3,$

  $\alpha_1 = 2, \quad \alpha_2 = 3, \quad \alpha_3 = 2,$

  $\beta_1 = 11.2838, \quad \beta_2 = 13.4382, \quad \beta_3 = 18.0541,$

  $E(\xi_1) = 10, \quad E(\xi_2) = 12, \quad E(\xi_3) = 16 \text{ [h]};$

- $F_{\eta_i}(x) = 1 - e^{-(x/\beta_{i+3})^{\alpha_{i+3}}}, \quad x \geqslant 0, \quad i = 1, 2, 3,$

  $\alpha_4 = 3, \quad \alpha_5 = 3, \quad \alpha_6 = 3,$

  $\beta_4 = 15.6779, \quad \beta_5 = 13.4382, \quad \beta_6 = 8.95877,$

  $E(\eta_1) = 14, \quad E(\eta_2) = 12, \quad E(\eta_3) = 8 \text{ [h]};$

- $F_{\zeta_i}(x) = 1 - e^{-\lambda_i x}, \quad x \geqslant 0, \quad i = 1, 2, 3,$

  $\lambda_1 = 0.0004, \quad \lambda_2 = 0.0008, \quad \lambda_3 = 0.0006,$

  $E(\zeta_1) = 2500, \quad E(\zeta_2) = 1250, \quad E(\zeta_3) = 2500 \text{ [h]};$

- $F_\kappa(x) = 1 - e^{-\gamma x}, \quad x \geqslant 0,$

  $\gamma = 0.04, \quad E(\kappa) = 25 \text{ [h]}.$

9.6.2 Calculated model parameters

We calculate the necessary parameters of the models. Using (9.8), we compute elements of the transition probability matrix P.

$$P = \begin{bmatrix} 0 & 0 & 0 & 0.9960 & 0 & 0 & 0.0040 \\ 0 & 0 & 0 & 0 & 0.9905 & 0 & 0.0095 \\ 0 & 0 & 0 & 0 & 0 & 0.9904 & 0.0096 \\ 0.2 & 0.4 & 0.4 & 0 & 0 & 0 & 0 \\ 0.2 & 0.4 & 0.4 & 0 & 0 & 0 & 0 \\ 0.2 & 0.4 & 0.4 & 0 & 0 & 0 & 0 \\ 0.2 & 0.4 & 0.4 & 0 & 0 & 0 & 0 \end{bmatrix}.$$

We compute the matrix of holding times expectations corresponding to P by applying a formula:

$$m_{ij} = E(T_{ij}) = \frac{1}{p_{ij}} \int_0^\infty t \, dQ_{ij}(t),$$

$$\left[E(T_{ij})\right] = \begin{bmatrix} 0 & 0 & 0 & 9.9890 & 0 & 0 & 6.3570 \\ 0 & 0 & 0 & 0 & 11.9848 & 0 & 6.7756 \\ 0 & 0 & 0 & 0 & 0 & 15.9581 & 10.1505 \\ 14.0 & 14.0 & 14.0 & 0 & 0 & 0 & 0 \\ 12.0 & 12.0 & 12.0 & 0 & 0 & 0 & 0 \\ 8.0 & 8.0 & 8.0 & 0 & 0 & 0 & 0 \\ 25.0 & 25.0 & 25.0 & 0 & 0 & 0 & 0 \end{bmatrix}.$$

We calculate the matrix of waiting times expectations using the formula

$$m_i = E(T_i) = \sum_{j=1}^7 p_{ij} m_{ij}.$$

As a result, we get

$$[E(T_i)] = \begin{bmatrix} 9.9746 & 11.9351 & 15.9027 & 14.0 & 12.0 & 8.0 & 25.0 \end{bmatrix}^T.$$

By solving (9.9) we get the stationary distribution of the embedded Markov chain:

$$\pi = \begin{bmatrix} 0.1 & 0.2 & 0.2 & 0.0996 & 0.1981 & 0.1981 & 0.0042 \end{bmatrix}^T.$$

Applying (9.10), we obtain the limit probability distribution of the states:

$$P_i = \begin{bmatrix} 0.0808 & 0.1934 & 0.2577 & 0.1130 & 0.1926 & 0.1283 & 0.0342 \end{bmatrix}^T.$$

In this case, a limiting (steady-state) availability (3.107) is given by

$$A = \sum_{j=1}^{6} P_j.$$

Hence,

$$A = 0.9658.$$

We calculate the matrix of the holding times second moments using an equality

$$\overline{m^2}_{ij} = E(T_{ij}^2) = \frac{1}{p_{ij}} \int_0^\infty t^2 dQ_{ij}(t),$$

$$\left[E(T_{ij}^2) \right] = \begin{bmatrix} 0 & 0 & 0 & 126.943 & 0 & 0 & 63.4194 \\ 0 & 0 & 0 & 0 & 162.646 & 0 & 67.129 \\ 0 & 0 & 0 & 0 & 0 & 324.39 & 161.983 \\ 220.867 & 220.867 & 220.867 & 0 & 0 & 0 & 0 \\ 163.022 & 163.022 & 163.022 & 0 & 0 & 0 & 0 \\ 72.453 & 72.453 & 72.453 & 0 & 0 & 0 & 0 \\ 1250.0 & 1250.0 & 1250.0 & 0 & 0 & 0 & 0 \end{bmatrix}.$$

Using the formula

$$\overline{m^2}_i = E(T_i^2) = \sum_{j=1}^{7} p_{ij} \overline{m^2}_{ij}$$

we calculate the matrix of second moments of waiting times

$$\left[E(T_i^2) \right] = \begin{bmatrix} 126.563 & 163.022 & 322.84 & 220.867 & 163.022 & 72.453 & 1250.0 \end{bmatrix}^{\mathrm{T}}.$$

We obtain the mean time to failure of the object by solving the matrix equation (9.15):

$$\begin{bmatrix} 0 & 0 & 0 & 0.9960 & 0 & 0 \\ 0 & 0 & 0 & 0 & 0.9905 & 0 \\ 0 & 0 & 0 & 0 & 0 & 0.9904 \\ 0.2 & 0.4 & 0.4 & 0 & 0 & 0 \\ 0.2 & 0.4 & 0.4 & 0 & 0 & 0 \\ 0.2 & 0.4 & 0.4 & 0 & 0 & 0 \end{bmatrix} \begin{bmatrix} E(\Theta_{17}) \\ E(\Theta_{27}) \\ E(\Theta_{37}) \\ E(\Theta_{47}) \\ E(\Theta_{57}) \\ E(\Theta_{67}) \end{bmatrix} = \begin{bmatrix} 9.9746 \\ 11.9351 \\ 15.9027 \\ 14.0 \\ 12.0 \\ 8.0 \end{bmatrix}.$$

The solution is

$$[E(\Theta_{i7})] = \begin{bmatrix} 2837.56 \\ 2821.93 \\ 2821.65 \\ 2838.94 \\ 2836.94 \\ 2832.94 \end{bmatrix}.$$

The mean time to the object failure is

$$E(T) = 0.2.56 + 0.4.93 + 0.4.65 = 2824.994 \ [\text{h}].$$

To compute the second moment of the time to failure we have to solve the following matrix equation:

$$
\begin{bmatrix}
0 & 0 & 0 & 126.943 & 0 & 0 \\
0 & 0 & 0 & 0 & 162.646 & 0 \\
0 & 0 & 0 & 0 & 0 & 324.39 \\
220.867 & 220.867 & 220.867 & 0 & 0 & 0 \\
163.022 & 163.022 & 163.022 & 0 & 0 & 0 \\
72.453 & 72.453 & 72.453 & 0 & 0 & 0
\end{bmatrix}
\begin{bmatrix}
E(\Theta_{17}^2) \\
E(\Theta_{27}^2) \\
E(\Theta_{37}^2) \\
E(\Theta_{47}^2) \\
E(\Theta_{57}^2) \\
E(\Theta_{67}^2)
\end{bmatrix}
$$

$$
=
\begin{bmatrix}
56,616.1 \\
67,517.4 \\
89,871.6 \\
79,319.3 \\
67,961.7 \\
45,271.5
\end{bmatrix}.
$$

The solution of the equation is

$$
\left[E(\Theta_{i7}^2) \right] =
\begin{bmatrix}
1.60599 \times 10^7 \\
1.59711 \times 10^7 \\
1.59694 \times 10^7 \\
1.60675 \times 10^7 \\
1.60562 \times 10^7 \\
1.60335 \times 10^7
\end{bmatrix}.
$$

The second moment of time to the object failure is

$$E(T^2) = 0.2.60599 \times 10^7 + 0.4.59711 \times 10^7 + 0.4.59694 \times 10^7$$

$$= 1.598818 \times 10^7 \ \left[\text{h}^2\right].$$

The standard deviation of the time to failure is

$$D(T) = \sqrt{1.598818 \times 10^7 - (2824.994)^2} = 2829.76 \ [\text{h}].$$

The formula (9.40) allows us to find the approximate reliability function. Finally, we have

$$R(t) \approx \exp\left[-0.0003208t\right].$$

The theory of the semi-Markov processes provides methods for calculating the reliability characteristics and parameters. The semi-Markov model of the multitask operation process allowed us to compute the limit distribution of the state, the mean time to failure, the standard deviation of the time to failure, and the approximate reliability function.

Semi-Markov failure rate process | 10

Abstract

In this chapter the reliability function is defined by the stochastic failure rate process with a nonnegative and right-continuous trajectories. Equations for the conditional reliability functions of an object are derived under the assumption that the failure rate is a semi-Markov process with an at most countable state space. A proper theorem is presented. The equations for Laplace transforms of conditional reliability functions for the finite space semi-Markov random walk failure rate process are presented in the chapter. The countable linear systems of equations for the appropriate Laplace transforms allow us to find the reliability functions for the Poisson and the Furry-Yule failure rate processes. Frequently, the random tasks and environmental conditions cause a random load of an object in its operation. The failure rate of an object, depending on the random load, may be the random process, too. This chapter presents the limit theorem concerning a failure rate under the assumption that it is a linear function of a random load process with an ergodic mean.

Keywords: Semi-Markov model, Stochastic failure rate process, Semi-Markov random walk failure rate process, Poisson failure rate processes, Furry-Yule failure rate processes

10.1 Introduction

Traditionally, a reliability function is defined by a failure rate that is a real function taking nonnegative real values. In this chapter, the reliability function is determined by the stochastic failure rate process with a nonnegative and right-continuous trajectories.

Equations for the conditional reliability functions of an element, under assumptions that the failure rate is a special semi-Markov process or a piecewise Markov process with a finite state space, was introduced by Kopocińska and Kopociński [56]. Kopocińska [57] has considered the reliability of a component with an alternating failure rate. The result from Ref. [56], for general semi-Markov process with the at most countable and for the nonnegative real state space, was extended by Grabski [32, 36, 37].

10.2 Reliability function with random failure rate

We suppose that the failure rate, denoted by $\{\lambda(t) : t \geqslant 0\}$, is a stochastic process with a nonnegative and right-continuous trajectories.

Definition 10.1. The reliability function of an object is defined as an expectation of the random process $\{\exp(-\int_0^t \lambda(u)du) : t \geqslant 0\}$:

$$R(t) = E\left[\exp\left(-\int_0^t \lambda(u)du\right)\right]. \tag{10.1}$$

Semi-Markov Processes: Applications in System Reliability and Maintenance. http://dx.doi.org/10.1016/B978-0-12-800518-7.00010-7

Example 10.1. Suppose that the failure rate is the stochastic process $\{\lambda(t) : t \geqslant 0\}$ given by $\lambda(t) = Ct$, where C is a nonnegative random variable having the exponential distribution with parameter $\beta > 0$:

$$P(C \leqslant u) = 1 - e^{-\beta u}, \quad t \geqslant 0.$$

A trajectory of the process $\{\exp(-\int_0^t \lambda(u)du), \quad t \geqslant 0\}$ is a real function $\xi(t) = \exp\left(-c\frac{t^2}{2}\right)$, where c is a value of the random variable C.

From Definition 10.1 we obtain

$$R(t) = E\left[\exp\left(-\int_0^t Cx\,dx\right)\right] = \int_0^\infty e^{(-\int_0^t ux\,dx)}\beta\,e^{-\beta u}du$$

$$= \beta \int_0^\infty e^{-ut^2/2}e^{-\beta u}du = \beta \int_0^\infty e^{-((t^2/2)+\beta)u}du = \frac{2\beta}{t^2 + 2\beta}, \quad t \geqslant 0.$$

$\triangle$

From the Fubini's theorem and the Jensen's inequality we immediately get the following result

Proposition 10.1. If

$$\int_0^t E[\lambda(u)]\,du < \infty$$

then

$$R(t) \geqslant \exp\left[-\int_0^t E[\lambda(u)]\,du\right]. \tag{10.2}$$

From that inequality it follows that an object reliability function with the random failure rate $\{\lambda(t) : t \geqslant 0\}$ is greater than or equal to the reliability function with the deterministic failure rate equal to the mean $\overline{\lambda}(t) = E[\lambda(t)]$.

Example 10.2. An expectation of the failure rate process $\{\lambda(t) : t \geqslant 0\}$ from Example 10.1 is

$$\overline{\lambda}(t) = E[Ct] = \frac{1}{\beta}t.$$

Hence,

$$\overline{R}(t) = \exp\left[-\int_0^t \overline{\lambda}(u)du\right] = \exp\left[-\int_0^t \frac{1}{\beta}u\,du\right] = \exp\left[-\frac{t^2}{2\beta}\right].$$

Those functions for $\beta = 0.1$ are shown in Figure 10.1 $\triangle$

10.3   Semi-Markov failure rate process

Now, we suppose that a random failure rate process $\{\lambda(t) : t \geqslant 0\}$ is a semi-Markov process on a discrete state space $S = \{\lambda_j : j \in J\}$, where $J = \{0, 1, 2, \ldots\}$ or $J = \{0, 1, 2, \ldots, r\}$, with a kernel $\mathbf{Q}(t) = [Q_{ij}(t) : i, j \in J]$. The function

$$R_i(t) = E\left[\exp\left(-\int_0^t \lambda(u)du\right)|\lambda(0) = \lambda_i\right], \quad i \in J \tag{10.3}$$

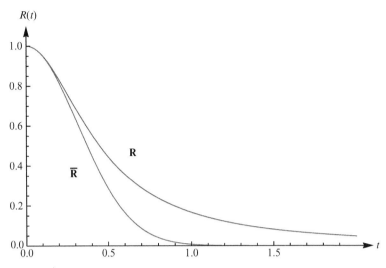

Figure 10.1 The reliability functions $R(t) = \frac{2\beta}{t^2 + 2\beta}$ and $\bar{R}(t) = \exp\left[-\frac{t^2}{2\beta}\right]$.

is said to be the *conditional reliability function* with a random failure rate $\{\lambda(t) : t \geqslant 0\}$.

Theorem 10.1 (Grabski [36]). If the failure rate function $\{\lambda(t) : t \geqslant 0\}$ is a regular semi-Markov process on a discrete state space $S = \{\lambda_j : j \in J\}$ with the kernel $\mathbf{Q}(t) = \left[Q_{ij}(t) : i, j \in J\right]$, then conditional reliability functions $R_i(t), i \in J$, satisfy the system of equations

$$R_i(t) = e^{-\lambda_i t}(1 - G_i(t)) + \int_0^t e^{-\lambda_i x} \sum_{j \in J} R_j(t - x) dQ_{ij}(x), \quad i \in J, \tag{10.4}$$

where

$$G_i(t) = \sum_{j \in J} Q_{ij}(t).$$

The solutions are unique in the class of the measurable and uniformly bounded functions.

Proof: [32, 56]. We apply the Laplace transform to solve that system of integral equations . Let

$$\tilde{R}_i(s) = \int_0^\infty e^{-st} R_i(t) dt, \quad \tilde{G}_i(s) = \int_0^\infty e^{-st} G_i(t) dt, \quad \tilde{q}_{ij}(s) = \int_0^\infty e^{-st} dQ_{ij}(t) dt.$$

Passing (10.4) to the Laplace transform, we obtain the system of linear equations

$$\tilde{R}_i(s) = \frac{1}{s + \lambda_i} - \tilde{G}_i(s + \lambda_i) + \sum_{j \in J} \tilde{R}_j(s) \tilde{q}_{ij}(s + \lambda_i), \quad i \in J. \tag{10.5}$$

This system is equivalent to the system of linear equations

$$\sum_{j \in J} (\delta_{ij} - \tilde{q}_{ij}(s + \lambda_i)) \tilde{R}_j(s) = \frac{1}{s + \lambda_i} - \tilde{G}_i(s + \lambda_i), \quad i \in J, \tag{10.6}$$

which in matrix notation has the form

$$(I - \tilde{q}_\Lambda(s)) \tilde{R}(s) = \tilde{H}(s), \tag{10.7}$$

where

$$\tilde{q}_\Lambda(s) = \left[\tilde{q}_{ij}(s + \lambda_i) : i, j \in J \right]$$

is the square matrix and

$$\left[\tilde{R}_i(s) : i \in J \right]', \quad \tilde{H}(s) = \left[\frac{1}{s + \lambda_i} - \tilde{G}_i(s + \lambda_i) : i \in J \right]'$$

are one-column matrices.

    Let

$$\tilde{R}_i(0^+) = \lim_{p \to 0^+} \tilde{R}_i(p), \quad p \in (0, \infty).$$

If this limit exists, then

$$\tilde{R}_i(0^+) = \int_0^\infty R_i(t) \mathrm{d}t = \bar{\mu}_i \tag{10.8}$$

is *the mean time to failure* of the object with the random failure rate $\{\lambda(t) : t \geqslant 0\}$, which starts from the state λ_i. Using the fact mentioned above, we obtain the system of linear equations for the mean times to failure, which in matrix notation takes the form

$$\left(I - \tilde{q}_\Lambda(0^+) \right) \bar{\mu} = \tilde{H}(0^+), \tag{10.9}$$

where

$$\tilde{q}_\Lambda(0^+) = \lim_{p \to 0^+} \left[\tilde{q}_{ij}(p + \lambda_i), i, j \in J \right] \tag{10.10}$$

is a square matrix and

$$\mu = \tilde{R}(0^+) = [\mu_i : i \in J]', \quad \tilde{H}(0^+) = \lim_{p \to 0^+} \tilde{H}(p)$$

are the corresponding one-column matrices.

    If there exists a finite second moment of the time to failure, then

$$\bar{\mu}_i^2 = -\tilde{R}_i'(0^+) = 2 \int_0^\infty t R(t) \mathrm{d}t.$$

Differentiating the system of linear equations (10.6) with respect to s, we get

$$\sum_{j\in J}\left[-\tilde{q}'_{ij}(\lambda_i+s)\tilde{R}_j(s)+(\delta_{ij}-\tilde{q}_{ij}(\lambda_i+s))\tilde{R}'_j(s)\right]$$

$$=-\frac{1+(\lambda_i+s)^2\tilde{G}'_i(\lambda_i+s)}{(\lambda_i+s)^2},\quad i\in J.$$

Hence,

$$\sum_{j\in J}(\delta_{ij}-\tilde{q}_{ij}(\lambda_i+s))\tilde{R}'_j(s)=-\frac{1+(\lambda_i+s)^2\tilde{G}'_i(\lambda_i+s)}{(\lambda_i+s)^2}$$

$$+\sum_{j\in J}\tilde{q}'_{ij}(\lambda_i+s)\tilde{R}_j(s),\quad i\in J.$$

Substituting $s=p\in(0,\infty)$ and letting $p\to 0^+$, we obtain

$$\left(I-\tilde{q}_\Lambda(0^+)\right)\bar{\mu^2}=\tilde{K}(0^+),\tag{10.11}$$

where

$$\bar{\mu^2}=-\lim_{p\to 0^+}\tilde{R}'(p)=\left[\bar{\mu^2}_i:i\in J\right]'$$

and

$$\tilde{K}(0^+)=\lim_{p\to 0^+}\left[\frac{1+(\lambda_i+p)^2\tilde{G}'_i(\lambda_i+p)}{(\lambda_i+p)^2}-\sum_{j\in J}\tilde{q}'_{ij}(\lambda_i+p)\bar{\mu}_j:\quad i\in J\right]'$$

are one-column matrices.

10.4 Random walk failure rate process

Let $\{\lambda(t):t\geqslant 0\}$ be a semi-Markov process with the state space $S=\{\lambda_0,\lambda_1,\ldots,\lambda_n\}$ and the kernel

$Q(t)$

$$=\begin{bmatrix} 0 & G_0(t) & 0 & 0 & \ldots & 0 & 0 \\ a_1G_1(t) & 0 & b_1G_1(t) & 0 & \ldots & 0 & 0 \\ 0 & a_2G_2(t) & 0 & b_2G_2(t) & 0 & \ldots & 0 & 0 \\ \ldots & \ldots & \ldots & \ldots & \ldots & \ldots & \ldots \\ \ldots & \ldots & \ldots & \ldots & \ldots & \ldots & \ldots \\ \ldots & \ldots & \ldots & \ldots & \ldots & \ldots & \ldots \\ 0 & 0 & 0 & 0 & a_{n-1}G_{n-1}(t) & 0 & b_{n-1}G_{n-1}(t) \\ 0 & 0 & 0 & 0 & 0 & G_n(t) & 0 \end{bmatrix},$$

$$\tag{10.12}$$

where $G_0(t), G_1(t), \ldots, G_n(t)$ denote the cumulative distribution functions with a nonnegative support $R_+ = [0, \infty)$ and $a_k > 0, b > 0, a_k + b_k = 1$, for $k = 1, \ldots, n-1$. This stochastic process $\{\lambda(t) : t \geqslant 0\}$ is called a *semi-Markov random walk* or the *semi-Markov birth and death process*. Suppose that the distributions $G_0(t), G_1(t), \ldots, G_n(t)$ are absolutely continuous with respect to the Lebesgue measure.

Let $\mathbf{p} = [p_0, p_1, \ldots, p_n]$ be an initial probability distribution of the process. Now, the matrices from (10.7) are

$I - \tilde{q}_\Lambda(s)$

$$
= \begin{bmatrix}
1 & \tilde{d}_0(s) & 0 & 0 & \ldots & 0 & 0 \\
a_1 \tilde{d}_1(s) & 1 & b_1 \tilde{d}_1(s) & 0 & \ldots & 0 & 0 \\
0 & a_2 \tilde{d}_2(s) & 1 & b_2 \tilde{d}_2(s) & 0 & \ldots & 0 \\
\ldots & \ldots & \ldots & \ldots & \ldots & \ldots & \ldots \\
\ldots & \ldots & \ldots & \ldots & \ldots & \ldots & \ldots \\
\ldots & \ldots & \ldots & \ldots & \ldots & \ldots & \ldots \\
0 & 0 & \ldots & 0 & a_{n-1}\tilde{d}_{n-1}(s) & 1 & b_{n-1}\tilde{d}_{n-1}(s) \\
0 & 0 & 0 & \ldots & 0 & \tilde{d}_n(s) & 1
\end{bmatrix},
$$

(10.13)

where

$$
\tilde{g}_i(s) = \int_0^\infty e^{-st} dG_i(t), \quad i = 0, 1, \ldots, n,
$$

$$
\tilde{d}_k(s) = \tilde{g}_k(s + \lambda_k), \quad i = 0, 1, \ldots, n,
$$

$$
\tilde{R}(s) = \begin{bmatrix} \tilde{R}_0(s) \\ \tilde{R}_1(s) \\ \vdots \\ \tilde{R}_n(s) \end{bmatrix}, \quad
\tilde{H}(s) = \begin{bmatrix} \frac{1}{s+\lambda_0} - \tilde{G}_0(s+\lambda_0) \\ \frac{1}{s+\lambda_1} - \tilde{G}_1(s+\lambda_1) \\ \vdots \\ \frac{1}{s+\lambda_1} - \tilde{G}_2(s+\lambda_2) \end{bmatrix}.
$$

The Laplace transform of the unconditional reliability function is

$$
\tilde{R}(s) = p_0 \tilde{R}_0(s) + p_1 \tilde{R}_1(s) + \cdots + p_n \tilde{R}_n(s).
$$

(10.14)

From (10.2), it follows that

$$
R(t) \geqslant \bar{R}(t) = \exp\left[-\int_0^t \bar{\lambda}(u) du\right],
$$

(10.15)

where

$$
\bar{\lambda}(u) = E[\lambda(t)] = \sum_{k=0}^n \lambda_k P_k(t),
$$

$$
P_k(t) = P(\lambda(t) = \lambda_k), \quad t \geqslant 0, \ k = 0, 1, \ldots, n.
$$

(10.16)

The limiting distribution of the states for the semi-Markov random walk is derived in (8.24)–(8.29). Therefore limiting probabilities

$$P_k = \lim_{t \to \infty} P_k(t) = \lim_{t \to \infty} P(\lambda(t) = \lambda_k), \quad k \in J \tag{10.17}$$

take the form

$$P_0 = \frac{m_0}{m_0 + \sum\limits_{j=1}^{n} \left[\prod\limits_{k=0}^{j} \frac{b_{k-1}}{a_k} \right] m_j}, \tag{10.18}$$

$$P_j = \frac{\prod\limits_{k=0}^{j} \frac{b_{k-1}}{a_k} m_j}{m_0 + \sum\limits_{j=1}^{n} \left[\prod\limits_{k=1}^{j} \frac{b_{k-1}}{a_k} \right] m_j}, \quad j = 1, \ldots, n.$$

Finally, for large t we obtain an approximate lower bound for the reliability function

$$R(t) \geqslant \exp\left[-\int_0^t \bar{\lambda}(u) \mathrm{d}u \right] \approx \exp\left[-\sum_{k=0}^{n} \lambda_k P_k t \right]. \tag{10.19}$$

Example 10.3. Let $\{u(t) : t \geqslant 0\}$ denote the random load of an object. We assume that this stochastic process is the semi-Markov random walk taking its values in the state space $U = \{u_i : i \in J = \{0, 1, \ldots, n\}$ and its kernel is defined by (10.12). Let

$$\lambda(t) = g(u(t)), \quad t \geqslant 0, \tag{10.20}$$

where $g(x), x \geqslant 0$ is an increasing and continuous real function taking values on $\mathbb{R}$. It is easy to notice that $\{\lambda(t) : t \geqslant t\}$ is the semi-Markov random walk with the state space $S = \{\lambda_i = g(\gamma_i) : i \in J = \{0, 1, \ldots, n\}$ and the same kernel (10.12). Assume that the load of the car engine is semi-Markov random walk $\{u(t) : t \geqslant 0\}$ with $n = 3$. The state space is $U = \{u_0, u_1, u_2, u_3\}$ and the semi-Markov kernel takes a form

$$\mathbf{Q}(t) = \begin{bmatrix} 0 & G_0(t) & 0 & 0 \\ a_1 G_1(t) & 0 & b_1 G_1(t) & 0 \\ 0 & a_2 G_2(t) & 0 & b_2 G_2(t) \\ 0 & 0 & G_3(t) & 0 \end{bmatrix}, \tag{10.21}$$

where $G_i(t), i = 0, 1, 2, 3$ are the cumulative distribution functions with a nonnegative support $R_+ = [0, \infty)$. Suppose that those distributions are absolutely continuous with respect to the Lebesgue measure.

Let $\mathbf{p} = [1, 0, 0, 0]$ be an initial probability distribution of the process. The random failure process $\{\lambda(t) : t \geqslant t\}$ is the semi-Markov random walk with the state space $S = \{\lambda_i = g(u_i) : i = 0, 1, 2, 3\}$ and the identical kernel. Now, the matrices from equation (10.7) are

$$I - \tilde{q}_\Lambda(s) = \begin{bmatrix} 1 & -\tilde{g}_0(s+\lambda_0) & 0 & 0 \\ -a_1\tilde{g}_1(s+\lambda_1) & 1 & -b_1\tilde{g}_1(s+\lambda_1) & 0 \\ 0 & -a_2\tilde{g}_2(s+\lambda_2) & 1 & -b_2\tilde{g}_2(s+\lambda_2) \\ 0 & 0 & -\tilde{g}_3(s+\lambda_3) & 1 \end{bmatrix},$$

(10.22)

where

$$\tilde{g}_i(s) = \int_0^\infty e^{-st} dG_i(t), \quad i = 0, 1, 2, 3,$$

(10.23)

$$\tilde{R}(s) = \begin{bmatrix} \tilde{R}_0(s) \\ \tilde{R}_1(s) \\ \tilde{R}_2(s) \\ \tilde{R}_3(s) \end{bmatrix}, \quad \tilde{H}(s) = \begin{bmatrix} \frac{1}{s+\lambda_0} - \tilde{G}_0(s+\lambda_0) \\ \frac{1}{s+\lambda_1} - \tilde{G}_1(s+\lambda_1) \\ \frac{1}{s+\lambda_2} - \tilde{G}_2(s+\lambda_2) \\ \frac{1}{s+\lambda_3} - \tilde{G}_3(s+\lambda_3) \end{bmatrix}.$$

(10.24)

The Laplace transform of the unconditional reliability function is

$$\tilde{R}(s) = \tilde{R}_0(s).$$

(10.25)

Suppose

$$G_0(t) = 1 - (1 + \alpha_0 t + 0.5\alpha_0^2 t^2)e^{-\alpha_0 t}, \quad t \geq 0,$$
$$G_1(t) = 1 - (1 + \alpha_1 t)e^{-\alpha_1 t}, \quad t \geq 0,$$
$$G_2(t) = 1 - e^{-\alpha_2 t}, \quad t \geq 0, \quad G_3(t) = 1 - e^{-\alpha_3 t}, \quad t \geq 0.$$

The corresponding Laplace transforms have the forms

$$\tilde{G}_0(s) = \frac{\alpha_0^3}{s(\alpha_0+s)^3}, \quad \tilde{G}_1(s) = \frac{\alpha_1^2}{s(\alpha_1+s)^2}, \quad \tilde{G}_2(s) = \frac{\alpha_2}{s(\alpha_2+s)},$$

$$\tilde{G}_3(s) = \frac{\alpha_3}{s(\alpha_3+s)},$$

$$\tilde{g}_0(s) = \frac{\alpha_0^3}{s(\alpha_0+s)^3}, \quad \tilde{g}_1(s) = \frac{(\alpha_1)^2}{(\alpha_1+s)^2}, \quad \tilde{g}_2(s) = \frac{\alpha_2}{\alpha_2+s},$$

$$\tilde{g}_3(s) = \frac{\alpha_3}{\alpha_3+s}.$$

Let

$$a_1 = 0.4, \quad b_1 = 1 - a_1, \ a_2 = 0.6, \quad b_2 = 1 - a_2,$$
$$\lambda_0 = 0, \quad \lambda_1 = 0.1, \quad \lambda_2 = 0.2, \quad \lambda_3 = 0.3,$$
$$\alpha_0 = 0.02, \ \alpha_1 = 0.04, \quad \alpha_2 = 0.06, \ \alpha_3 = 0.04.$$

Using the MATHEMATICA computer program, we have solved (10.7). We have obtained the conditional reliability function $R_0(t)$ as the inverse Laplace transform of

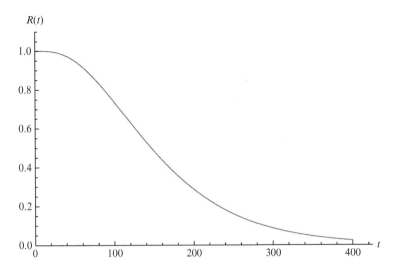

Figure 10.2 The reliability function.

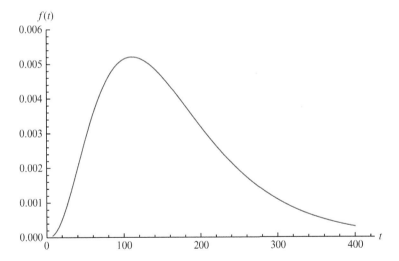

Figure 10.3 The probability density function.

$\tilde{R}_0(s)$. In this case, the function $R_0(t)$ is equal to the unconditional reliability function $R(t)$. This function and a corresponding probability density function are shown in Figures 10.2 and 10.3.

We can easily compute a mean time to failure

$$E(T) = \tilde{R}_0(s)|_{s \to 0^+} = 164.873.$$

$\triangle$

10.5   Alternating failure rate process

For $n = 1$, the semi-Markov random walk is a two-states semi-Markov process that is called an *alternating process*. The alternating failure rate process was discussed by Kopocińska [57]. The above-mentioned process denoted by $\{\lambda(t) : t \geqslant 0\}$, takes values in the states space $S = \{\lambda_0, \lambda_1\}$. The alternating failure rate process is defined by a kernel matrix

$$\boldsymbol{Q}(t) = \begin{bmatrix} 0 & G_0(t) \\ G_1(t) & 0 \end{bmatrix} \tag{10.26}$$

and an initial distribution $\boldsymbol{p} = [p_0, p_1]$. Now, the matrices from (10.7) take the form

$$\boldsymbol{I} - \tilde{\boldsymbol{q}}_\Lambda(s) = \begin{bmatrix} 1 & -\tilde{g}_0(s + \lambda_0) \\ -\tilde{g}_1(s + \lambda_1) & 1 \end{bmatrix}, \tag{10.27}$$

where

$$\tilde{g}_0(s) = \tilde{q}_{01}(s) = \int_0^\infty e^{-st} dG_0(0), \quad \tilde{g}_1(s) = \tilde{q}_{10}(s) = \int_0^\infty e^{-st} dG_1(0), \tag{10.28}$$

$$\tilde{\boldsymbol{R}}(s) = \begin{bmatrix} \tilde{R}_0(s) \\ \tilde{R}_1(s) \end{bmatrix}, \quad \tilde{\boldsymbol{H}}(s) = \begin{bmatrix} \frac{1}{s+\lambda_0} - \tilde{G}_0(s + \lambda_0) \\ \frac{1}{s+\lambda_1} - \tilde{G}_1(s + \lambda_1) \end{bmatrix}. \tag{10.29}$$

A solution of those systems of equations is

$$\tilde{R}_0(s) = \frac{\frac{1}{s+\lambda_0} - \tilde{G}_0(s + \lambda_0) + \tilde{g}_0(s + \lambda_0)\left[\frac{1}{s+\lambda_1} - \tilde{G}_1(s + \lambda_1)\right]}{1 - \tilde{g}_0(s + \lambda_0)\tilde{g}_1(s + \lambda_1)}, \tag{10.30}$$

$$\tag{10.31}$$

$$\tilde{R}_1(s) = \frac{\frac{1}{s+\lambda_1} - \tilde{G}_1(s + \lambda_1) + \tilde{g}_1(s + \lambda_1)\left[\frac{1}{s+\lambda_0} - \tilde{G}_0(s + \lambda_0)\right]}{1 - \tilde{g}_0(s + \lambda_0)\tilde{g}_1(s + \lambda_1)}. \tag{10.32}$$

An unconditional reliability function is

$$\tilde{R}(s) = p_0 \tilde{R}_0(s) + p_1 \tilde{R}_1(s). \tag{10.33}$$

Example 10.4. We assume that an initial distribution and a kernel of the process are $\boldsymbol{p} = [p_0, p_1]$,

$$\boldsymbol{Q}(t) = \begin{bmatrix} 0 & 1 - (1 + \beta t)e^{-\beta t} \\ 1 - e^{-\alpha t} & 0 \end{bmatrix}, \quad \text{where } \alpha, \beta, t \geqslant 0.$$

The CDFs of waiting times in the states λ_0, λ_1 are

$$G_0(t) = 1 - (1 + \beta t)e^{-\beta t}, \quad G_1(t) = 1 - e^{-\alpha t}, \quad t \geqslant 0.$$

The corresponding Laplace transforms are

$$\tilde{G}_0(s) = \frac{\beta^2}{s(\beta+s)^2}, \quad \tilde{G}_1(s) = \frac{\alpha}{s(\alpha+s)}, \quad \tilde{g}_0(s) = \frac{\beta^2}{(\beta+s)^2}, \quad \tilde{g}_1(s) = \frac{\alpha}{\alpha+s}.$$

We calculate the conditional reliability function for the parameters

$$\lambda_0 = 0, \quad \lambda_1 = 0.2, \quad \alpha = 0.01, \quad \beta = 0.1, \quad p_0 = 0, \quad p_1 = 1.$$

Substituting those functions into (10.7), we get

$$\tilde{R}_1(s) = \frac{\frac{1}{s+0.2} - \frac{0.01}{(s+0.2)(s+0.21)} + \frac{0.01}{s+0.21}\left[\frac{1}{s} - \frac{0.01}{s(s+0.1)^2}\right]}{1 - \frac{0.0001}{(s+0.1)^2(s+0.21)}}.$$

Using the MATHEMATICA computer program, we obtain a reliability function as an inverse transform

$$R_1(t) = 1.25\,e^{-0.2t} - 0.495913\,e^{-0.137016t} + 0.245913\,e^{-0.0729844t}, \quad t \geqslant 0.$$

The function $R_1(t)$ is equal to unconditional reliability function $R(t)$. A corresponding probability density function is

$$f(t) = 0.25e^{-0.2t} - 0.067948e^{-0.137016t} + 0.0179478e^{-0.0729844t}, \quad t \geqslant 0.$$

$$\triangle$$

10.6 Poisson failure rate process

Assume that the random failure rate $\{\lambda(t) : t \geqslant 0\}$ is a Poisson process with parameter $\lambda > 0$. This process is a semi-Markov process on the counting state space $S = \{0, 1, 2, \ldots\}$ defined by the initial distribution $\mathbf{p}(0) = [1, 0, 0, \ldots]$ and the kernel

$$\mathbf{Q}(t) = \begin{bmatrix} 0 & G_0(t) & 0 & 0 & 0 & \cdots \\ 0 & 0 & G_1(t) & 0 & 0 & \cdots \\ 0 & 0 & 0 & G_2(t) & 0 & \cdots \\ 0 & 0 & 0 & 0 & G_3(t) & 0 \\ \cdot & \cdot & \cdot & \cdot & \cdot & \cdot \end{bmatrix}, \tag{10.34}$$

where $G_i(t) = 1 - e^{-\lambda t}, t \geqslant 0, \quad i = 0, 1, 2, \ldots.$

Theorem 10.2 (Grabski [36]). If the random failure rate $\{\lambda(t) : t \geqslant 0\}$ is the Poisson process with parameter $\lambda > 0$, then the reliability function defined by (10.1) is

$$R(t) = \exp\left[-\lambda\left(t - 1 + e^{-t}\right)\right]. \tag{10.35}$$

Proof: [36].

Recall the well known fact that, if $\{\lambda(t) : t \geqslant 0\}$ is the Poisson process with the parameter $\lambda > 0$, then

$$E[\lambda(t)] = \lambda t, \quad t \geqslant 0.$$

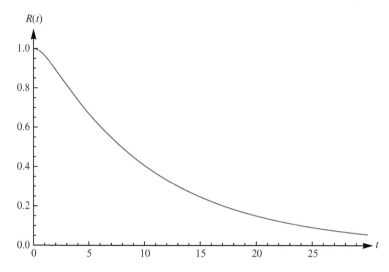

Figure 10.4 The reliability functions $R(t)$ and $\bar{R}(t)$.

From inequality (10.2), we get

$$R(t) \geqslant \exp\left(-\frac{\lambda}{2}t^2\right).$$

Let

$$\hat{R}(t) = \exp\left(-\frac{\lambda}{2}t^2\right).$$

For comparison, the reliability function and the lower bound function $\hat{R}(t)$ with $\lambda = 0.1$ are shown in Figure 10.4.

Proposition 10.2.

1. The density function of the time to failure for the Poisson failure rate process has the form

$$f(t) = \lambda e^{-\left[\lambda\left(t-1+e^{-t}\right)\right]}\left(1 - e^{-t}\right). \tag{10.36}$$

2. The hazard rate function corresponding to the reliability function for the Poisson failure rate process is

$$h(t) = \lambda\left(1 - e^{-t}\right). \tag{10.37}$$

Note that $\lim_{t\to\infty} h(t) = \lambda$. This means that for large t the reliability function (10.35) is approximately equal to the exponential reliability function.

The density function (10.36), with parameter $\lambda = 0.1$, and the hazard rate function (10.37) are shown in Figures 10.5 and 10.6.

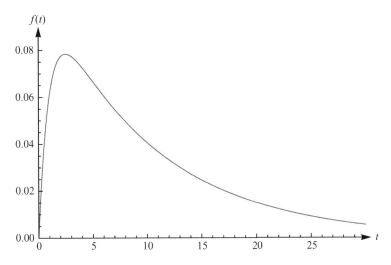

Figure 10.5 The density function for the Poisson failure rate process.

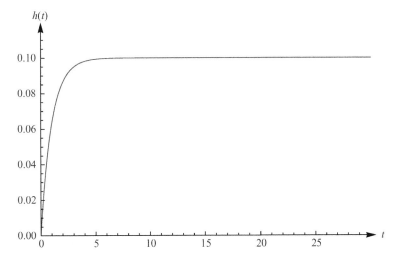

Figure 10.6 The hazard rate function for the Poisson failure rate process.

10.7   Furry-Yule failure rate process

Assume that the random failure rate $\{\lambda(t) : t \geqslant 0\}$ is the Furry-Yule process. The Furry-Yule process with the parameter $\lambda > 0$ is the semi-Markov process on the counting state space $S = \{0, 1, 2, \ldots\}$ defined by the initial distribution $\mathbf{p}(0) = [1, 0, 0, \ldots]$ and the kernel

$$Q(t) = \begin{bmatrix} 0 & G_0(t) & 0 & 0 & 0 & \cdots \\ 0 & 0 & G_1(t) & 0 & 0 & \cdots \\ 0 & 0 & 0 & G_2(t) & 0 & \cdots \\ 0 & 0 & 0 & 0 & G_3(t) & 0 \\ \cdot & \cdot & & & \cdot & \cdot & \cdot \end{bmatrix}, \tag{10.38}$$

where $G_i(t) = 1 - e^{-(i+1)\lambda t}, t \geqslant 0, i = 0, 1, 2, \ldots.$

Theorem 10.3 (Grabski [37]). If the random failure rate $\{\lambda(t) : t \geqslant 0\}$ is the Furry-Yule process with the parameter $\lambda > 0$, then the reliability function defined by (10.1) is

$$R(t) = \frac{(\lambda + 1)e^{-\lambda t}}{1 + \lambda e^{-(\lambda+1)t}}. \tag{10.39}$$

Proof: [37].

Well-known equalities for the differentiable reliability function come to the following conclusion.

Proposition 10.3. A density function $f(t), t \geqslant 0$ and a hazard rate function $h(t), t \geqslant 0$ corresponding to the reliability function (10.39) are

$$f(t) = \frac{\lambda(\lambda + 1)e^{-\lambda t}(1 - e^{-(\lambda+1)t})}{(1 + \lambda e^{-(\lambda+1)t})^2}, \tag{10.40}$$

$$h(t) = \frac{\lambda\left[1 - e^{-(\lambda+1)t}\right]}{(1 + \lambda e^{-(\lambda+1)t})}. \tag{10.41}$$

10.8 Failure rate process depending on random load

Very often the tasks and environmental conditions are randomly changeable and they cause a random load of an object in its operation process. We suppose that the failure rate of the object depends on its random load. Let $\{u(t) : t \geqslant 0\}$ denote the random load of an object defined on at most countable state space U, with a nonnegative and right-continuous trajectories. Suppose that the failure rate of an object depends on the random load process

$$\lambda(t) = g(u(t)), \quad t \geqslant 0, \tag{10.42}$$

where $g(x), x \geqslant 0$ is an increasing and continuous real function taking values on $\mathbb{R}_+$. It is easy to notice that: *If $\{u(t) : t \geqslant 0\}$ is the semi-Markov process taking its values on the state space $U = \{u_i : i \in J = \{0, 1, \ldots, n\}\}$, then the process $\{\lambda(t) : t \geqslant t\}$ is the semi-Markov process with the same kernel.* Now we assume that a failure rate is a linear function of a random load process with an ergodic mean.

Theorem 10.4 (Grabski [32]). Let $\{u(t) : t \geqslant 0\}$ be a random load process with an ergodic mean, that is,

$$\lim_{T \to \infty} \frac{1}{T} \int_0^T u(x)\mathrm{d}x = E\left[u(t)\right] = \bar{u}.$$

If

$$\lambda(t) = \varepsilon u(t) \tag{10.43}$$

then

$$\lim_{\varepsilon \to 0} R\left(\frac{t}{\varepsilon}\right) = \exp(-\bar{u}t). \tag{10.44}$$

Proof: [32, 56].

For small ε, we obtain

$$R(x) \approx \exp(-\varepsilon \bar{u}x). \tag{10.45}$$

Now, we formulate another version of Theorem 10.4, assuming that random load is a semi-Markov process on the discrete state space.

Theorem 10.5. If the random load $\{u(t) : t \geqslant 0\}$ is an ergodic semi-Markov process on the discrete state space, defined by the kernel $\mathbf{Q}(t) = \left[Q_{ij}(t) : i,j \in J\right]$ and

$$\lambda(t) = \varepsilon u(t), \tag{10.46}$$

then

$$\lim_{\varepsilon \to 0} R\left(\frac{t}{\varepsilon}\right) = \exp\left(-\bar{u}t\right), \tag{10.47}$$

where

$$\bar{u} = \frac{\sum_{i \in J} u_i m_i \pi_i}{\sum_{i \in S} m_i \pi_i}, \tag{10.48}$$

$$m_i = \int_0^\infty [1 - G_i(t)]\, \mathrm{d}t = \int_0^\infty \left[1 - \sum_{j \in J} Q_{ij}(t)\right] \mathrm{d}t$$

and $\pi_i, i \in J$ is the stationary distribution of the embedded Markov chain $\{u(\tau_n) : n \in N_0\}$, with the probability transition matrix

$$\mathbf{P} = \left[p_{ij} : i,j \in J\right], \quad p_{ij} = Q_{ij}(\infty).$$

Proof: [32, 56].

For small ε, we obtain

$$R(x) \approx \exp(-\varepsilon \bar{u}x). \tag{10.49}$$

10.9 Conclusions

In many practical situations the randomly changeable environmental conditions cause random load of an object and it implies the random failure rate of it. For the reliability function defined by a random failure rate we obtained an interesting property: the reliability function with the random failure rate is greater than or equal to the reliability function with the deterministic failure rate equal to the mean of the corresponding random failure rate. A main problem discussed here is the reliability function defined by the semi-Markov failure rate process. For the semi-Markov failure rate we have derived equations that allow us to obtain the conditional reliability functions. Applying the Laplace transformation for the introduced system of renewal equations for an at most countable states space, we have obtained the reliability function for the random walk failure rate process, for the Poisson and Furry-Yule failure rate processes. Moreover, we have derived the lower bounds for the considered reliability functions. If an object random load process is ergodic and the corresponding failure rate process is the linear function of it with a small parameter, then the reliability function is approximately exponential. It seems to be possible to extend the results presented here on the continuous time nonhomogeneous semi-Markov process (see, e.g., Papadopoulou and Vassiliou [81]).

Simple model of maintenance

Abstract

Preventive maintenance is conducted to keep a device working and extend its life. In this chapter we present a simple semi-Markov model of preventive maintenance. The constructed model allows us to formulate a problem of optimizing the time to preventive maintenance. The theorem formulates and proves a sufficient condition that an optimal time exists for preventive service.

Keywords: Semi-Markov model, Preventive maintenance, Limiting probabilities of state, Optimization

11.1 Introduction

We can determine preventive maintenance as maintenance of devices, machines, or systems before a failure occurs. The main goal of maintenance is to avoid or to mitigate the consequences of an object failure. Very often, the time and cost of repair are greater than the time and cost of preventive maintenance. Preventive maintenance is conducted to keep a device working and/or extend its life. Preventive maintenance activities include, for example, partial or complete overhauls at specified periods, oil changes, and replacement or repair of worn elements before they cause a system failure. The ideal preventive maintenance program would prevent all devices from failing before this happens. There are many probabilistic models that enable us to choose an optimal policy of preventive maintenance. Gertsbakh [24] have studied Semi-Markov models of object operation with states that included preventive maintenance. Based on these models, the author has formulated and solved many kinds of problems concerning preventive maintenance. Quite different preventive maintenance problems have been considered by Beker [5], Girtler and Grabski [27], Harriaga [41], and Knopik [67].

We assume that the preventive maintenance treatments are designed to restore the full working order of the object. Moreover, we suppose that the moment the failed object repair begins is the instant of its failure, while the start of the preventive service is executed according the principle: start a preventive maintenance when the object has worked without damage by the time T. We will present a stochastic model of the object operation process, which allows us to formulate an optimization problem that deals with choosing the optimal time T. As a model of operation process, we accept three-state semi-Markov processes. Limiting the distribution of the process enables us to construct a criterion function with values denoting the average incomes earned per unit time in a long operating life. From the presented theorem, it follows that the solution of the considered optimization problem depends on the expected value of an object time to failure, expectation of the repair time, the expected value of duration of

Semi-Markov Processes: Applications in System Reliability and Maintenance. http://dx.doi.org/10.1016/B978-0-12-800518-7.00011-9

the preventive maintenance, and properly measured income and cost of operation. The problem considered in this chapter is some modification of Grabski's results presented in Ref. [30].

11.2   Description and assumptions

We assume that time to the object (machine, device) failure ζ is a nonnegative random variable governed by PDF $f_\zeta(\cdot)$ with a finite positive expected value $\bar{\zeta} = E(\zeta)$. A damaged object is repaired at once. Repair of the object recovers it completely and lasts for a random time γ. We assume that γ is a nonnegative random variable defined by CDF $F_\gamma(t) = P(\gamma \leqslant t)$ with a positive expected value $\bar{\gamma} = E(\gamma)$. A preventive maintenance policy is conducted with respect to the working object. Preventive maintenance is performed after expiration of the time η from the end of the previous preventive maintenance or repair. We assume that η is a random variable with a single-point distribution determined by CDF

$$F_\eta(t) = \begin{cases} 0 & \text{for } t \leqslant T \\ 1 & \text{for } t > T \end{cases} . \tag{11.1}$$

We assume that preventive maintenance completely recovers the technical object state. We describe the duration of every preventive service by a nonnegative random variable κ with CDF $F_\kappa(t) = P(\kappa \leqslant t)$ and finite positive expectation $\bar{\kappa} = E(\kappa)$. We suppose that at the moment $t = 0$ the object's work begins and at every renewal instant begins a successive cycle of the object operation, which is determined by the independent copies of the independent random variables $\zeta, \gamma, \eta, \kappa$. The operating object brings a units of income on a unit of time. The emergency maintenance of the object after its failure brings losses. We suppose that a repair cost per unit of time is b. The cost of the preventive maintenance per unit of time is c. We assume $a > 0$, $b \geqslant 0, c \geqslant 0$.

11.3   Model

Accordingly to the description and assumptions, the model of the object operation process may be a semi-Markov process. To construct a suitable model, we introduce the following states:

  1—Working state of the object.
  2—Emergency maintenance after failure.
  3—Preventive maintenance.

To define the semi-Markov model $\{X(t) : t \geqslant 0\}$ completely for a preventive maintenance problem, we have determine a kernel of the process

$$\mathbf{Q}(t) = [Q_{ij}(t) : i, j \in S], \quad t \geqslant 0,$$

and its initial distribution

$$p_i^{(0)} = P(X(0) = i), \quad i \in S.$$

The kernel of the SM process $\{X(t) : t \geqslant 0\}$ takes the form

$$Q(t) = \begin{bmatrix} 0 & Q_{12}(t) & Q_{13}(t) \\ Q_{21}(t) & 0 & 0 \\ Q_{31}(t) & 0 & 0 \end{bmatrix},$$

where

$$Q_{12}(t) = P(\zeta \leqslant t, \eta > \zeta) = \int_0^t [1 - F_\eta(x)]f_\zeta(x)\mathrm{d}x \qquad (11.2)$$

$$= \begin{cases} F_\zeta(t) & \text{for } t \leqslant T \\ F_\zeta(T) & \text{for } t > T \end{cases},$$

$$Q_{13}(t) = P(\tau \leqslant t, \zeta > \tau) = \int_0^t [1 - F_\zeta(x)]\mathrm{d}F_\eta(x) \qquad (11.3)$$

$$= \begin{cases} 0 & \text{for } t \leqslant T \\ 1 - F_\zeta(T) & \text{for } t > T \end{cases},$$

$$Q_{21}(t) = P(\gamma \leqslant t) = F_\gamma(t), \qquad (11.4)$$

$$Q_{31}(t) = P(\kappa \leqslant t) = F_\kappa(t). \qquad (11.5)$$

An initial distribution is

$$p_i = P(X(0) = i) = \begin{cases} 1 & \text{for } i = 1 \\ 0 & \text{for } i = 2, 3 \end{cases}. \qquad (11.6)$$

Therefore, the semi-Markov model has been completely defined.

11.4 Characteristics of operation process

To formulate the problem of optimizing the time to the object preventive maintenance we have to evaluate some necessary characteristics of the process. The probability transition matrix of an embedded Markov chain $\{X(\tau_n); n \in \mathbb{N}_0\}$ has the form

$$P = \begin{bmatrix} 0 & p_{12} & p_{13} \\ 1 & 0 & 0 \\ 1 & 0 & 0 \end{bmatrix}, \qquad (11.7)$$

where

$$p_{12} = \lim_{t \to \infty} Q_{12}(t) = F_\zeta(T), \qquad (11.8)$$

$$p_{13} = \lim_{t \to \infty} Q_{13}(t) = 1 - F_\zeta(T) = R_\zeta(T). \qquad (11.9)$$

We get a stationary distribution

$$\pi = [\pi_1, \pi_2, \pi_3]$$

of the embedded Markov chain by solving a system of equations (4.74)

$$\begin{aligned} \pi_2 + \pi_3 &= \pi_1, \\ \pi_1 p_{12} &= \pi_2, \\ \pi_1 p_{13} &= \pi_3, \\ \pi_1 + \pi_2 + \pi_3 &= 1. \end{aligned} \qquad (11.10)$$

The probabilities

$$\pi_1 = \frac{1}{1 + p_{12} + p_{13}} = \frac{1}{2},$$

$$\pi_2 = \frac{p_{12}}{1 + p_{12} + p_{13}} = \frac{F_\zeta(T)}{2}, \tag{11.11}$$

$$\pi_3 = \frac{p_{13}}{1 + p_{12} + p_{13}} = \frac{R_\zeta(T)}{2}$$

are the only solutions. Expected values of the random variables $T_i, i \in S$ representing duration of the states are

$$E(T_1) = \int_0^\infty t \, d[Q_{12}(t) + Q_{13}(t)] = \int_0^T t f_\zeta(t) dt + T[R_\zeta(T)], \tag{11.12}$$

$$E(T_2) = \int_0^\infty t \, dQ_{21}(t) = \int_0^\infty t \, dF_\gamma(t) = \bar{\gamma}, \tag{11.13}$$

$$E(T_3) = \int_0^\infty t \, dQ_{31}(t) = \int_0^\infty t \, dF_\kappa(t) = \bar{\kappa}. \tag{11.14}$$

From (3.73), we obtain a limiting distribution of the semi-Markov model $\{X(t) : t \geqslant 0\}$

$$P_1 = \frac{\int_0^T t f_\zeta(t) dt + TR_\zeta(T)}{M(T)}, \quad P_2 = \frac{[1 - R_\zeta(T)]\bar{\gamma}}{M(T)}, \quad P_3 = \frac{R_\zeta(T)\bar{\kappa}}{M(T)}, \tag{11.15}$$

where

$$M(T) = \int_0^T t f_\zeta(t) dt + TR_\zeta(T) + [1 - R_\zeta(T)]\bar{\gamma} + R_\zeta(T)\bar{\kappa}. \tag{11.16}$$

11.5   Problem of time to preventive service optimization

As we know, a stochastic process $\{K_j(t) : t \geqslant 0\}$ given by the rule

$$K_j(t) = \int_0^t I_{\{j\}}(X(u)) du, \quad j \in S \tag{11.17}$$

represents a cumulative time spent in state j by the SM process $\{X(t) : t \geqslant 0\}$ during its execution in an interval $[0, t]$. A quantity

$$L(t) = aE[K_1(t)] - bE[K_2(t)] - cE[K_3(t)] \tag{11.18}$$

denotes a mean profit from the object operation in the interval $[0, t]$. A number

$$K = \lim_{t \to \infty} \frac{L(t)}{t} \tag{11.19}$$

represents the income per unit of time that comes from the operation of the object for a long time. Because

$$\lim_{t \to \infty} \frac{E[K_j(t)]}{t} = P_j, \tag{11.20}$$

then

$$K = aP_1 - bP_2 - cP_3. \tag{11.21}$$

Let

$$H_\zeta(T) = \int_0^T t f_\zeta(t) dt. \tag{11.22}$$

According to rule (11.21), we have

$$K(T) = \frac{a[H_\zeta(T) + TR_\zeta(T)] - b[1 - R_\zeta(T)]\bar{\gamma} - cR(T)\bar{\kappa}}{M(T)}. \tag{11.23}$$

The function $K(T), T \geqslant 0$ describes dependence of the unit profit on the object operating time T to the moment of beginning preventive maintenance. This function allows us to formulate a problem of optimization.

Find the number $T^* > 0$ such that

$$K(T^*) = \max_{T > 0}\{K(T)\}. \tag{11.24}$$

Theorem 11.1 contains sufficient conditions of existence for the solution of this problem.

Theorem 11.1. Let

$$\lambda_\zeta(t) = \frac{f_\zeta(t)}{R_\zeta(t)}, \quad r = \bar{\zeta} \lim_{t \to \infty} \lambda_\zeta(t), \quad q = \lim_{t \to \infty} t f_\zeta(t), \tag{11.25}$$

$$B = (c - b)\bar{\gamma}\bar{\kappa} < 0, \quad C = (a + c)\bar{\kappa} > 0, \quad D = (a + b)\bar{\gamma} > 0,$$

$$A = C - D < 0, \quad E = \left(1 - \frac{1}{r}\right)D - \frac{B}{\bar{\zeta}} - Aq.$$

If the time to failure ζ is a random variable with continuous PDF $f_\zeta(t)$ for $t \in (0, \infty)$ and

$$-B \lambda(0^+) < C < E, \tag{11.26}$$

then there exists number T^* for which the function $K(T), T \in (0, \infty)$ given by (11.23) takes the maximum value. The number T^* is the solution of the equation

$$[AH_\zeta(T) + B] \lambda_\zeta(T) + ATf_\zeta(T) + AR_\zeta(T) + D = 0. \tag{11.27}$$

Proof: The criterion function has the form

$$K(T) = \frac{a[H_\zeta(T) + TR_\zeta(T)] - b[1 - R_\zeta(T)]\bar{\gamma} - cR(T)\bar{\kappa}}{M(T)}, \tag{11.28}$$

where

$$H_\zeta(T) = \int_0^T t f_\zeta(t) \mathrm{d}t \tag{11.29}$$

and

$$M(T) = \int_0^T t f_\zeta(t) \mathrm{d}t + T R_\zeta(T) + [1 - R_\zeta(T)]\bar{\gamma} + R_\zeta(T)\bar{\kappa}. \tag{11.30}$$

The derivative of the criterion function has the form

$$K'(T) = \frac{R(T)\left[A H_\zeta(T) + B\right]\lambda_\zeta(T) + A T f_\zeta(T) + A R_\zeta(T) + D}{[M(T)]^2}. \tag{11.31}$$

For each $T \in [0, \infty), R(T) > 0$ and $M(T) > 0$. So, it is sufficient to examine the expression

$$W(T) = \left[A H_\zeta(T) + B\right]\lambda_\zeta(T) + A T f_\zeta(T) + A R_\zeta(T) + D. \tag{11.32}$$

Notice that

$$W(0^+) = -B\lambda(0) + A + D = C - B\lambda(0^+). \tag{11.33}$$

From inequalities (11.26) we have $C > -B\lambda(0^+) > 0$. This means that $W(0^+) > 0$. On the other hand, under the assumption that $C < E$ we have $W(\infty) = \lim_{T \to \infty} W(T) = E < 0$. From the last inequality and the continuity of the function $W(T)$ it follows that there exists the number $T_0 > 0$ such that $W(T_0) < 0$. From the Darboux property of continuous functions there exists a number of $T^* \in (0, T_0)$ such that $w(T^*) = 0$. Thus, the theorem has been proved. □

For $a = 1, b = 0, c = 0$ the function

$$K(T) = \frac{H_\zeta(T) + T R_\zeta(T)}{M(T)} \tag{11.34}$$

denotes steady-state availability of the object. From this theorem, we obtain the following conclusion.

Proposition 11.1. If the time to failure ζ is a random variable with continuous PDF $f_\zeta(t)$ for $t \in (0, \infty)$ and

$$\bar{\kappa} < \left(1 - \frac{1}{r}\right)\bar{\gamma}, \tag{11.35}$$

then there exists number T^* for which the function $K(T), T \in (0, \infty)$ given by (11.34) takes the maximum value. The number T^* is the root of the equation

$$A H_\zeta(T)\lambda_\zeta(T) + A T f_\zeta(T) + A R_\zeta(T) + D = 0. \tag{11.36}$$

where $A = \bar{\kappa} - \bar{\gamma}$ and $D = \gamma$.
Equation (11.36) is equivalent to the following equation:

$$\left[H_\zeta(T)\right]\lambda_\zeta(T) + T f_\zeta(T) + R_\zeta(T) = \frac{\bar{\gamma}}{\bar{\gamma} - \bar{\kappa}}. \tag{11.37}$$

We can see that the solution of the optimization problem exists if a mean time of preventive maintenance is accordingly smaller than a mean time of the object repair. A parameter r given by (11.25) depends on the object failure rate.

11.6   Example

Suppose that the time to failure ζ has Erlang distribution of second order with CDF

$$F_\zeta(t) = \begin{cases} 1 - (1 + \lambda t)e^{-\lambda t} & \text{dla } t > 0 \\ 0 & \text{dla } t \leqslant 0 \end{cases}. \tag{11.38}$$

Assume that the working object (machine, device) brings a [unit/h] of income. Suppose that, a mean time of repair is $\overline{\gamma}$ [h] and its cost is b [unit/h], while the mean time of preventive maintenance is $\overline{\kappa}$ [h] and its cost is c [unit/h].

We find the optimal operating time to preventive maintenance T^* for the criterion function $K(T), T \in [0, \infty)$ determining by (11.23). First, we have to calculate a parameter r. According to (11.25), we have

$$\lambda_\zeta(t) = \frac{\lambda^2 t}{1 + \lambda t}, \quad t \geqslant 0, \quad \overline{\zeta} = \frac{2}{\lambda}, \quad r = \overline{\zeta} \lim_{t \to \infty} \lambda(t) = 2. \tag{11.39}$$

Note that

$$q = \lim_{t \to \infty} t\lambda^2 t e^{-\lambda t} = 0.$$

Hence, according to (11.26), we have

$$E = (1 - \frac{1}{r})D - \frac{B}{\overline{\zeta}}.$$

We evaluate the optimal operating time to preventive maintenance by solving equation (11.27) with unknown parameter T. In this case, the equation takes the form

$$W(T) = 0, \tag{11.40}$$

where

$$W(T) = \left[AH_\zeta(T) + B \right] \lambda_\zeta(T) + ATf_\zeta(T) + AR_\zeta(T) + D, \tag{11.41}$$

where

$$H_\zeta(T) = \int_0^T \lambda^2 t^2 e^{-\lambda t} dt = \frac{2 - (2 + 2\lambda T + \lambda^2 T^2)e^{-\lambda T}}{\lambda},$$

$$\lambda_\zeta(T) = \frac{\lambda^2 T}{1 + \lambda T}, \quad f_\zeta(T) = \lambda^2 T e^{-\lambda T}, \quad R(T) = (1 + \lambda T)e^{-\lambda T},$$

$$B = (b - c)\overline{\gamma}\overline{\kappa}, \quad C = (a + c)\overline{\kappa}, \quad D = (a + b)\overline{\gamma}, \quad A = C - D.$$

A failure rate defined by (11.39) is an increasing function satisfying condition $\lambda(0) = 0$. In this case, inequalities (11.26) take the form

$$0 < C < E = \frac{D}{2} - \frac{B}{\zeta}. \tag{11.42}$$

To solve (11.40), we use the MATHEMATICA program applying procedure

$\text{FindRoot}[W(x) == 0, \{x, 50\}]$,

where the function $W(T), T > 0$ is given by (11.41). The parameters and solutions are presented in Table 11.1. Notice that in each case, $c < b$ and $\bar{\kappa} < \bar{\gamma}$. It means that a unit cost of a preventive service is less than a unit cost of an emergency service, and an expectation of a duration of the preventive maintenance time is less than an expectation of the emergency maintenance duration. This table shows that all inequalities (11.26) are satisfied.

Figure 11.1 shows a graphical solution of the equation. The number $T^* = 137.49$ [h] is the optimal working time to preventive maintenance under the assumption that $\lambda = 0.01$ and the rest of the parameters are equal to the parameters presented in Table 11.1.

The criterion function $K(T), T > 0$ is shown in Figure 11.2.

Table 11.1 Parameters and results

λ	a	b	c	$\bar{\gamma}$	$\bar{\kappa}$	ζ	B	C	E	T^*
0.01	80	20	5	20	5	200.00	−1500	425	1007.5	137.49
0.008	80	20	5	20	5	250.00	−1500	425	1006.0	172.53
0.006	80	20	5	20	5	333.33	−1500	425	1004.5	230.94

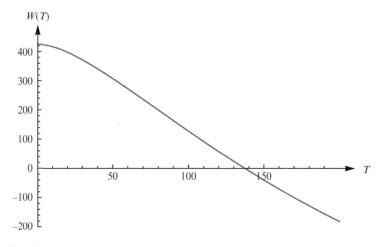

Figure 11.1 The function $W(T), T > 0$.

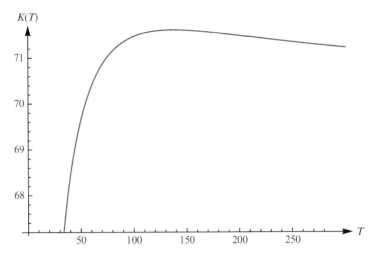

Figure 11.2 The criterion function $K(T), T > 0$ for the Erlang PDF.

Table 11.2 Parameters and results

λ	$\bar{\gamma}$	$\bar{\kappa}$	$\bar{\zeta}$	B	C	E	T^*
0.01	20	5	200.00	0	5	10	173.55
0.008	20	5	250.00	0	5	10	216.94
0.006	20	5	333.33	0	5	10	289.26

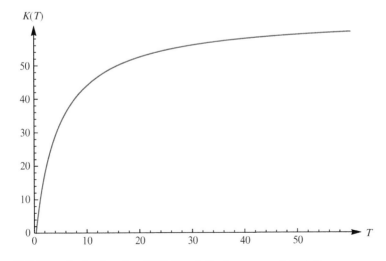

Figure 11.3 The criterion function $K(T), T > 0$, for the exponential PDF.

For $a = 1, \quad b = 0, \quad c = 0$, a criterion function $K(T), T > 0$ denotes a limiting availability of the object. For the same other parameters that are shown in Table 11.1, we obtain the optimal working times to preventive maintenance. Parameters and results are presented in Table 11.2.

We should discuss the case of the exponential distribution. In this case, the criterion function takes the form

$$K(T) = \frac{\frac{a}{\lambda}\left(1 - e^{-\lambda T}\right) - b\gamma(1 - e^{-\lambda,T}) - c\kappa e^{-\lambda T}}{\frac{1}{\lambda}\left(1 - e^{-\lambda T}\right) + \gamma(1 - e^{-\lambda,T}) + \kappa e^{-\lambda T}}. \tag{11.43}$$

The inequalities (11.26) from Theorem 11.1 are not fulfilled. There is no solution of (11.27). Thus, there is no solution to the optimization problem. This means that in the case of the exponential distribution, execution of any preventive maintenance is unprofitable. The function $K(T), T > 0$ for parameters presented in the first row of the Table 11.1 is shown in Figure 11.3.

Semi-Markov model of system component damage

12

Abstract

The semi-Markov model of system component damage is discussed in this chapter. Models that deal with unrepairable systems are presented here. The multistate reliability functions and corresponding expectations, second moments, and standard deviations are evaluated for the presented cases of the component damage. The so-called inverse problem for a simple exponential model of damage is presented in the chapter.

Keywords: Semi-Markov process, Multistate system, Multistate reliability functions, Semi-Markov model of damage

12.1 Semi-Markov model of multistate object

Markov and semi-Markov processes for modeling multistate systems are applied in many different reliability problems [4, 32, 52, 60, 72–74, 90, 97–99].

We will consider an object with finite sets of the ordered reliability states $S = \{0, 1, \ldots, n\}$, where the state 0 is the worst while the state n is the best. We can treat object as one-component system or a component of the many components system. We suppose that the probabilistic model of reliability evolution of the object damage is a stochastic process $\{X(t) : t \geqslant 0\}$, taking values in the set of states $S = \{0, 1, \ldots, n\}$. We suppose that the trajectories of the process are the right-continuous functions. We also assume that a flow graph describing the process state changes is a coherent subgraph of the graph shown in Figure 12.1.

The semi-Markov process has all the features mentioned above. Therefore, we may assume that a damage process can be modeled by SM process $\{X(t) : t \geqslant 0\}$ with a state space $S = \{0, 1, \ldots, n\}$ determined by a kernel $\mathbf{Q}(t) = [Q_{ij}(t) : i, j \in S], t \geqslant 0$, and initial distribution $\mathbf{p} = [p_i(0) : i \in S]$.

Let

$$P_{iB}(t) = P(\forall u \in [0, t], \quad X(u) \in B | X(0) = i), \quad i \in B \subset S. \tag{12.1}$$

This denotes that the process that starts from state i during all the time from the interval $[0, t]$ occupies the states belonging to a subset B. We will apply a theorem that is some conclusion of Theorem coming from [72].

Theorem 12.1. Functions $P_{iB}(t), i \in B \subset S$ satisfy the system of integral equations

$$P_{iB}(t) = 1 - G_i(t) + \sum_{j \in B} \int_0^t P_{jB}(t - x) dQ_{ij}(x), \quad i \in B. \tag{12.2}$$

Semi-Markov Processes: Applications in System Reliability and Maintenance. http://dx.doi.org/10.1016/B978-0-12-800518-7.00012-0

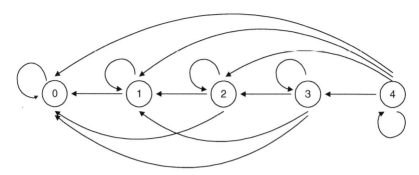

Figure 12.1 A general flow graph of a damage process.

We should mention that these equations are derived in Section 3.4. Using Laplace transformation, we obtain the system of linear equation

$$\tilde{P}_{iB}(s) = \frac{1}{s} - \tilde{G}_i(s) + \sum_{j \in B} \tilde{q}_{ij}(s)\tilde{P}_{jB}(s), \quad i \in B, \tag{12.3}$$

where

$$\tilde{P}_{iB}(s) = \int_0^\infty e^{-st} P_{iB}(t)dt. \tag{12.4}$$

If B is a subset of working states, then the function

$$R_i(t) = P_{iB}(t), \quad i \in B \subset S \tag{12.5}$$

is the reliability function of a system with initial state $i \in B$ at $t = 0$.

Proposition 12.1. Conditional reliability functions satisfy the system of integral equations

$$R_i(t) = 1 - G_i(t) + \sum_{j \in B} \int_0^t R_j(t - x)dQ_{ij}(x), \quad i \in B. \tag{12.6}$$

If T is a random variable denoting a lifetime of the system component, then

$$R_i(t) = P(T > t | X(0) = i), \quad i \in B \tag{12.7}$$

and

$$R_i(t, s) = P(T > t + s | T > t, X(0) = i) = \frac{R_i(t + s)}{R_i(t)}, \quad i \in B. \tag{12.8}$$

12.2   General Semi-Markov model of damage process

We suppose that the technical state of the system component is described by the semi-Markov process which is defined by the renewal kernel

$$Q(t) = \begin{bmatrix} Q_{00}(t) & 0 & 0 & \cdots & 0 \\ Q_{10}(t) & Q_{11}(t) & 0 & \cdots & 0 \\ Q_{20}(t) & Q_{21}(t) & Q_{22}(t) & \cdots & 0 \\ Q_{30}(t) & Q_{31}(t) & Q_{32}(t) & \cdots & 0 \\ \cdots & \cdots & \cdots & \cdots & 0 \\ Q_{n0}(t) & Q_{n1}(t) & \cdots & Q_{nn-1}(t) & Q_{nn}(t) \end{bmatrix}. \tag{12.9}$$

A corresponding flow graph for $n = 4$ is shown in Figure 12.1. Let

$$T_{[l]} = \inf\{t : X(t) \in A_{[l]}\}, \tag{12.10}$$

where

$$A_{[l]} = \{0, \ldots, l-1\} \quad \text{and} \quad A'_{[l]} = S - A_{[l]} = \{l, \ldots, n\}.$$

The function

$$\Phi_{i[l]}(t) = P(T_{[l]} \leqslant t | X(0) = i), \quad i \in A'_{[l]} \tag{12.11}$$

represents the cumulative distribution function (CDF) of the first passage time from the state $i \in A'_{[l]}$ to the subset $A_{[l]}$ for the $\{X(t) : t \geqslant 0\}$. If $X(0) = n$, then the random variable $T_{[l]}$ represents the lifetime of one component system in the subset $A'_{[l]}$. A corresponding reliability function is

$$R_{n[l]}(t) = P(T_{[l]} > t | X(0) = n) = 1 - \Phi_{nA_{[l]}}(t). \tag{12.12}$$

On the other hand,

$$P(T_{[l]} > t | X(0) = n) = P(\forall u \in [0, t] X(u) \in A'_{[l]} | X(0) = n). \tag{12.13}$$

In this case, we have

$$P(T_{[l]} > t | X(0) = n) = P(X(t) \in A'_{[l]} | X(0) = n). \tag{12.14}$$

Applying (12.3), we obtain a system of linear equations for Laplace transform of reliability functions.

$$\tilde{R}_{i[l]}(s) = \frac{1}{s} - \tilde{G}_i(s) + \sum_{j \in A'_{[l]}} \tilde{q}_{ij}(s)\tilde{R}_{j[l]}(s), \quad i \in A'_{[l]}, \tag{12.15}$$

where

$$\tilde{G}_i(s) = \int_0^\infty e^{-st} G_i(t)dt, \quad \tilde{R}_{i[l]}(s) = \int_0^\infty e^{-st} R_{i[l]}(t)dt \tag{12.16}$$

are the Laplace transforms of the functions $G_i(t), R_{i[l]}(t), t \geqslant 0$. Passing to a matrix form, we get

$$\left(I - \tilde{q}_{A'_{[l]}}(s)\right) \tilde{R}_{A'_{[l]}}(s) = \tilde{G}_{A'_{[l]}}(s), \tag{12.17}$$

where

$$I = [\delta_{ij} : i, j \in A'_{[l]}]$$

is the unit matrix and

$$\tilde{\mathbf{q}}_{A'_{[l]}}(s) = [\tilde{q}_{ij}(s) : i, j \in A'_{[l]}],$$

$$\tilde{\mathbf{G}}_{A'_{[l]}}(s) = \frac{1}{s}\left[1 - \sum_{j \in S}\tilde{q}_{ij}(s) : i \in A'_{[l]}\right]^{\mathrm{T}},$$

$$\tilde{\mathbf{R}}_{A'_{[l]}}(s) = \left[R_{i[l]} : i \in A'_{[l]}\right]^{\mathrm{T}}.$$

A vector function

$$\tilde{\mathbf{R}}(s) = \left[\tilde{R}_{n[0]}(s), \tilde{R}_{n[1]}(s), \dots, \tilde{R}_{n[n]}(s)\right] \tag{12.18}$$

is the Laplace transform of the multistates reliability function of the object.

Example 12.1. Let $S = \{0, 1, 2, 3\}$. Hence,

$$\begin{aligned}
A_{[1]} &= \{0\}, & A'_{[1]} &= \{1, 2, 3\}, \\
A_{[2]} &= \{0, 1\}, & A'_{[2]} &= \{2, 3\}, \\
A_{[3]} &= \{0, 1, 2\}, & A'_{[3]} &= \{3\}.
\end{aligned} \tag{12.19}$$

For $l = 1$, the matrices from (12.17) take the form

$$\mathbf{I} - \tilde{\mathbf{q}}_{A'_{[l]}}(s) = \begin{bmatrix} 1 - \tilde{q}_{11}(s) & 0 & 0 \\ -\tilde{q}_{21}(s) & 1 - \tilde{q}_{22}(s) & 0 \\ -\tilde{q}_{31}(s) & -\tilde{q}_{32}(s) & 1 - \tilde{q}_{33}(s) \end{bmatrix}, \tag{12.20}$$

$$\tilde{\mathbf{G}}_{A'_{[l]}}(s) = \frac{1}{s}\begin{bmatrix} 1 - \tilde{q}_{10}(s) - \tilde{q}_{11}(s) \\ 1 - \tilde{q}_{20}(s) - \tilde{q}_{21}(s) - \tilde{q}_{22}(s) \\ 1 - \tilde{q}_{30} - \tilde{q}_{31}(s) - \tilde{q}_{32}(s) - \tilde{q}_{33}(s) \end{bmatrix}. \tag{12.21}$$

We are interested in the element $\tilde{R}_{3[1]}(s)$ of the solution.

$$\tilde{\mathbf{R}}_{A'[l]}(s) = \begin{bmatrix} \tilde{R}_{1[1]}(s) \\ \tilde{R}_{2[1]}(s) \\ \tilde{R}_3[1](s) \end{bmatrix}. \tag{12.22}$$

Its Laplace transform is

$$\tilde{R}_{3[1]}(s) = \frac{\tilde{u}_3(s)}{s(1 - \tilde{q}_{11}(s))(1 - \tilde{q}_{22}(s))(1 - \tilde{q}_{33}(s))},$$

where

$$\begin{aligned}
\tilde{u}_3(s) = {} & 1 - \tilde{q}_{11}(s) - \tilde{q}_{22}(s) + \tilde{q}_{11}(s)\tilde{q}_{22}(s) - \tilde{q}_{30}(s) \\
& + \tilde{q}_{11}(s)\tilde{q}_{30(s)} + \tilde{q}_{22}(s)q_{30}(s) - \tilde{q}_{11}(s)\tilde{q}_{22}(s)\tilde{q}_{30}(s) \\
& - \tilde{q}_{10}(s)\tilde{q}_{31}(s) + \tilde{q}_{10}(s)\tilde{q}_{22}(s)\tilde{q}_{31}(s) - \tilde{q}_{20}(s)\tilde{q}_{32}(s) \\
& + \tilde{q}_{11}(s)\tilde{q}_{20}(s)\tilde{q}_{32}(s) - \tilde{q}_{10}(s)\tilde{q}_{21}(s)\tilde{q}_{32}(s) - \tilde{q}_{33}(s) \\
& + \tilde{q}_{11}(s)q_{33}(s) + \tilde{q}_{22}(s)\tilde{q}_{33}(s) - \tilde{q}_{11}(s)\tilde{q}_{22}(s)\tilde{q}_{33}(s).
\end{aligned}$$

For $l = 2$, the matrices from (12.17) take the form

$$\mathbf{I} - \tilde{\mathbf{q}}_{A'_{[2]}}(s) = \begin{bmatrix} 1 - \tilde{q}_{22}(s) & 0 \\ -\tilde{q}_{32}(s) & 1 - \tilde{q}_{33}(s) \end{bmatrix}.$$

By solving (12.17), we get

$$\tilde{R}_{3[2]}(s) = \frac{\tilde{u}_2(s)}{s(1 - \tilde{q}_{11}(s))(1 - \tilde{q}_{22}(s))},$$

where

$$\tilde{u}_2(s) = 1 - \tilde{q}_{22}(s) - \tilde{q}_{30}(s) - \tilde{q}_{33}(s) - \tilde{q}_{31}(s)$$
$$+\tilde{q}_{22}(s)\tilde{q}_{31}(s) - \tilde{q}_{20}(s)\tilde{q}_{32}(s) - \tilde{q}_{21}(s)\tilde{q}_{32}(s)$$
$$+\tilde{q}_{22}(s)\tilde{q}_{33}(s)) + \tilde{q}_{22}(s)\tilde{q}_{30}(s).$$

For $l = 3$, the matrices from (12.17) are

$$\mathbf{I} - \tilde{q}_{A'_{[3]}}(s) = \left[\; 1 - \tilde{q}_{33}(s) \; \right],$$

$$\tilde{\mathbf{G}}_{A'_{[3]}}(s) = \frac{1}{s}\left[\; 1 - \tilde{q}_{30} - \tilde{q}_{31}(s) - \tilde{q}_{32}(s) - \tilde{q}_{33}(s) \; \right].$$

Now, the solution of (12.17) is

$$\tilde{R}_{3[3]}(s) = \frac{1 - \tilde{q}_{30} - \tilde{q}_{31}(s) - \tilde{q}_{32}(s) - \tilde{q}_{33}(s)}{s(1 - \tilde{q}_{33}(s))}.$$

Mostly, the elements $Q_{ii}(t), i = 1, 2, \ldots, n$ are equal to 0. Let us suppose that

$$Q(t) = \begin{bmatrix} Q_{00}(t) & 0 & 0 & 0 \\ Q_{10}(t) & 0 & 0 & 0 \\ Q_{20}(t) & Q_{21}(t) & 0 & 0 \\ Q_{30}(t) & Q_{31}(t) & Q_{32}(t) & 0 \end{bmatrix}. \quad (12.23)$$

Hence, we obtain

$$\tilde{R}_{3[1]}(s) = \frac{\tilde{u}_3(s)}{s}, \quad (12.24)$$

where

$$\tilde{u}_3(s) = 1 - \tilde{q}_{30}(s) - \tilde{q}_{10}(s)\tilde{q}_{31}(s)$$
$$-\tilde{q}_{20}(s)\tilde{q}_{32}(s) - \tilde{q}_{10}(s)\tilde{q}_{21}(s)\tilde{q}_{32}(s),$$

$$\tilde{R}_{3[2]}(s) = \frac{\tilde{u}_2(s)}{s}, \quad (12.25)$$

where

$$\tilde{u}_2(s) = 1 - \tilde{q}_{30}(s) - \tilde{q}_{31}(s)$$
$$-\tilde{q}_{20}(s)\tilde{q}_{32}(s) - \tilde{q}_{21}(s)\tilde{q}_{32}(s),$$

and

$$\tilde{R}_{3[3]}(s) = \frac{1 - \tilde{q}_{30} - \tilde{q}_{31}(s) - \tilde{q}_{32}(s)}{s}. \quad (12.26)$$

The Laplace transform of the multistate reliability function of that object is

$$\tilde{\mathbf{R}}(s) = \left[\tilde{R}_{3[0]}(s), \tilde{R}_{3[1]}(s), \tilde{R}_{3[2]}(s), \tilde{R}_{3[3]}(s) \right].$$

12.3   Multistate model of two kinds of failures

We assume that the failures are caused by wear or some random events. With positive probabilities only the state changes from k to $k-1$ or from k to 0 are possible (Figure 14.2). The time of state change from a state k to $k-1, k=1,\ldots,n$ because of wear is assumed to be a nonnegative random variable η_k with a PDF $f_k(x), x \geqslant 0$. Time to a total failure (state 0) of the object in state k is a nonnegative random variable ζ_k exponentially distributed with parameter λ_k. Under those assumptions, the stochastic process $\{X(t) : t \geqslant 0\}$ describing the reliability state of the component is the semi-Markov process with a state space $S = \{0, 1, \ldots, n\}$ and a kernel

$$
\mathbf{Q}(t) =
\begin{bmatrix}
Q_{00}(t) & 0 & 0 & \ldots & 0 \\
Q_{10}(t) & 0 & 0 & \ldots & 0 \\
Q_{20}(t) & Q_{21}(t) & 0 & \ldots & 0 \\
Q_{30}(t) & 0 & Q_{32}(t) & \ldots & 0 \\
\ldots & \ldots & \ldots & \ldots & 0 \\
Q_{n0}(t) & 0 & \ldots & Q_{nn-1}(t) & 0
\end{bmatrix},
\tag{12.27}
$$

where

$$
Q_{kk-1}(t) = P(\eta_k \leqslant t, \zeta_k > \eta_k) = \int_0^t e^{-\lambda_k x} f_k(x)\mathrm{d}x,
\tag{12.28}
$$

$$
Q_{k0}(t) = P(\zeta_k \leqslant t, \eta_k > \zeta_k) = \int_0^t \lambda_k e^{-\lambda_k x}[1 - F_k(x)]\mathrm{d}x
\tag{12.29}
$$

for $k = 1, \ldots, n$.

To explain such a model, we assume that $n = 3$ and we suppose that random variables $\eta_k, k = 1, 2, 3$ have gamma distributions with parameters $\alpha_k = 1, 2, \ldots$ and $\beta_k > 0$ with PDF

$$
f_k(x) = \frac{\beta_k^{\alpha_k} x^{\alpha_k-1} e^{-\beta_k x}}{(\alpha_k - 1)!}.
\tag{12.30}
$$

In this case, the matrix (12.27) has the form

$$
\mathbf{Q}(t) =
\begin{bmatrix}
Q_{00}(t) & 0 & 0 & 0 \\
Q_{10}(t) & 0 & 0 & 0 \\
Q_{20}(t) & Q_{21}(t) & 0 & 0 \\
Q_{30}(t) & 0 & Q_{32}(t) & 0
\end{bmatrix}.
\tag{12.31}
$$

Notice that this matrix is equal to matrix (12.23) from Example 12.1 with $Q_{31}(t) = 0$. Therefore, we can apply equalities (12.24)–(12.26) to calculate components of a multistate reliability function. Finally, we obtain the Laplace transforms

$$
\tilde{R}_{3[1]}(s) = \frac{\tilde{u}_3(s)}{s},
\tag{12.32}
$$

$$\tilde{u}_3(s) = 1 - \tilde{q}_{30}(s) - \tilde{q}_{20}(s)\tilde{q}_{32}(s) - \tilde{q}_{10}(s)\tilde{q}_{21}(s)\tilde{q}_{32}(s),$$

$$\tilde{R}_{3|2|}(s) = \frac{\tilde{u}_2(s)}{s}, \tag{12.33}$$

$$\tilde{u}_2(s) = 1 - \tilde{q}_{30}(s) - \tilde{q}_{20}(s)\tilde{q}_{32}(s) - \tilde{q}_{21}(s)\tilde{q}_{32}(s),$$

and

$$\tilde{R}_{3|3|}(s) = \frac{1 - \tilde{q}_{30} - \tilde{q}_{32}(s)}{s}, \tag{12.34}$$

where

$$\tilde{q}_{10}(s) = \frac{\beta_1^2}{(s + \beta_1 + \lambda_1)^2} + \frac{\lambda_1(s + 2\beta_1 + \lambda_1)}{(s + \beta_1 + \lambda_1)^2},$$

$$\tilde{q}_{21}(s) = \frac{\beta_2^2}{(s + \beta_2 + \lambda_2)^2}, \quad \tilde{q}_{20}(s) = \frac{\lambda_2(s + 2\beta_2 + \lambda_2)}{(s + \beta_2 + \lambda_2)^2}, \tag{12.35}$$

$$\tilde{q}_{32}(s) = \frac{\beta_3^2}{(s + \beta_3 + \lambda_3)^2}, \quad \tilde{q}_{30}(s) = \frac{\lambda_3(s + 2\beta_3 + \lambda_3)}{(s + \beta_3 + \lambda_3)^2}.$$

For a numerical example, we fix

$$\alpha_1 = 2, \quad \beta_1 = 0.04, \quad \lambda_1 = 0.004,$$
$$\alpha_2 = 2, \quad \beta_2 = 0.02, \quad \lambda_2 = 0.002, \tag{12.36}$$
$$\alpha_3 = 2, \quad \beta_3 = 0.01, \quad \lambda_3 = 0.001.$$

Substituting function (12.35) with the parameters above to (12.32)–(12.34), we obtain

$$\tilde{R}_{3|1|}(s) = \frac{\tilde{w}_1(s)}{(0.021 + s)^2(0.032 + s)^2(0.044 + s)^2},$$

where

$$\tilde{w}_1(s) = 1.59533 10^{-7} + 0.000014s + 0.000626s^2 + 0.015224s^3 + 0.193s^4 + s^5,$$

and

$$\tilde{R}_{3|2|}(s) = \frac{0.000066784 + 0.004048s + 0.105s^2 + s^3}{(0.021 + s)^2(0.032 + s)^2},$$

$$\tilde{R}_{3|3|}(s) = \frac{0.041 + s}{(0.021 + s)^2}.$$

Using the inverse Laplace transformation, we obtain the reliability functions

$$R_{3|1|}(t) = 52.6698\, e^{-0.044t} + 58.277\, e^{-0.032t} - 109.947\, e^{-0.021t} +$$
$$+ 0.189036\, e^{-0.044t}t + 1.17355\, e^{-0.032t}t + 0.509862\, e^{-0.021t}t, \tag{12.37}$$

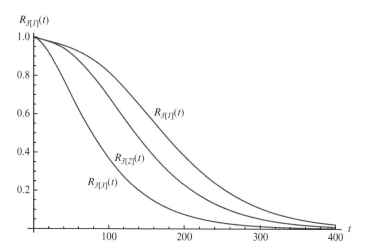

Figure 12.2 The l level reliability functions.

$$R_{3[2]}(t) = 21.3373\, e^{-0.032t} - 20.3373\, e^{-0.021t} +$$
$$+0.0991736\, e^{-0.032t} t + 0.155537\, e^{-0.021t} t, \tag{12.38}$$

$$R_{3[3]}(t) = e^{-0.021t}(1 + 0.02t). \tag{12.39}$$

These reliability functions are shown in Figure 12.2.

We calculate the corresponding expectations, second moments, and standard deviations of an l level system lifetime using the following formulas:

$$m_1[l] = [E[T_{[l]}|X(0) = 3] = \lim_{s \to 0} \tilde{R}_{3[l]}(s), \quad l = 1, 2, 3,$$
$$m_2[l] = E[T_{[l]}^2|X(0) = 3] = -2 \lim_{s \to 0} [\tilde{R}'_{3[l]}(s)],$$
$$\sigma[l] = \sqrt{m_2[l] - [m_1[l]]^2}.$$

For the given parameters, we get

$$m_1[1] = 182.48, \quad m_1[2] = 147.89, \quad m_1[3] = 92.97,$$
$$m_2[1] = 41960.1, \quad m_2[2] = 28727.2, \quad m_2[3] = 13173.5,$$
$$\sigma[1] = 93.06, \qquad \sigma[2] = 82.80, \qquad \sigma[3] = 67.30.$$

12.4 Inverse problem for simple exponential model of damage

We suppose that only the state changes from k to $k - 1$, $k = 1, 2, \ldots, n$, are possible with the positive probabilities. Now, the stochastic process $\{X(t) : t \geqslant 0\}$ describing the reliability state of the object is the semi-Markov process with a state space $S = \{0, 1, \ldots, n\}$ and a kernel

$$\mathbf{Q}(t) = \begin{bmatrix} Q_{00}(t) & 0 & 0 & \cdots & 0 \\ Q_{10}(t) & 0 & 0 & \cdots & 0 \\ 0 & Q_{21}(t) & 0 & \cdots & 0 \\ 0 & 0 & Q_{32}(t) & \cdots & 0 \\ \cdots & \cdots & \cdots & \cdots & 0 \\ 0 & 0 & \cdots & Q_{nn-1}(t) & 0 \end{bmatrix}. \tag{12.40}$$

From (12.17), we obtain the Laplace transforms of the multistate reliability function components.

$$\tilde{R}_{n[1]}(s) = \frac{1 - \tilde{q}_{10}(s)\tilde{q}_{21}(s)\ldots\tilde{q}_{nn-1}(s)}{s}, \tag{12.41}$$

$$\tilde{R}_{n[2]}(s) = \frac{1 - \tilde{q}_{21}(s)\tilde{q}_{32}(s)\ldots\tilde{q}_{nn-1}(s)}{s},$$

$$\vdots$$

$$\tilde{R}_{n[n-1]}(s) = \frac{1 - \tilde{q}_{n-1n-2}(s)\tilde{q}_{nn-1}(s)}{s},$$

$$\tilde{R}_{n[n]}(s) = \frac{1 - \tilde{q}_{nn-1}(s)}{s},$$

The multistate reliability function is called exponential if all its components (except of $R_{n[0]}(t)$) are exponential functions [52]

$$\mathbf{R}(t) = \left[1, e^{-\lambda_1 t}, e^{-\lambda_2 t}, \ldots, e^{-\lambda_n t}\right],$$

where

$$0 < \lambda_1 < \lambda_2 < \cdots < \lambda_n.$$

This means that

$$\tilde{R}_{n[l]}(s) = \frac{1}{s + \lambda_l}, \quad l = 1, 2, \ldots, n.$$

We set a problem. Find the elements

$$Q_{kk-1}(t), \quad k = 1, 2, \ldots, n$$

of a semi-Markov kernel for which the system of equations (12.41) is fulfilled. For computing these functions, we have to solve the following system of equations:

$$\frac{1}{s + \lambda_1} = \frac{1 - \tilde{q}_{10}(s)\tilde{q}_{21}(s)\ldots\tilde{q}_{nn-1}(s)}{s},$$

$$\frac{1}{s + \lambda_2} = \frac{1 - \tilde{q}_{21}(s)\tilde{q}_{32}(s)\ldots\tilde{q}_{nn-1}(s)}{s},$$

$$\vdots$$

$$\frac{1}{s + \lambda_{n-1}} = \frac{1 - \tilde{q}_{n-1n-2}(s)\tilde{q}_{nn-1}(s)}{s},$$

$$\frac{1}{s + \lambda_n} = \frac{1 - \tilde{q}_{nn-1}(s)}{s}.$$

The solution takes the form

$$\tilde{q}_{kk-1}(s) = \frac{\lambda_k (s + \lambda_{k+1})}{(s + \lambda_k) \lambda_{k+1}}, \quad k = 1, 2, \ldots, n - 1, \tag{12.42}$$

$$\tilde{q}_{nn-1}(s) = \frac{\lambda_n}{s + \lambda_n}. \tag{12.43}$$

We obtain the elements $Q_{kk-1}(t), k = 1, 2, \ldots, n$ of the kernel (12.40) as the inverse Laplace transforms of the functions

$$\tilde{Q}_{kk-1}(s) = \frac{\tilde{q}_{kk-1}(s)}{s}, \quad k = 1, 2, \ldots, n. \tag{12.44}$$

A CDF of a waiting time in state k for the kernel (12.40) is equal to a function $Q_{kk-1}(t)$, for $k = 1, 2, \ldots, n$.

$$G_k(t) = P(T_k \leqslant t) = Q_{kk-1}(t), \quad k = 1, 2, \ldots, n \tag{12.45}$$

Applying the results (12.42) and (12.43), we get

$$G_k(t) = L^{-1} \left[\frac{\lambda_k (s + \lambda_{k+1})}{s (s + \lambda_k) \lambda_{k+1}} \right] \tag{12.46}$$

$$= \begin{cases} 0 & \text{for } t < 0, \\ 1 - \left(1 - \frac{\lambda_k}{\lambda_{k+1}}\right) e^{-\lambda_k t} & \text{for } t \geqslant 0, \end{cases} \quad k = 1, \ldots, n - 1, \tag{12.47}$$

and

$$G_n(t) = L^{-1} \left[\frac{\lambda_n}{s(s + \lambda_n)} \right] = \begin{cases} 0 & \text{for } t < 0, \\ 1 - e^{-\lambda_n t} & \text{for } t \geqslant 0. \end{cases} \tag{12.48}$$

Therefore, now we can formulate the following theorem.

Theorem 12.2. For the multistate exponential reliability function

$$\mathbf{R}(t) = [1, e^{-\lambda_1 t}, e^{-\lambda_2 t}, \ldots, e^{-\lambda_n t}], \tag{12.49}$$

where

$$0 < \lambda_1 < \lambda_2 < \cdots < \lambda_n, \tag{12.50}$$

the CDF of the waiting time T_k of the semi-Markov process defined by the kernel (12.40) is

$$G_k(t) = \begin{cases} 0 & \text{for } t < 0, \\ 1 - \left(1 - \frac{\lambda_k}{\lambda_{k+1}}\right) e^{-\lambda_k t} & \text{for } t \geqslant 0, \end{cases} \quad k = 1, \ldots, n-1, \tag{12.51}$$

$$G_n(t) = \begin{cases} 0 & \text{for } t < 0, \\ 1 - e^{-\lambda_n t} & \text{for } t \geqslant 0. \end{cases} \tag{12.52}$$

From Theorem 12.2 it follows that the probability distributions for the random variables $T_k, k = 1, 2, \ldots, n-1$ are a mixture of discrete and absolutely continuous distributions

$$G_k(t) = p G_k^{(d)}(t) + q G_k^{(c)}(t), \quad k = 1, \ldots, n-1, \tag{12.53}$$

where

$$p = \frac{\lambda_k}{\lambda_{k+1}}, \quad q = 1 - \frac{\lambda_k}{\lambda_{k+1}}, \tag{12.54}$$

$$G_k^{(d)}(t) = \begin{cases} 0 & \text{for } t < 0, \\ 1 & \text{for } t \geqslant 0, \end{cases} \quad G_n^{(c)}(t) = \begin{cases} 0 & \text{for } t < 0, \\ 1 - e^{-\lambda_k t} & \text{for } t \geqslant 0. \end{cases} \tag{12.55}$$

From the above-mentioned theorem, it follows that

$$P(T_k = 0) = \frac{\lambda_k}{\lambda_{k+1}}, \quad k = 1, \ldots, n-1. \tag{12.56}$$

This means that a sequence of state changes $(n, n-1, \ldots, 1, 0)$ with waiting times $(T_n > 0, T_{n-1} = 0, \ldots, T_1 = 0)$ is possible. The probability of these sequences of events is

$$P(T_n > 0, T_{n-1} = 0, \ldots, T_1 = 0) = \frac{\lambda_{n-1}}{\lambda_n} \frac{\lambda_{n-2}}{\lambda_{n-1}} \cdots \frac{\lambda_1}{\lambda_2} = \frac{\lambda_1}{\lambda_n}. \tag{12.57}$$

Finally, in this case a value of n-level time to failure is

$$T_{n[n]} = t_n + 0 + \cdots + 0 = t_n.$$

12.5 Conclusions

Constructed multistate semi-Markov models allow us to find the reliability characteristics and parameters of unrepairable systems. Multistate reliability functions and corresponding their expectations, second moments, and standard deviations are calculated for presented models. Solutions of equations that follow from semi-Markov processes theory are obtained by using Laplace transformation. Some interesting conclusions follow from presented Theorem 12.1 concerning the multistate exponential reliability function. A sequence of state changes $(n, n-1, \ldots, 1, 0)$ with waiting times $T_n > 0, T_{n-1} = 0, \ldots, T_1 = 0$ is possible. The probability of these events sequences is

$$P(T_n > 0, T_{n-1} = 0, \ldots, T_1 = 0) = \frac{\lambda_1}{\lambda_n}.$$

Multistate systems with SM components

13

Abstract

This chapter deals with unrepairable and repairable multistate monotone systems consisting of multistate components that are modeled by the semi-Markov processes. In the case of an unrepairable system, the multistate reliability functions of the system components and the whole system are discussed. In the case of a repairable system, the probability distribution and the limiting distribution of the system reliability state are investigated. All concepts and models presented are illustrated by simple numerical examples.

Keywords: Semi-Markov process, Multistate system, Multistate reliability functions, Binary representation

13.1 Introduction

Many papers are devoted to the reliability of multistate monotone systems (MMSs) [4, 35, 52, 59, 73, 74]. Many others concern the semi-Markov models of multistate systems [32, 60, 72]. Some results of investigation of the MMS with components modeled by the independent semi-Markov processes (SMPs) are presented in the chapter. The chapter is organized as follows. The second section contains a basic notation, concepts, and assumptions. Particularly, it deals with a system structure and a concept of a MMS. The third section is devoted to the unrepairable system components. We assume that the states of system components are modeled by the independent SMPs. Some characteristics of a SMP are used as reliability characteristics of the system components. In the next section, the binary representation of the MMSs is discussed. The concept of a minimal path vector to level 1 is crucial to these considerations. The multistate reliability functions of the system components and the whole system are discussed in the section. The last section provides the semi-Markov model of the renewable multistate system. The probability distribution and the limiting distribution of the systems reliability state are computed. The concepts and models presented are illustrated by some numerical examples.

13.2 Structure of the system

Consider a system consisting of n components with the index set $C = \{1, \ldots, n\}$. We suppose that $S_k = \{0, 1, \ldots, z_k\}$, $k \in C$ is the set of the states of the component k, and $S = \{0, 1, \ldots, s\}$ is the set of the system states. All the states are ordered. States of the

Semi-Markov Processes: Applications in System Reliability and Maintenance. http://dx.doi.org/10.1016/B978-0-12-800518-7.00013-1

system (a component k) denote successive levels of the object's technical condition from the perfect functioning level z (z_k) to the complete failure level 0. Therefore, the state 0 is the worst and the state z, (z_k) is the best.

The function

$$\psi : S_1 \times \cdots \times S_n \to S$$

is called the *system structure function*.

If the system structure function is nondecreasing in each argument and

$$\psi(0,\ldots,0) = 0, \quad \psi(z_1,\ldots,z_n) = z,$$

then it is said to be *monotone*. Formally, a multistate system is represented by a sequence of symbols $(C, S, S_1, \ldots, S_n, \psi)$. If the system structure function is monotone, the system is called a MMS. We assume that the systems considered in this chapter are MMS. The state of a component k at fixed instant t may be described by the random variable $X_k(t)$ taking its value in S_k. The random vector $\mathbf{X}(t) = (X_1(t),\ldots,X_n(t))$ represents the states of all system components at fixed moment t. The state of the system at the fixed instant t is completely defined by the states of components through the system structure function ψ

$$\mathbf{Y}(t) = \psi(\mathbf{X}(t)). \tag{13.1}$$

If the parameter t runs the interval $[0, \infty)$, all random variables mentioned above become random processes. Therefore, $\{\mathbf{Y}(t) : t \in [0, \infty)\}$ is a stochastic process with the state space $S = \{0, 1, \ldots, z\}$. The process determines a reliability state of the system.

13.3   Reliability of unrepairable system components

We suppose that the reliability states of system components are described by the independent SMPs $\{X_k(t) : t \geqslant 0\}$, $k \in C$. Unfortunately, the random process $\{\mathbf{Y}(t) : t \geqslant 0\}$, $\mathbf{Y}(t) = \psi(X_1(t),\ldots,X_n(t))$, taking its values from the set $S = \{0, 1, \ldots, z\}$, which describes the system reliability state at time $t \in [0, \infty)$ is not a SMP. We have at least two ways to analyze the reliability of the multistate system. The first one is based on the extension of the process $\{\mathbf{Y}(t) : t \geqslant 0\}$ to a SMP by construction of the superposition of independent Markov renewal processes associated with the semi-Markov processes $\{X_k(t) : t \geqslant 0\}$, $k \in C$ [61]. This way needs more advanced mathematical concepts, which go beyond the scope of this book. The second way consists of calculating the reliability characteristics of the multistate system based on the characteristics of its independent components. In this chapter, we apply the second way.

We suppose that the SMP representing the reliability state of the component k is determined by the following kernel:

$$
\begin{bmatrix}
Q_{00}^{(k)}(t) & 0 & 0 & \cdots & 0 \\
Q_{10}^{(k)}(t) & 0 & 0 & \cdots & 0 \\
Q_{20}^{(k)}(t) & Q_{21}^{(k)}(t) & 0 & \cdots & 0 \\
Q_{30}^{(k)}(t) & Q_{31}^{(k)}(t) & Q_{32}^{(k)}(t) & \cdots & 0 \\
\cdots & \cdots & \cdots & \cdots & 0 \\
Q_{z_k0}^{(k)}(t) & Q_{z_k1}^{(k)}(t)\cdots & \cdots & Q_{z_kz_{k-1}}^{(k)}(t) & 0
\end{bmatrix}. \tag{13.2}
$$

Note that this matrix is a submatrix of the kernel (12.9) in which $Q_{ii}(t) = 0$, $i = 1,\ldots,n$. This means that the process determined by (13.2) may be treated as the process of kth component damage.

Let

$$
T_{[l]}^{(k)} = \inf\{t : X_k(t) \in A_{[l]}^{(k)}\} \tag{13.3}
$$

where

$$
A_{[l]}^{(k)} = \{0,\ldots,l-1\} \quad \text{and} \quad A_{[l]}'^{(k)} = S - A_{[l]}^{(k)} = \{l,\ldots,z_k\}.
$$

The function

$$
\Phi_{i[l]}^{(k)}(t) = P(T_{[l]}^{(k)} \leqslant t | X(0) = i), \quad i \in A_{[l]}'^{(k)} \tag{13.4}
$$

represents the cumulative distribution function (CDF) of the first passage time from the state $i \in A_{[l]}'^{(k)}$ to the subset $A_{[l]}^{(k)}$ for the process $\{X_k(t) : t \geqslant 0\}$. If $X(0) = s_k$, then the random variable $T_{[l]}^{(k)}$ represents the l-level lifetime of the component k. A corresponding reliability function is

$$
R_{z_k[l]}^{(k)}(t) = P(T_{[l]}^{(k)} > t | X(0) = z_k) = 1 - \Phi_{z_k[l]}^{(k)}(t). \tag{13.5}
$$

From Chapter 3 we know that the Laplace-Stieltjes transforms of the CDFs $\Phi_{i[l]}^{(k)}(t)$, $i \in A_{[l]}'^{(k)}$ satisfy the integral system of equations

$$
\tilde{\varphi}_{i[l]}^{(k)}(s) = \sum_{j\in A_{[l]}'^{(k)}} \tilde{q}_{ij}^{(k)}(s) + \sum_{j\in A_k'(u)} \tilde{q}_{ij}^{(k)}(s)\tilde{\varphi}_{j[u]}^{(k)}(s), \quad i \in A_{[l]}'^{(k)}, \tag{13.6}
$$

where

$$
\tilde{\varphi}_{i[l]}^{(k)}(s) = \int_0^\infty e^{-st} d\Phi_{i[l]}^{k)}(t), \quad \tilde{q}_{ij}^{(k)}(s) = \int_0^\infty e^{-st} dQ_{ij}^{(k)}(t).
$$

The Laplace transform

$$
\tilde{R}_{z_k[l]}^{(k)}(s) = \int_0^\infty e^{-st} R_{z_k[l]}^{(k)}(t)
$$

of the kth component reliability function to level l is given by the formula

$$
\tilde{R}_{s_k[l]}^{(k)}(s) = \frac{1 - \tilde{\varphi}_{k[u]}^{(k)}(s)}{s}. \tag{13.7}
$$

On the other hand,

$$R_{z_k[l]}^{(k)}(t) = P(T_{k[l]} > t | X(0) = z_k) = P(\forall u \in [0, t], \ X_k(u) \in A'_{k[l]} | X(0) = z_k).$$
(13.8)

As components of the system are unrepairable, then we have

$$R_{z_k[l]}^{(k)}(t) = P(\forall u \in [0, t], \ X_k(u) \in A_{[l]}^{\prime(k)} | X(0) = z_k) = P(X_k(t) \in A_{[l]}^{\prime(k)} | X(0) = z_k).$$
(13.9)

Finally, we get

$$R_{z_k[l]}^{(k)}(t) = \sum_{j=l}^{z_k} P(X(t) = j | X(0) = z_k) = \sum_{j=l}^{z_k} P_{z_k j}^{(k)}(t).$$
(13.10)

Applying (12.2) we obtain a linear system of equations for the Laplace transforms of the reliability functions to level l for the system components

$$\tilde{R}_{i[l]}^{(k)}(s) = \frac{1}{s} - \tilde{G}_i^{(k)}(s) + \sum_{j \in A_{[l]}^{\prime(k)}} \tilde{q}_{ij}^{(k)}(s) \tilde{R}_{j[l]}^{(k)}(s), \quad i \in A_{[l]}^{\prime(k)},$$
(13.11)

where

$$\tilde{G}_i^{(k)}(s) = \int_0^\infty e^{-st} G_i^{(k)}(t) dt, \quad \tilde{R}_{i[l]}^{(k)}(s) = \int_0^\infty e^{-st} R_{i[l]}^{(k)}(t) dt$$
(13.12)

are the Laplace transforms of the functions $G_i^{(k)}(t)$ and $R_{i[l]}^{(k)}(t)$, $t \geqslant 0$. Passing to the matrix notation, we get

$$\left(I - \tilde{q}_{[l]}^{(k)}(s) \right) \tilde{R}_{[l]}^{(k)}(s) = \tilde{G}_{[l]}^{(k)}(s).$$
(13.13)

The function

$$R_{[l]}^{(k)}(t) = P(T_{[l]}^{(k)} > t) = P(X_k(t) \in A_{[l]}^{\prime(k)})$$
(13.14)

means the reliability function to level l of a kth system component.

Definition 13.1. The vector function

$$\boldsymbol{R}^{(k)}(t) = [R_{[0]}^{(k)}(t), R_{[1]}^{(k)}(t), \ldots, R_{[z_k]}^{(k)}(t)]$$
(13.15)

is said to be the multistate reliability function of the kth component of the system. Notice that

$$A_{[0]}^{\prime(k)} = S_k \supset A_{[1]}^{\prime(k)} \supset \cdots \supset A_{[s_k]}^{\prime(k)}.$$

From the well-known property of probability, we have

$$1 = P(X_k(t) \in S_k) \geqslant P(X_k(t) \in A_{[1]}^{\prime(k)}) \geqslant \cdots \geqslant P(X_k(t) \in A_{[z_k]}^{\prime(k)}).$$

This means that

$$1 = R_{[0]}^{(k)}(t) \geqslant R_{[1]}^{(k)}(t) \geqslant \cdots \geqslant R_{[z_k]}^{(k)}(t).$$
(13.16)

Equation (13.13) enables us to calculate the Laplace transform of the multistates reliability function of the kth component.

$$\tilde{\boldsymbol{R}}^{(k)}(s) = \left[\tilde{R}_{z_k[0]}^{(k)}(s), \tilde{R}_{z_k[1]}^{(k)}(s), \dots, \tilde{R}_{z_k[z_k]}^{(k)}(s) \right].$$ (13.17)

Its inverse Laplace transform is equal to the vector function (13.15).

13.4 Binary representation of MMSs

A vector $\boldsymbol{y} = (y_1, y_2, \dots, y_n) \in S_1 \times \cdots \times S_n$ is called a *path vector to level (of level)* l of the MMS if $\psi(\boldsymbol{y}) \geq l$.

The path vector $\boldsymbol{y}$ is said to be a *minimal path vector to level* l if, in addition, the inequality $\boldsymbol{x} < \boldsymbol{y}$ implies $\psi(\boldsymbol{x}) < l$. The inequality $\boldsymbol{x} < \boldsymbol{y}$ means that $x_i \leq y_i$ for $i = 1, \dots, n$ and $x_i < y_i$ for some i. We denote the set of all minimal path vectors to level l by U_l, $l = 1, \dots, z(z_k)$ and $U_0 = \{\boldsymbol{0}\}$, where $\{\boldsymbol{0}\} = (0, 0, \dots, 0)$.

In reliability analysis of the MMSs, we may use their binary representation. This approach was presented among other in papers by Block and Savits [10] and Korczak [59]. We define the binary random variables $\{X_{kr}(t) : t \geq 0\}$, $k \in C$, $r \in S_k$:

$$X_{kr}(t) = \begin{cases} 1 & \text{dla } X_k(t) \geq r \\ 0 & \text{dla } X_k(t) < r. \end{cases}$$ (13.18)

We determine the system level indicators ψ_j, $j \in \{1, \dots, z\}$:

$$\psi_j(\boldsymbol{X}(t)) = \begin{cases} 1 & \text{for } \psi(\boldsymbol{X}(t)) \geq j \\ 0 & \text{for } \psi(\boldsymbol{X}(t)) < j. \end{cases}$$ (13.19)

We will use symbols $\coprod$ and $\sqcup$ introduced by Barlow and Proshan [7], which denote the binary operations:

$$\coprod_{i=1}^{n} x_k = 1 - \prod_{i=1}^{n}(1 - x_k), \quad x_k \in \{0, 1\}$$

$$x_1 \sqcup x_2 = 1 - (1 - x_1)(1 - x_2), \quad x_1, x_2 \in \{0, 1\}.$$

From (13.18) and (13.19), and the definition of the minimal path vectors we obtain a following binary representation of the stochastic process describing evolution of the MMS:

$$\psi_l(\boldsymbol{X}(t)) = \coprod_{\boldsymbol{y} \in U_l} \prod_{k \in C, \, y_k > 0} X_{ky_k}(t) = 1 - \prod_{\boldsymbol{y} \in U_l} \left(1 - \prod_{k \in C, y_k > 0} X_{ky_k}(t) \right).$$ (13.20)

Example 13.1. Consider a multistate system $(C, S, S_1, S_2, S_3, \psi)$, where $C = \{1, 2, 3\}$, $S = \{0, 1, 2\}$, $S_1 = \{0, 1, 2\}$, $S_2 = \{0, 1, 2\}$, $S_3 = \{0, 1\}$, and the system structure function is determined by the formulas

$$\psi(x) = 0 \text{ for } x = (x_1, x_2, x_3) \in D_0,$$
$$\psi(x) = 1 \text{ for } x = (x_1, x_2, x_3) \in D_1, \tag{13.21}$$
$$\psi(x) = 2 \text{ for } x = (x_1, x_2, x_3) \in D_2,$$

where

$$
\begin{aligned}
D_0 &= \{(0,0,0), (0,0,1)(0,1,0), (1,0,0), (1,1,0), \\
&\quad (2,0,0), (2,2,0), (0,2,0), (1,2,0), (2,1,0)\}, \\
D_1 &= \{(1,0,1), (0,1,1), (1,1,1), (2,0,1), (0,2,1)\}, \\
D_2 &= \{(2,1,1), (1,2,1), (2,2,1)\}.
\end{aligned} \tag{13.22}
$$

First, we have to determine the set U_l of all minimal path vectors to the level 1 for $l = 1, 2$. We take under consideration the set D_1. The vector $y = (1,0,1)$ is a minimal path vector to level 1, because according to definition $\psi(y) = 1 \geqslant 1$ and there exists a vector $x = (0,0,1)$ such that $x < y$ and $\psi(x) = 0 < 1$. The vector $y = (0,2,1)$ is not a minimal path vector to level 1, because $\psi(y) = 1 \geqslant 1$ and for $x = (0,1,1)$ we have $x < y$ and $\psi(x) = 1$. Also, the vector $y = (2,0,1)$ is not a minimal path vector to level 1, because $\psi(y) = 1 \geqslant 1$ and for $x = (1,0,1)$ is $x < y$ and $\psi(x) = 1$.

Analyzing all vectors from D_1, we get a set of the minimal path vectors of the level $l = 1$, which is denoted as

$$U_1 = \{(1,0,1), (0,1,1)\}. \tag{13.23}$$

In the similar manner, we get

$$U_2 = \{(2,1,1), (1,2,1)\}. \tag{13.24}$$

From (13.20), we have

$$\psi_l(x) = \coprod_{y \in U_l} \prod_{k \in C, y_k > 0} x_{ky_k} = 1 - \prod_{y \in U_l} \left(1 - \prod_{k \in C, y_k > 0} x_{k y_k}\right). \tag{13.25}$$

Applying this equality, we have

$$\psi_1(x) = x_{11} x_{31} \coprod x_{21} x_{31} = x_{11} x_{31} + x_{21} x_{31} - x_{11} x_{21} x_{31}. \tag{13.26}$$

In a similar way, using an equality

$$x_{kr} x_{k_p} = x_{k \max\{r,p\}},$$

we get

$$\psi_2(x) - x_{12} x_{21} x_{31}) + x_{11} x_{22} x_{31} - x_{12} x_{22} x_{31}. \tag{13.27}$$

13.5 Reliability of unrepairable system

We suppose that the SMPs $\{X_1(t) : t \geqslant 0\}, \ldots, \{X_n(t) : t \geqslant 0\}$ are independent. A stochastic process $\{Y(t) : t \geqslant 0\}$,

$$Y(t) = \psi(X(t)) = \psi(X_1(t), \ldots, X_n(t)), \tag{13.28}$$

taking its values in a state space $S = \{0, 1, \ldots, z\}$ describes a reliability state for $t \in [0, \infty)$. It is not a SMP. Let $A'_{[l]} = \{l, l+1 \ldots, z\}$ and $A_{[l]} = S - A'_{[l]} = \{0, 1 \ldots, l-1\}$. A random variable

$$T_{[l]} = \inf\{t : S(t) \in A_{[l]}\} \tag{13.29}$$

denotes the time to failure to level (of level) l of the system. A reliability function to level 1 of the system is determined by the rule

$$R_{[l]}(t) = P(T_{[l]} > t). \tag{13.30}$$

We have at least two ways of calculating it. The first one consists of applying distributions of the processes that describe the reliability evolution of the system components. The l level reliability function of the system may be computed according to the rule

$$R_{[l]}(t) = \sum_{j \in A'_{[l]}} P_j(t), \tag{13.31}$$

where

$$P_j(t) = P(S(t) = j) = P(X(t) \in D_j) = \sum_{(x_1, \ldots, x_n) \in D_j} P^1_{x_1}(t) \ldots P^n_{x_n}(t), \tag{13.32}$$

$$D_j = \psi^{-1}(j).$$

The second way leads through the computation of the components of reliability functions to level 1. Applying (13.20), we have

$$R_{[l]}(t) = E[\psi_l(X(t))] = 1 - \prod_{y \in U_l} \left(1 - \prod_{k \in C, y_k > 0} E\left[X_{k y_k}(t)\right] \right). \tag{13.33}$$

Definition 13.2. The vector function

$$R(t) = \left[1, R_{[1]}(t), \ldots, R_{[z]}(t)\right] \tag{13.34}$$

is called the multistate reliability function of the system.
The vector

$$m = \left[1, m_{[1]}, \ldots, m_{[z]}\right], \tag{13.35}$$

$$m_{[l]} = \int_0^\infty R_{[l]}(t), \quad l = 1, \ldots, z$$

is said to be the *multistate mean time to failure of the system.*

13.6 Numerical illustrative example

To explain and illustrate the concepts presented above, we will construct a simple reliability model of the multistate system with the semi-Markov components.

Example 13.2. We assume that the multistate reliability system consists of three components, the reliability evolution of which are modeled by independent SMPs $\{X_1(t) : t \geqslant 0\}$, $\{X_2(t) : t \geqslant 0\}$, $\{X_3(t) : t \geqslant 0\}$ with the state spaces $S_1 = S_2 = \{0, 1, 2\}$, $S_3 = \{0, 1\}$. We assume that the kernels of processes 1 and 2 are the same:

$$\boldsymbol{Q}^{(k)}(t) = \begin{bmatrix} Q_{00}^{(k)}(t) & 0 & 0 \\ Q_{10}^{(k)}(t) & 0 & 0 \\ Q_{20}^{(k)}(t) & Q_{21}^{(k)}(t) & 0 \end{bmatrix}, \tag{13.36}$$

where

$$Q_{00}^{(k)}(t) = 1 - e^{-\alpha t},$$

$$Q_{10}^{(k)}(t) = 1 - (1 + \beta t)e^{-\beta t}, \tag{13.37}$$

$$Q_{20}^{(k)}(t) = a[1 - (1 + \gamma t)e^{-\gamma t}], \quad Q_{21}^{(k)}(t) = b[1 - (1 + \gamma t)e^{-\gamma t}],$$

$$t \geqslant 0, \ \alpha > 0, \ \beta, \gamma, \ a, \ b > 0, \ a + b = 1.$$

Suppose that the initial distributions are

$$P(X^{(k)}(0) = 2) = 1, \quad k = 1, 2.$$

Assume that a kernel of the last process is of the form

$$\boldsymbol{Q}^{(3)}(t) = \begin{bmatrix} Q_{00}^{(3)}(t) & 0 \\ Q_{10}^{(3)}(t) & 0 \end{bmatrix} = \begin{bmatrix} 1 - e^{-\kappa t} & 0 \\ 1 - (1 + \lambda t)e^{-\lambda t} & 0 \end{bmatrix}, \tag{13.38}$$

where

$$t \geqslant 0, \quad \kappa > 0, \ \lambda > 0.$$

The initial distribution of the process is $P\{X^{(3)}(0) = 1\} = 1$. We also suppose that the system structure is the same as in Example 13.1. It means that the structure function is defined by (13.21). To find distributions of those processes, we have to solve the systems of equations (13.16) for their Laplace transforms. The inverse Laplace transformation of equations solutions follows to the demanding functions.

According to rule (13.32), we obtain the distribution of the system states

$$P_0(t) = \sum_{(x_1, x_2, x_3) \in D_0} P_{x_1}^{(1)}(t) P_{x_2}^{(2)}(t) P_{x_3}^{(3)}(t), \tag{13.39}$$

$$P_1(t) = \sum_{(x_1, x_2, x_3) \in D_1} P_{x_1}^{(1)}(t) P_{x_2}^{(2)}(t) P_{x_3}^{(3)}(t), \tag{13.40}$$

$$P_2(t) = \sum_{(x_1, x_2, x_3) \in D_2} P_{x_1}^{(1)}(t) P_{x_2}^{(2)}(t) P_{x_3}^{(3)}(t). \tag{13.41}$$

Applying (13.31), we obtain the multistate reliability function of the system.

$$\boldsymbol{R}(t) = [R_{[1]}(t),\ R_{[2]}(t)] = [P_1(t) + P_2(t), P_2(t)].$$

Now we illustrate the second way to calculate the multistate system reliability function.

The second method of computing the multistate system reliability function needs to calculate the reliability functions of its components to level l. Applying (13.33), we have

$$R_{[1]}(t) = E[\psi_1(\boldsymbol{X}(t))] = E[X_{11}(t)\, X_{31}(t)] + E[X_{21}(t)\, X_{31}(t)]$$
$$- E[X_{11}(t)\, X_{22}(t), X_{31}(t)]. \tag{13.42}$$

Hence, using the independence of the processes discussed here we get the reliability function of the system to level 1:

$$R_{[1]}(t) = R_{[1]}^{(1)}(t)\, R_{[1]}^{(3)}(t) + R_{[1]}^{(2)}(t)\, R_{[1]}^{(3)}(t) - R_{[1]}^{(1)}(t)\, R_{[1]}^{(2)}(t)\, R_{[1]}^{(3)}(t) \tag{13.43}$$

In the same way, according to (13.27) we have

$$R_{[2]}(t) = R_{[2]}^{(1)}(t)\, R_{[1]}^{(2)}(t)\, R_{[1]}^{(3)}(t) + R_{[1]}^{(1)}(t)\, R_{[2]}^{(2)}(t)\, R_{[1]}^{(3)}(t) - R_{[2]}^{(1)}(t)\, R_{[2]}^{(2)}(t)\, R_{[1]}^{(3)}(t). \tag{13.44}$$

The reliability functions of the component to level $l = 1, 2$ from (13.43) and (13.44) we evaluate using (13.13).

In this case $S_k = \{0, 1, 2\}, k = 1, 2$. Hence,

$$A_{[1]} = \{0\}, \qquad A'_{[1]} = \{1, 2\},$$
$$A_{[2]} = \{0, 1\}, \qquad A'_{[2]} = \{2\}. \tag{13.45}$$

For $l = 1$, the matrices from (13.13) take the form

$$\boldsymbol{I} - \tilde{\boldsymbol{q}}_{A'_{[l]}}^{(k)}(s) = \begin{bmatrix} 1 & 0 \\ -\tilde{q}_{21}^{(k)}(s) & 1 \end{bmatrix}, \tag{13.46}$$

$$\tilde{\boldsymbol{G}}_{A'_{[l]}}^{(k)}(s) = \frac{1}{s} \begin{bmatrix} 1 - \tilde{q}_{10}^{(k)}(s) \\ 1 - \tilde{q}_{20}^{(k)}(s) - \tilde{q}_{21}^{(k)}(s) \end{bmatrix}. \tag{13.47}$$

The element $\tilde{R}_{2[1]}^{(k)}(s)$ of the solution of (13.46) is

$$\tilde{R}_{2[1]}^{(k)}(s) = \frac{1 - \tilde{q}_{20}^{(k)}(s) - \tilde{q}_{21}^{(k)}(s)\tilde{q}_{10}^{(k)}(s)}{s}. \tag{13.48}$$

For $l = 2$, the matrices from (13.13) take the form

$$\boldsymbol{I} - \tilde{\boldsymbol{q}}_{A'_{[2]}}^{(k)}(s) = \begin{bmatrix} 1 \end{bmatrix},$$

$$\tilde{\boldsymbol{G}}_{A'_{[l]}}^{(k)}(s) = \frac{1}{s} \begin{bmatrix} 1 - \tilde{q}_{20}^{(k)}(s) - \tilde{q}_{21}^{(k)}(s) \end{bmatrix}. \tag{13.49}$$

The solution of (13.13) is

$$\tilde{R}_{2[2]}^{(k)}(s) = \frac{1 - \tilde{q}_{20}^{(k)}(s) - \tilde{q}_{21}^{(k)}(s)}{s}. \tag{13.50}$$

In (13.10), the interval transition probabilities are used

$$P_{ij}^{(k)}(t) = P(X_k(t) = j | X_k(0) = i), \quad j \in S_k$$

and *distribution of a SMP*

$$P_j^{(k)}(t) = P(X_k(t) = j), \quad j \in S_k.$$

If $P(X_k(0) = s_k) = 1$, then $P_{z_kj}^{(k)}(t) = P_j^{(k)}(t)$. The Laplace-Stieltjes transforms of the kernel elements (13.37) are

$$\tilde{q}_{10}^{(k)}(s) = \frac{\beta^2}{(s + \beta)^2}, \quad \tilde{q}_{20}^{(k)}(s) = \frac{a\gamma^2}{(s + \gamma)^2}, \quad \tilde{q}_{21}^{(k)}(s) = \frac{b\gamma^2}{(s + \gamma)^2} \tag{13.51}$$

for $k = 1, 2$. For parameters

$$\alpha = 0.1, \ \beta = 0.02, \ \gamma = 0.01, \ a = 0.2, \ b = 0.8, \ \eta = 0.01, \ \kappa = 0.1,$$

the solutions for the Laplace transforms given by (13.48) and (13.50) are

$$\tilde{R}_{2[1]}^{(k)}(s) = \frac{0.0000832 + 0.00488s + 0.12s^2 + s^3}{(0.02 + s)^2(0.04 + s)^2}, \tag{13.52}$$

$$\tilde{R}_{2[2]}^{(k)}(s) = \frac{08 + s}{(0.04 + s)^2} \quad \text{for } k = 1, 2. \tag{13.53}$$

We get the reliability functions of the system components as the inverse Laplace transforms of these functions. Thus, we obtain

$$R_{[1]}^{(k)}(t) = 4.2e^{-0.04t} - 3.2e^{-0.02t}0.04e^{-0.04t} + 0.064e^{-0.02t}t \tag{13.54}$$

$$R_{[2]}^{(k)}(t) = 0.04e - 0.04t(25 + t) \quad \text{for } k = 1, 2. \tag{13.55}$$

For $k = 3$, we have

$$R_{[1]}^{(3)}(t) = 1 - Q_{10}^{(3)}(t) = (1 + \lambda t)e^{-\lambda t}.$$

Using equalities (13.43) and (13.44), we obtain elements of the multistate reliability function of the system:

$$R_{[1]}(t) = 2e^{-0.01t}(1 + 0.01t)(4.2e^{-0.04t} - 3.2e^{-0.02t}$$

$$+ 0.04e^{-0.04t}t + 0.064e^{-0.02t}t) - e^{-0.01t}(1 + 0.01t)(4.2e^{-}0.04t \tag{13.56}$$

$$- 3.2e^{-0.02t} + 0.04e^{0.04t}t + 0.064e^{-0.02t}t)^2,$$

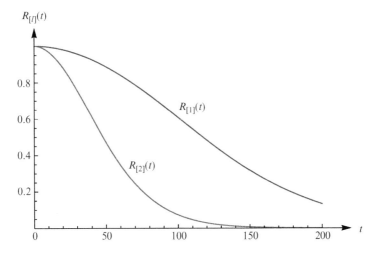

Figure 13.1 The reliability functions to level $l = 1, 2$.

$$R_{[2]}(t) = -0.0016e^{-0.09t}(1 + 0.01t)(25. + t)^2$$

$$+ 0.08e^{-0.05t}(1 + 0.01t)(25 + t)(4.2e^{-0.04t}$$ (13.57)

$$- 3.2e^{-0.02t} + 0.04e^{-0.04t}t + 0.064e^{-0.02t}t).$$

The functions are shown in Figure 13.1.
The multistate reliability function may be written as a vector function

$$\boldsymbol{R}(t) = \left[1, R_{[1]}(t), R_{[2]}(t)\right].$$ (13.58)

$\triangle$

13.7 Renewable multistate system

Assume that the SMP $\{X_k(t) : t \geqslant 0\}$, describing the reliability state of the system's component k, is defined by the following kernel:

$$
\begin{bmatrix}
0 & 0 & 0 & \cdots & 0 & Q^{(k)}_{0z_k}(t) \\
Q^{(k)}_{10}(t) & 0 & 0 & \cdots & 0 & Q^{(k)}_{1z_k}(t) \\
Q^{(k)}_{20}(t) & Q^{(k)}_{21}(t) & 0 & \cdots & 0 & Q^{(k)}_{2z_k}(t) \\
Q^{(k)}_{30}(t) & Q^{(k)}_{31}(t) & Q^{(k)}_{32}(t) & \cdots & 0 & Q^{(k)}_{3z_k}(t) \\
\cdots & \cdots & \cdots & \cdots & \cdots & \cdots \\
Q^{(k)}_{z_k0}(t) & Q^{(k)}_{z_k1}(t) \cdots & & \cdots & \cdots & Q^{(k)}_{z_k\,z_k-1}(t) & 0
\end{bmatrix}
$$ (13.59)

This means that a full renewal of the component is possible for all reliability states. Moreover, we assume that for all $k \in C$

$$P_{z_k}^{(k)}(0) = 1, \ k \in C \quad \text{and} \quad P_{z(0)} = 1, \ j \in S.$$

We know that the random vector process $\{\mathbf{X}(t) = (X_1(t), \ldots, X_n(t)) : t \geq 0\}$ represents the states of all system components at any time $t \geq 0$.

A function $P_j^{(k)}(t) = P(X_k(t) = j)$, $t \geq 0$ represents a probability of the state $j \in S_k$ at the moment t for the kth component, while the function $P_j(t) = P(S(t) = j)$, $t \geq 0$ denotes probability of the state $j \in S$ at the moment t for the multistate system. The states probabilities of the system with independent components are completely determined by the probabilities of components states through the system structure function ψ. Recall the important rule (13.32):

$$P_j(t) = P(S(t) = j) = P(\mathbf{X}(t) \in D_j) = \sum_{(x_1,\ldots,x_n) \in D_j} P_{x_1}^1(t) \ldots P_{x_n}^n(t), \tag{13.60}$$

$$D_j = \psi^{-1}(j).$$

To calculate probabilities $P_j^{(k)}(t)$, $j \in S_k$, first we have to compute the interval transition probabilities $P_{ij}^{(k)}(t)$, $i, j \in S_k$. From Chapter 3, we know that the Laplace transforms of them fulfill the system of equations

$$\tilde{P}_{ij}^{(k)}(s) = \delta_{ij}[\frac{1}{s} - \tilde{G}_i^{(k)}(s)] + \sum_{r \in S_k} \tilde{q}_{ir}^{(k)}(s)\tilde{P}_{rj}^{(k)}(s), \quad i, j \in S, \tag{13.61}$$

where

$$\tilde{P}_{ij}^{(k)}(s) = \int_0^\infty e^{-st} P_{ij}^{(k)}(t) dt, \quad \tilde{q}_{ir}^{(k)}(s) = \int_0^\infty e^{-st} dQ_{ir}^{(k)}(t),$$

$$\tilde{G}_i(s) = \int_0^\infty e^{-st} G_i^{(k)}(t) dt, \quad G_i^{(k)}(t) = \sum_{j \in S} Q_{ij}^{(k)}(t).$$

In matrix notation, we have

$$\tilde{\boldsymbol{P}}^{(k)}(s) = \left(\frac{1}{s}\boldsymbol{I} - \tilde{\boldsymbol{G}}^{(k)}(s)\right) + \tilde{\boldsymbol{q}}^{(k)}(s)\tilde{\boldsymbol{P}}^{(k)}(s), \tag{13.62}$$

where

$$\tilde{\boldsymbol{P}}^{(k)}(s) = [\tilde{P}_{ij}^{(k)}(s) : i, j \in S], \quad \tilde{\boldsymbol{G}}^{(k)}(s) = [\delta_{ij}\tilde{G}_i^{(k)}(s) : i, j \in S],$$
$$\tilde{\boldsymbol{q}}^{(k)}(s) = [\tilde{q}_{ij}^{(k)}(s) : i, j \in S], \quad \boldsymbol{I} = [\delta_{ij} : i, j \in S].$$

Hence,

$$\tilde{\boldsymbol{P}}^{(k)}(s) = (\boldsymbol{I} - \tilde{\boldsymbol{q}}^{(k)}(s))^{-1}\left(\frac{1}{s}\boldsymbol{I} - \tilde{\boldsymbol{G}}^{(k)}(s)\right). \tag{13.63}$$

According to assumptions concerning the initial distribution of the process, we have

$$P_j^{(k)}(t) = P_{z_k j}^{(k)}(t), \quad j \in S_k.$$

Applying Theorem 3.6, we are able to evaluate the limiting distributions for SMPs $\{X_k(t) : t \geqslant 0\}$, $k \in C$. If for $k \in C$ and $\in S_k$

$$0 < m_i^{(k)} = \int_0^\infty [1 - G_i^{(k)}(x)]dx < \infty, \quad G_i^{(k)}(x) = \sum_{j \in S_k} Q_{ij}^{(k)}(t),$$

then for $k = 1, 2$ there exists limiting distribution

$$P_{ij}^{(k)} = \lim_{t \to \infty} P_{ij}^{(k)}(t) = P_j^{(k)} = \lim_{t \to \infty} P_j^{(k)}(t) \tag{13.64}$$

and

$$P_j^{(k)} = \frac{\pi_j^{(k)} m_j^{(k)}}{\sum_{i \in S} \pi_i^{(k)} m_i^{(k)}}, \tag{13.65}$$

where the stationary distributions $\pi_i^{(k)}$, $i \in S_k$, $k \in C$, of the embedded Markov chains $\{X_k(\tau_n) : n \in \mathbb{N}_0\}$, $k \in C$ satisfy the systems of equations

$$\sum_{i \in S_k} \pi_i^{(k)} p_{ij}^{(k)} = \pi_j^{(k)}, \quad j \in S_k, \quad \sum_{j \in S} \pi_j^{(k)} = 1, \tag{13.66}$$

where $p_{ij}^{(k)} = \lim_{t \to \infty} Q_{ij}^{(k)}(t)$.

Example 13.3. Like in Example 13.2 we assume that the multistate reliability system consists of three components of the reliability evolution that are modeled by independent SMPs $\{X_1(t) : t \geqslant 0\}$, $\{X_2(t) : t \geqslant 0\}$, $\{X_3(t) : t \geqslant 0\}$ with the state spaces $S_1 = S_2 = \{0, 1, 2\}$, $S_3 = \{0, 1\}$. Now we assume that the kernels are

$$\boldsymbol{Q}^{(k)}(t) = \begin{bmatrix} 0 & 0 & Q_{02}^{(k)}(t) \\ Q_{10}^{(k)}(t) & 0 & Q_{12}^{(k)}(t) \\ Q_{20}^{(k)}(t) & Q_{21}^{(k)}(t) & 0 \end{bmatrix}, \tag{13.67}$$

where

$$Q_{02}^{(k)}(t) = 1 - (1 + \mu t)e^{-\mu t},$$
$$Q_{10}^{(k)}(t) = a_1[1 - e^{-\beta t}], \qquad Q_{12}^{(k)}(t) = b_1[1 - (1 + \nu t)e^{-\nu t}] \tag{13.68}$$
$$Q_{20}^{(k)}(t) = b_2[1 - (1 + \alpha t)e^{-\alpha t}], \qquad Q_{21}^{(k)}(t) = a_2(1 - e^{-\alpha t})$$

for

$$k = 1, 2, \quad t \geqslant 0, \quad \alpha > 0, \quad \beta > 0, \quad \mu > 0, \quad \nu > 0,$$
$$a_1 > 0, \quad b_1 > 0, \quad a_1 + b_1 = 1, \quad a_2 > 0, \quad b_2 > 0, \quad a_2 + b_2 = 1.$$
$$\tag{13.69}$$

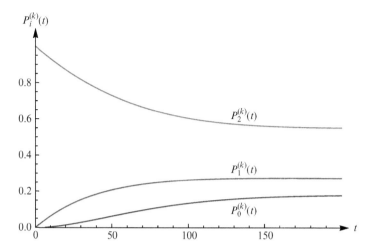

Figure 13.2 The distribution of the first and the second components states.

The Laplace-Stieltjes transforms of these functions are

$$\tilde{q}_{02}^{(k)}(s) = \frac{\mu^2}{(s+\mu)^2}, \quad \tilde{q}_{10}^{(k)}(s) = \frac{a_1\beta}{s+\beta}, \quad \tilde{q}_{12}^{(k)}(s) = \frac{b_1 v^2}{(s+v)^2},$$

$$\tilde{q}_{20}^{(k)}(s) = \frac{a_2\alpha^2}{(s+\alpha)^2}, \quad \tilde{q}_{21}^{(k)}(s) = \frac{b_2\alpha}{s+\alpha}. \tag{13.70}$$

The initial distributions are

$$P(X^{(k)}(0) = 2) = 1, \quad k = 1, \, 2.$$

For components $k = 1, 2$, the inverse transforms of the solution of (13.62) allows us to obtain the distribution of the system components states. Figure 13.2 shows the functions that form this distribution for the following parameters:

$$k = 1, 2, \quad t \geqslant 0, \quad \alpha = 0.008, \quad \beta = 0.012, \quad \mu = 0.04, \quad v = 0.08,$$
$$a_1 = 0.9, \quad b_1 = 0.1, \quad a_2 = 0.12, \, b_2 = 0.88. \tag{13.71}$$

We can calculate the limiting distribution of these processes much easier using the formula (13.65). The transition probability matrix of an embedded Markov chain for considered processes is

$$\mathbf{P}^{(k)} = \begin{bmatrix} 0 & 0 & 0.12 \\ 0.9 & 0 & 0.1 \\ 0.12 & 0.88 & 0 \end{bmatrix}. \tag{13.72}$$

The expected values of the waiting times of states are

$$m_1^{(k)} = 150, \quad m_2^{(k)} = 77.5, \quad m_3^{(k)} = 14.$$

The numbers

$$\pi_1^{(k)} = 0.3267, \quad \pi_2^{(k)} = 0.3152, \quad \pi_3^{(k)} = 0.3581$$

are the solution of the equations system (13.66). Applying (13.65), we finally obtain

$$P_1^{(k)} = 0.1797, \quad P_2^{(k)} = 0.2687, \quad P_3^{(k)} = 0.5516, \quad k = 1, 2.$$

A kernel of the last process $\{X_3(t) : t \geqslant 0\}$ is assumed to be

$$\mathbf{Q}^{(3)}(t) = \begin{bmatrix} 0 & Q_{01}^{(3)}(t) \\ Q_{10}^{(3)}(t) & 0 \end{bmatrix}, \tag{13.73}$$

where

$$Q_{10}^{(3)}(t) = 1 - e^{-\lambda t}, \quad Q_{01}^{(3)}(t) = 1 - (1 + \kappa t)e^{-\kappa t}$$

for

$$t \geqslant 0, \quad \lambda > 0 \quad \kappa > 0.$$

The Laplace-Stieltjes transforms of these functions are

$$\tilde{q}_{01}^{(3)}(s) = \frac{\kappa^2}{(s + \kappa)^2}, \quad \tilde{q}_{10}^{(3)}(s) = \frac{\lambda}{s + \lambda}. \tag{13.74}$$

The initial distribution of the process is assumed to be

$$P\{X^{(3)}(0) = 1\} = 1.$$

Moreover, we suppose that

$$\lambda = 0.009, \quad \kappa = 0.02. \tag{13.75}$$

We obtain the distribution of the process states in a similar way. The functions

$$\tilde{P}_0^{(3)}(s) = \frac{0.00072 + 0.009s}{0.00232s + 0.089s^2 + s^3}, \quad \tilde{P}_1^{(3)}(s) = \frac{(0.04 + s)^2}{s(0.00232 + 0.089s + s^2)} \tag{13.76}$$

are the solution of (13.62). Functions forming the third component's distribution are shown in Figure 13.3.

Applying the rule

$$P_i = \lim_{s \to 0} s \tilde{P}_i^{(3)}(s), \quad i = 0, 1 \tag{13.77}$$

to (13.76), we get the limiting distribution of the third component states

$$P_0^{(3)} = 0.3103, \quad P_1^{(3)} = 0.6897. \tag{13.78}$$

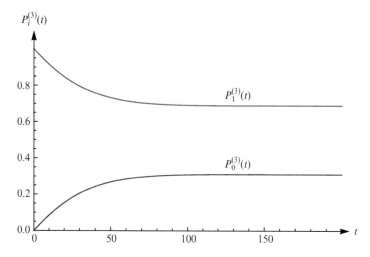

Figure 13.3 The distribution of the third component states.

From (13.39), we obtain the rule that enables computing the limiting distribution of the system reliability states:

$$P_0 = \sum_{(x_1,x_2,x_3)\in D_0} P^{(1)}_{x_1} P^{(2)}_{x_2} P^{(3)}_{x_3},$$ (13.79)

$$P_1 = \sum_{(x_1,x_2,x_3)\in D_1} P^{(1)}_{x_1} P^{(2)}_{x_2} P^{(3)}_{x_3},$$ (13.80)

$$P_2 = \sum_{(x_1,x_2,x_3)\in D_2} P^{(1)}_{x_1} P^{(2)}_{x_2} P^{(3)}_{x_3},$$ (13.81)

where the sets D_0, D_1, D_2 are determined in Example 13.1 (13.12). Substituting appropriate numerical values, we obtain probabilities of the system reliability states for large t

$$P_0 = 0.3326, \quad P_1 = 0.2531, \quad P_3 = 0.4143.$$

△

13.8   Conclusions

In many real-life situations the binary models seem to be not sufficient for describing reliability of the system, because in addition to "down" state (0) and "up" state (1) the system may be capable on different levels, from perfect functioning to complete failure. Then the multistate models are more adequate. The decomposition method of the multistate unrepairable system to binary systems allows us to apply well-known methods of classical reliability theory in multinary cases. SMPs are very useful as reliability models of the multistate system components. The SMP theory provides some concepts and theorems that enable us to construct the appropriate probability models of the multistate reliability system. Unfortunately, all these models are constructed under the assumption of independence of processes describing the reliability of the system components.

Semi-Markov maintenance nets

14

Abstract

The semi-Markov models of functioning maintenance systems, which are called maintenance nets are presented in this chapter. Elementary maintenance operations form the states of the SM model. Some concepts and results of the semi-Markov process theory provide the possibility of computing important characteristics and parameters of the maintenance process. Two semi-Markov models of maintenance nets are discussed in the chapter.

Keywords: Semi-Markov model, Maintenance nets, Maintenance operation, Profit from maintenance operation

14.1 Introduction

Using a technical object (machine, device) demands many kinds of maintenance operation such as current service, periodic preventive service, service paramount (repairs, replacement of subsystems), emergency service (repair, replacement), and so on. The quantitative (mathematical) description of the specific maintenance working process should take into account many parameters. These parameters relate to the number and type of work stations, the number of skilled professionals, the type of diagnostic equipment, tools and needed supplies, the type of spare parts, and the scope of performed operations. This type of maintenance operation, like repair of the machine, can be realized by exchanging the damaged system (module, component), adjustment, or object renewal in other ways. Until repairs are necessary appropriate tools, diagnostic equipment, spare parts, and many other materials are needed. Each maintenance process needs professional repairmen with specific qualifications. Every maintenance operation is performed by a scheme that depends on its type.

Semi-Markov maintenance nets were introduced by Silvestrov [90]. In Ref. [38], semi-Markov models of simple maintenance nets were presented.

14.2 Model of maintenance net

Figure 14.1 presents a diagram of the exemplary maintenance system.

14.2.1 States of maintenance operation

The states of a process depend on the type of the object maintenance. In the overall scheme, the states of the process should describe the kind of workstation, scope of activities at each station, and the quality of service determined by the condition of

Semi-Markov Processes: Applications in System Reliability and Maintenance. http://dx.doi.org/10.1016/B978-0-12-800518-7.00014-4

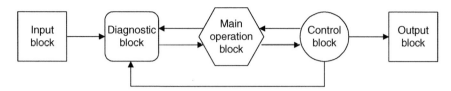

Figure 14.1 Diagram of the maintenance system.

the object after its execution. If we would like to take into account the economic aspect of the maintenance process, then assumptions of the model should also include information about the cost of all maintenance operation states.

We start construction of the model by determining the states corresponding to Figure 14.1. We divide the state space S into disjoint subset $E_1, E_2, \ldots, E_6$.

- E_1—a subset of waiting for service states whose elements are the states as follows: state 1, which is an object waiting due to the occupancy of the repair post; state 2, which is an object waiting due to the absence of spare parts.
- E_2—a subset of the preparation states for basic maintenance operations concerning a filed component. With access to repair of the damaged component, disassembly of a few other components is often needed. During this operation, *often turns* out that it is necessary to wait for the spare parts, which are missing at the service point at this moment. We assume here that the subset contains only one state 3, which means achieving access to the component requiring repair.
- E_3—a subset of diagnosing states containing states 4 and 5 where state 4 denotes a state the effect of which is an accurate identification of the object's technical condition, while state 5 is a state of diagnosing the results of which is incorrect identification of the object's technical condition or an unidentified object condition.
- E_4—a subset of the main maintenance operation. In this set of states, adjustments, replacements, and repairs of the object components are performed. Each state from this subset determines some range of activities. The activities consist of actions that are performed one after the another, or that are taken simultaneously by several repairmen or operators. In this simple model, we assume that the main maintenance operation is comprised of two stages, which correspond to the states 6 and 7. Note that we could determine the subset of a state in a completely different way.
- E_5—a subset of the inspection states. In this set of states, an inspection of the quality of the made maintenance operation and the technical condition of the object is performed. The maintenance process takes value a of 8 if the maintenance operation is done properly and the object is fit; otherwise, we say that the state of the process is 9. Of course, we can imagine other states belonging to a subset E_5: a state relies on the fact that the inspection does not detect the object failure or the state in which the fit object is identified by the inspection as an unfit one.
- E_6—a subset of final states. If the inspection gives positive results, then it follows the last state is 10, which denotes the object waiting for reuse.

Concluding, the states of the maintenance process are:

1. An object waiting due to the occupancy of the service post
2. An object waiting due to the absence of spare parts

3. Achieving access to the component that requires repair
4. Diagnosing which effect is an accurate identification of state
5. Diagnosing which results are an incorrect identification of state
6. The first stage of a main maintenance operation
7. The second stage of a main maintenance operation
8. Checking the technical object state with positive effect
9. Checking the technical object state with negative effect
10. The object waiting for reuse

Thus, the states of the maintenance operation form the state space

$$S = E_1 \cup E_2 \cup \ldots \cup E_6 = \{1, 2, \ldots, 10\}.$$

The possible state changes follow according to the flow graph shown in Figure 14.2.

14.2.2 Model of maintenance operation

An adequate stochastic model of the maintenance operation seems to be a semi-Markov process (SMP) with a set of states $S = \{1, 2, \ldots, 10\}$. The SMP kernel corresponding to the graph shown in Figure 14.2 has the form

$$Q(t) = \begin{bmatrix} 0 & 0 & Q_{13}(t) & 0 & 0 & 0 & 0 & 0 & 0 & 0 \\ 0 & 0 & Q_{23}(t) & 0 & 0 & 0 & 0 & 0 & 0 & 0 \\ 0 & Q_{32}(t) & 0 & Q_{34}(t) & Q_{35}(t) & Q_{36}(t) & 0 & 0 & 0 & 0 \\ 0 & Q_{42}(t) & 0 & 0 & 0 & Q_{46}(t) & 0 & 0 & 0 & 0 \\ 0 & 0 & 0 & 0 & 0 & Q_{56}(t) & 0 & 0 & 0 & 0 \\ 0 & Q_{62}(t) & Q_{63}(t) & 0 & 0 & 0 & Q_{67}(t) & 0 & 0 & 0 \\ 0 & 0 & 0 & 0 & 0 & 0 & 0 & Q_{78}(t) & Q_{79}(t) & 0 \\ 0 & 0 & 0 & 0 & 0 & 0 & 0 & 0 & 0 & Q_{8\,10}(t) \\ 0 & 0 & 0 & 0 & 0 & Q_{96}(t) & 0 & 0 & 0 & 0 \\ 0 & 0 & 0 & 0 & 0 & 0 & 0 & 0 & 0 & Q_{10\,10}(t) \end{bmatrix}.$$

$$(14.1)$$

We know that a transition probability from a maintenance state i to state j is

$$p_{ij} = \lim_{t \to \infty} Q_{ij}(t).$$

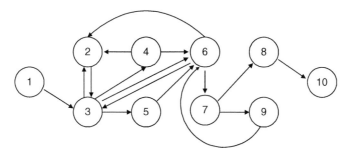

Figure 14.2 Flow graph of the maintenance operation.

The transition probability matrix $P = [p_{ij} : i,j \in S]$ of the SM maintenance process is

$$
P =
\begin{bmatrix}
0 & 0 & 1 & 0 & 0 & 0 & 0 & 0 & 0 & 0 \\
0 & 0 & 1 & 0 & 0 & 0 & 0 & 0 & 0 & 0 \\
0 & p_{32} & 0 & p_{34} & p_{35} & p_{36} & 0 & 0 & 0 & 0 \\
0 & p_{42} & 0 & 0 & 0 & p_{46} & 0 & 0 & 0 & 0 \\
0 & 0 & 0 & 0 & 0 & 1 & 0 & 0 & 0 & 0 \\
0 & p_{62} & p_{63} & 0 & 0 & 0 & p_{67} & 0 & 0 & 0 \\
0 & 0 & 0 & 0 & 0 & 0 & 0 & p_{78} & p_{79} & 0 \\
0 & 0 & 0 & 0 & 0 & 0 & 0 & 0 & 0 & 1 \\
0 & 0 & 0 & 0 & 0 & 1 & 0 & 0 & 0 & 0 \\
0 & 0 & 0 & 0 & 0 & 0 & 0 & 0 & 0 & 1
\end{bmatrix}
\tag{14.2}
$$

14.2.3 Characteristics of maintenance operation

A function

$$
G_i(t) = \sum_{j=1}^{10} Q_{ij}(t)
\tag{14.3}
$$

is cumulative distribution function (CDF) of a random variable T_i, which denotes an unconditional waiting time of ith state of the object maintenance.

A duration of the whole maintenance operation counting from beginning of state 1 to the end of state 10 is represented by a random variable Θ_{110}. Generally, the random variable Θ_{iA} denotes a first passage time from the state i to a subset of maintenance states A. The Laplace-Stieltjes transform of CDFs $\Phi_{i\,10}(t)$, $i = 1, \ldots, 10$ of random variables Θ_{i10}, $i = 1, \ldots, 10$ satisfy the linear system of equations (3.18), where transforms $\tilde{q}_{ij}(s)$ are known, while transforms $\tilde{\varphi}_{i10}(s)$, $i = 1, \ldots, 10$ are unknown. In matrix notation, the system of equations is equivalent to a matrix equation (3.21):

$$
\left(I - \tilde{q}_{A'}(s)\right)\tilde{\phi}_{A'}(s) = \tilde{b}(s),
\tag{14.4}
$$

where

$$
I - \tilde{q}_{A'}(s) =
\begin{bmatrix}
1 & 0 & -\tilde{q}_{13}(s) & 0 & 0 & 0 & 0 & 0 & 0 \\
0 & 1 & -\tilde{q}_{23}(s) & 0 & 0 & 0 & 0 & 0 & 0 \\
0 & -\tilde{q}_{32}(s) & 1 & -\tilde{q}_{34}(s) & -\tilde{q}_{35}(s) & -\tilde{q}_{36}(s) & 0 & 0 & 0 \\
0 & -\tilde{q}_{42}(s) & 0 & 1 & 0 & -\tilde{q}_{46}(s) & 0 & 0 & 0 \\
0 & 0 & 0 & 0 & 1 & -\tilde{q}_{56}(s) & 0 & 0 & 0 \\
0 & -\tilde{q}_{62}(s) & -\tilde{q}_{63}(s) & 0 & 0 & 1 & -\tilde{q}_{67}(s) & 0 & 0 \\
0 & 0 & 0 & 0 & 0 & 0 & 1 & -\tilde{q}_{78}(s) & -\tilde{q}_{79}(s) \\
0 & 0 & 0 & 0 & 0 & 0 & 0 & 1 & 0 \\
0 & 0 & 0 & 0 & 0 & -\tilde{q}_{96}(s) & 0 & 0 & 1
\end{bmatrix},
$$

$$
\tag{14.5}
$$

$$\tilde{\boldsymbol{\varphi}}_{A'}(s) = [\tilde{\varphi}_{i\,10}(s) : i \in A' = \{1, \ldots, 9\}]^{\mathrm{T}},$$

$$\tilde{\boldsymbol{b}}(s) = [0, 0, 0, 0, 0, 0, 0, \tilde{\varphi}_{8\,10}(s), 0]^{\mathrm{T}}.$$

We find the expected values of the random variables $\Theta_{i\,10}$, $i \in A' = \{1, \ldots, 9\}$ by solving an (3.25):

$$(\boldsymbol{I} - \boldsymbol{P}_{A'})\overline{\boldsymbol{\Theta}}_{A'} = \overline{\boldsymbol{T}}_{A'}, \tag{14.6}$$

where

$$\boldsymbol{P}_{A'} = [p_{ij} : i, j \in A'], \quad \overline{\boldsymbol{\Theta}}_{A'} = [E(\Theta_{iA}) : i \in A']^{\mathrm{T}}, \quad \overline{\boldsymbol{T}}_{A'} = [E(T_i) : i \in A']$$

and $\boldsymbol{I}$ is the unit matrix. We get the second moments as a solution of (3.26):

$$(\boldsymbol{I} - \boldsymbol{P}_{A'})\overline{\boldsymbol{\Theta}}_{A'}^2 = \boldsymbol{B}_A, \tag{14.7}$$

where

$$\boldsymbol{P}_{A'} = [p_{ij} : i, j \in A'], \quad \overline{\boldsymbol{\Theta}}_{A'} = [E(\Theta_{iA}^2) : i \in A']^{\mathrm{T}},$$
$$\boldsymbol{B}_A = [b_{iA} : i \in A']^{\mathrm{T}}, \quad b_{iA} = E(T_i^2) + 2\sum_{k \in A'} p_{ik}E(T_{ik})E(\Theta_{kA}).$$

14.2.4 Numerical example

We determine the elements of the matrix $\boldsymbol{Q}(t)$ which define semi-Markov model of describing the maintenance process.

- $Q_{13}(t) = G_1(t)$, where $G_1(t)$ is the CDF of a random variable T_1, denoting the time of an object waiting due to an occupancy of a proper maintenance stand. We suppose that the value of this random variable expectation is $\bar{g}_1 = E(T_1) = 0.25\,[h]$ and the value of its standard deviation is $\sigma_1 = D(T_1) = 0.15\,[h]$.
- $Q_{23}(t) = G_2(t)$, where $G_2(t)$ is the CDF of a random variable T_2, which denotes the time of an object waiting due to the absence of spare parts. We determine the expected value $\bar{g}_2 = E(T_2) = 0.55\,[h]$ and the standard deviation $\sigma_2 = D(T_2) = 0.50\,[h]$,
- $Q_{32}(t) = p_{32}G_3(t)$, $Q_{34}(t) = p_{34}G_3(t)$, $Q_{35}(t) = p_{35}G_3(t)$, $Q_{36}(t) = p_{36}G_3(t)$ where $G_3(t)$ is the CDF of a random variable T_3, which means achieving access to the component requiring a repair. A number p_{32} means a transition probability from state 3 to state 2, which denotes a state of an object waiting due to the absence of spare parts. A number p_{34} is a transition probability from state 3 to state 4 denoting a state the effect of which is an accurate identification of the object's technical condition; p_{35} means a transition probability from state 3 to state 5, denoting a state of diagnosing results of which is an incorrect or unidentified diagnosis; and p_{36} denotes a transition probability from state 3 to state 6, which means the first main maintenance operation. We suppose that an expected value of the waiting time T_3 is $\bar{g}_3 = E(T_3) = 0.25\,[h]$, while the standard deviation is $\sigma_3 = D(T_3) = 0.1\,[h]$. We fix $p_{32} = 0.12, p_{34} = 0.42, ; p_{35} = 0.01, p_{36} = 0.45$.
- $Q_{42}(t) = p_{42}G_4(t)$, $Q_{46}(t) = p_{46}G_4(t)$, where $G_4(t)$ is the CDF of the random variable T_4, which denotes a waiting time of diagnosis the effect of which is an accurate identification. We assume $\bar{g}_4 = E(T_4) = 0.2\,[h]$, $p_{42} = 0.08$, $p_{46} = 0.92$.

- $Q_{56}(t) = p_{56}G_5(t)$, where $G_5(t)$ is the CDF of a random variable T_5, denoting the time of diagnosing results for which identification of the object's technical state is incorrect or fuzzy. We assume $\bar{g}_5 = E(T_5) = 0.2\,[h]$ and $\sigma_5 = D(T_5) = 0.1\,[h]$.
- $Q_{62}(t) = p_{62}G_6(t)$, $Q_{63}(t) = p_{63}G_6(t)$, $Q_{67}(t) = p_{67}G_6(t)$, where $G_6(t)$ is the CDF of a random variable T_6, denoting duration of the first step of the main maintenance operation. We suppose $\bar{g}_6 = E(T_6) = 1.33\,[h]$ and $\sigma_6 = D(T_6) = 0.2\,[h]$, $p_{62} = 0.08$, $p_{63} = 0.12$, $p_{67} = 0.80$.
- $Q_{78}(t) = p_{78}G_7(t)$, $Q_{79}(t) = p_{79}G_7(t)$, where $G_7(t)$ is the CDF of a random variable T_7, denoting a duration of the second step of the main maintenance operation. We fix an expectation and a standard deviation of the random variable as $\bar{g}_7 = E(T_7) = 1.2\,[h]$ and $\sigma_7 = D(T_7) = 0.1\,[h]$. Transition probabilities of the embedded Markov chain from state 7 are $p_{78} = 0.98$ and $p_{79} = 0.02$.
- $Q_{8\,10}(t) = G_8(t)$, where $G_8(t)$ is the CDF of a random variable T_8, denoting a waiting time of inspection with a positive result. We determine its expectation and a standard deviation as $\bar{g}_8 = E(T_8) = 0.2\,[h]$ and $\sigma_8 = D(T_8) = 0.1\,[h]$.
- $Q_{96}(t) = G_9(t)$, where $G_9(t)$ is the CDF of a random variable T_9, denoting a waiting time of inspection with a negative result. We suppose that an expectation and a standard deviation of the random variable are $\bar{g}_9 = E(T_9) = 0.2\,[h]$ and $\sigma_9 = D(T_9) = 0.1\,[h]$. From state 9, an object passes to state 6, which denotes the first step of the main maintenance operation.

The matrices from equations (14.6) and (14.7) take the form

$$
I - P_{A'} =
\begin{bmatrix}
1 & 0 & -1 & 0 & 0 & 0 & 0 & 0 & 0 \\
0 & 0 & 1 & -1 & 0 & 0 & 0 & 0 & 0 \\
0 & -0.12 & 0 & -0.42 & -0.01 & -0.45 & 0 & 0 & 0 \\
0 & -0.08 & 0 & 1 & 0 & -0.92 & 0 & 0 & 0 \\
0 & 0 & 0 & 0 & 1 & -1 & 0 & 0 & 0 \\
0 & -0.08 & -0.12 & 0 & 0 & 1 & -0.8 & 0 & 0 \\
0 & 0 & 0 & 0 & 0 & 0 & 1 & -0.98 & -0.02 \\
0 & 0 & 0 & 0 & 0 & 0 & 0 & 1 & 0 \\
0 & 0 & 0 & 0 & 0 & -1 & 0 & 0 & 1
\end{bmatrix},
$$

$$
\overline{\Theta}_{A'} =
\begin{bmatrix}
E(\Theta_{1\,10}) \\
E(\Theta_{2\,10}) \\
E(\Theta_{3\,10}) \\
E(\Theta_{4\,10}) \\
E(\Theta_{5\,10}) \\
E(\Theta_{6\,10}) \\
E(\Theta_{7\,10}) \\
E(\Theta_{8\,10}) \\
E(\Theta_{9\,10})
\end{bmatrix},
\quad
\overline{T}_{A'} =
\begin{bmatrix}
0.25 \\
0.55 \\
0.25 \\
0.2 \\
0.2 \\
1.33 \\
1.2 \\
0.1 \\
0.9
\end{bmatrix},
\quad
\overline{\Theta}^2_{A'} =
\begin{bmatrix}
E(\Theta^2_{1\,10}) \\
E(\Theta^2_{2\,10}) \\
E(\Theta^2_{3\,10}) \\
E(\Theta^2_{4\,10}) \\
E(\Theta^2_{5\,10}) \\
E(\Theta^2_{6\,10}) \\
E(\Theta^2_{7\,10}) \\
E(\Theta^2_{8\,10}) \\
E(\Theta^2_{9\,10})
\end{bmatrix},
\quad
B_A =
\begin{bmatrix}
2.65946 \\
4.64332 \\
1.99446 \\
1.39485 \\
1.33886 \\
6.84199 \\
1.88306 \\
0.02 \\
6.61985
\end{bmatrix}.
$$

Using the procedure LinearSolve$[m, c]$ of the MATHEMATICA computer program, we get solutions of those equations

$$
\begin{bmatrix}
E(\Theta_{1\,10}) \\
E(\Theta_{2\,10}) \\
E(\Theta_{3\,10}) \\
E(\Theta_{4\,10}) \\
E(\Theta_{5\,10}) \\
E(\Theta_{6\,10}) \\
E(\Theta_{7\,10}) \\
E(\Theta_{8\,10}) \\
E(\Theta_{9\,10})
\end{bmatrix}
=
\begin{bmatrix}
3.96893 \\
4.26893 \\
3.71893 \\
3.50588 \\
3.42214 \\
3.22214 \\
1.38044 \\
0.1 \\
4.12214
\end{bmatrix}
,
\qquad
\begin{bmatrix}
E(\Theta_{1\,10}^2) \\
E(\Theta_{2\,10}^2) \\
E(\Theta_{3\,10}^2) \\
E(\Theta_{4\,10}^2) \\
E(\Theta_{5\,10}^2) \\
E(\Theta_{6\,10}^2) \\
E(\Theta_{7\,10}^2) \\
E(\Theta_{8\,10}^2) \\
E(\Theta_{9\,10}^2)
\end{bmatrix}
=
\begin{bmatrix}
18.84 \\
20.82 \\
16.18 \\
14.35 \\
13.61 \\
12.27 \\
2.28 \\
0.02 \\
18.89
\end{bmatrix}
.
$$

The results presented in Table 14.1 are computed for transition probabilities equal to

$$p_{62} = 0.08, \quad p_{63} = 0.12, \quad p_{67} = 0.80, \quad p_{78} = 0.98, \quad p_{79} = 0.02.$$

We change some transition probabilities. We increase a transition probability from state 6, which means the first stage of the main maintenance operation to state 2, denoting an object waiting due to an absence of spare parts, and we also degrease a transition probability from state 7, denoting the second stage of main maintenance operation to state 8, which means an inspection with positive effect. For the changed transition probabilities

$$p_{62} = 0.12, \quad p_{63} = 0.12, \quad p_{67} = 0.75, \quad p_{78} = 0.90, \quad p_{79} = 0.10$$

we have other results, which are shown in Table 14.2.

Comparing both tables we conclude that the expected value of the object whole maintenance operation time is essentially greater than in the second case. In the first case, we have $E(\Theta_{1\,10}) = 3.97\,[h]$ and in the second case $E(\Theta_{1\,10}) = 4.47\,[h]$. For the standard deviations, we have the same rule—in the first case we have $D(\Theta_{1\,10}) = 1.76\,[h]$ while in the second one $E(\Theta_{1\,10}) = 2.37\,[h]$.

Table 14.1 **Parameters of random variables** $\Theta_{i\,10}, i = 1, \ldots, 9$

State i	1	2	3	4	5	6	7	8	9
$E(\Theta_{i10})\,[h]$	3.97	4.27	3.72	3.50	3.42	3.22	1.38	0.1	4.12
$D(\Theta_{i10})\,[h]$	1.76	1.61	1.53	1.43	1.38	1.37	0.61	0.1	1.38

Table 14.2 **Parameters of random variables** $\Theta_{i\,10}, i = 1, \ldots, 9$

State i	1	2	3	4	5	6	7	8	9
$E(\Theta_{i10})\,[h]$	4.47	4.77	4.22	4.00	3.92	3.72	1.75	0.1	4.62
$D(\Theta_{i10})\,[h]$	2.37	2.26	2.21	2.14	2.10	2.10	1.52	0.1	2.10

14.2.5 Income from maintenance operation

The economic reward structure for the continuous-time SMP was introduced by Howard [45]. For the model presented above, we offer another approach. To compute the income from the maintenance operation, we construct a semi-Markov model with the same shape of the kernel. To underline differences, instead of the parameter t denoting time, we will use the parameter x denoting a profit; we replace waiting time T_i with a random variable C_i, which denotes a gain in a state i; we replace the SMP $\{X(t) : t \geqslant 0\}$ with the process $\{U(x) : x \geqslant 0\}$; and the first passage time from a state i to a state 10, which is signed as Θ_{i10}, we substitute with the random variable Z_{i10}, which means an income from the maintenance operation under the assumption that i is an initial state. We apply the model for the same numerical example, assuming that

$$E(C_1) = 0, \quad E(C_2) = 0, \quad E(C_3) = 30.6, \quad E(C_4) = 12.8, \quad E(C_5) = 2.4,$$
$$E(C_6) = 75.2, \quad E(C_7) = 88.23, \quad E(C_8) = 44.85, \quad E(C_9) = 0[\$]$$

and

$$D(C_1) = 0, \quad D(C_2) = 0, \quad D(C_3) = 7.99, \quad D(C_4) = 2.13, \quad D(C_5) = 1.04,$$
$$D(C_6) = 15.20, \quad E(C_7) = 16.96, \quad D(C_8) = 9.21, \quad E(C_9) = 0[\$].$$

We suppose the same transition probability matrix P (14.2). The matrices from (14.6) and (14.7), allowing us to calculate the expectation and standard deviation of the profit from the maintenance operation, take the form

$$I - P_{A'} = \begin{bmatrix}
1 & 0 & -1 & 0 & 0 & 0 & 0 & 0 & 0 \\
0 & 0 & 1 & -1 & 0 & 0 & 0 & 0 & 0 \\
0 & -0.12 & 0 & -0.42 & -0.01 & -0.45 & 0 & 0 & 0 \\
0 & -0.08 & 0 & 1 & 0 & -0.92 & 0 & 0 & 0 \\
0 & 0 & 0 & 0 & 1 & -1 & 0 & 0 & 0 \\
0 & -0.08 & -0.12 & 0 & 0 & 1 & -0.8 & 0 & 0 \\
0 & 0 & 0 & 0 & 0 & 0 & 1 & -0.98 & -0.02 \\
0 & 0 & 0 & 0 & 0 & 0 & 0 & 1 & 0 \\
0 & 0 & 0 & 0 & 0 & -1 & 0 & 0 & 1
\end{bmatrix},$$

$$\overline{Z}_{A'} = \begin{bmatrix}
E(Z_{1\,10}) \\
E(Z_{2\,10}) \\
E(Z_{3\,10}) \\
E(Z_{4\,10}) \\
E(Z_{5\,10}) \\
E(Z_{6\,10}) \\
E(Z_{7\,10}) \\
E(Z_{8\,10}) \\
E(Z_{9\,10})
\end{bmatrix}, \quad
\overline{C}_{A'} = \begin{bmatrix}
0 \\
0 \\
30.6 \\
12.8 \\
2.4 \\
75.2 \\
88.2 \\
44.85 \\
0
\end{bmatrix}, \quad
\overline{Z}_{A'}^2 = \begin{bmatrix}
E(Z_{1\,10}^2) \\
E(Z_{2\,10}^2) \\
E(Z_{3\,10}^2) \\
E(Z_{4\,10}^2) \\
E(Z_{5\,10}^2) \\
E(Z_{6\,10}^2) \\
E(Z_{7\,10}^2) \\
E(Z_{8\,10}^2) \\
E(Z_{9\,10}^2)
\end{bmatrix}, \quad
B_A = \begin{bmatrix}
0 \\
0 \\
1000.26 \\
172.25 \\
8.6 \\
5886.04 \\
8067.08 \\
2096.43 \\
0
\end{bmatrix}.$$

Table 14.3 **Parameters of random variables** Z_{i10}, $i = 1, \ldots, 7$

State i	1	2	3	4	5	6	7
$E(Z_{i10})$ [$]	307.98	307.98	307.98	281.65	267.85	265.45	155.11
$D(Z_{i10})$ [$]	112.96	112.96	112.96	112.1	111.35	111.37	77.35

The solutions of those equations are

$$
\begin{bmatrix}
E(Z_{1\,10}) \\
E(Z_{2\,10}) \\
E(Z_{3\,10}) \\
E(Z_{4\,10}) \\
E(Z_{5\,10}) \\
E(Z_{6\,10}) \\
E(Z_{7\,10}) \\
E(Z_{8\,10}) \\
E(Z_{9\,10})
\end{bmatrix}
=
\begin{bmatrix}
307.981 \\
307.981 \\
307.981 \\
281.65 \\
267.848 \\
265.448 \\
155.11 \\
44.85 \\
265.448
\end{bmatrix},
\quad
\begin{bmatrix}
E(Z_{1\,10}^2) \\
E(Z_{2\,10}^2) \\
E(Z_{3\,10}^2) \\
E(Z_{4\,10}^2) \\
E(Z_{5\,10}^2) \\
E(Z_{6\,10}^2) \\
E(Z_{7\,10}^2) \\
E(Z_{8\,10}^2) \\
E(Z_{9\,10}^2)
\end{bmatrix}
=
\begin{bmatrix}
107,612 \\
107,612 \\
107,612 \\
91,893.3 \\
84,140.9 \\
82,858.1 \\
30,042.6 \\
2096.43 \\
82,858.1
\end{bmatrix}.
$$

The expected values and standard deviation of random variables $Z_{i\,10}$, $i = 1, \ldots, 9$ are shown in Table 14.3.

We see that the total expected income from the maintenance operation with the initial state $i = 1$ is $E(Z_{1\,10}) = 307.98$ [$] and $D(Z_{1\,10}) = 112.96$ [$] is the standard deviation corresponding to it.

We should mention that we can use this approach only for a semi-Markov model with an absorbing state but not in a general case.

14.3 Model of maintenance net without diagnostics

Now we will construct and analyze a model of the maintenance net without diagnostics. Figure 14.3 shows a diagram of that kind of maintenance system.

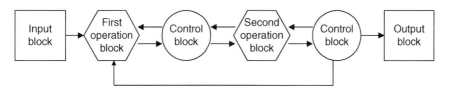

Figure 14.3 Flow graph of the maintenance operation.

14.3.1   Model of maintenance operation

We start the construction of the model from determining the states of the maintenance operation:

1. An object waiting for the operation to begin
2. The first stage of a maintenance operation
3. Control of a maintenance quality
4. The second stage of a maintenance operation
5. Checking the object's technical condition
6. The object waiting for reuse

The possible state changes of the maintenance process are shown in Figure 14.4. A SMP with a set of states $S = \{1, 2, 3, 4, 5, 6\}$ is an appropriate stochastic model of the maintenance operation without diagnostics. The SMP kernel corresponding to the graph shown in Figure 14.4 has the form

$$\mathbf{Q}(t) = \begin{bmatrix} 0 & Q_{12}(t) & 0 & 0 & 0 & 0 \\ 0 & 0 & Q_{23}(t) & 0 & 0 & 0 \\ 0 & Q_{32}(t) & 0 & Q_{34}(t) & 0 & 0 \\ 0 & 0 & 0 & 0 & Q_{45}(t) & 0 \\ 0 & Q_{52}(t) & 0 & Q_{54}(t) & 0 & Q_{56}(t) \\ 0 & 0 & 0 & 0 & 0 & Q_{66}(t) \end{bmatrix}. \tag{14.8}$$

We assume that

$$Q_{ij}(t) = p_{ij}G_i(t), \quad i, j \in S = \{1, \ldots, 6\},$$

where $G_i(t)$ is the CDF of a random variable T_i, which denotes a waiting time in a state i. According to the well-known rule

$$p_{ij} = \lim_{t \to \infty} Q_{ij}(t),$$

the transition probability matrix $\mathbf{P} = [p_{ij} : i, j \in S]$ of the SM maintenance process is

$$\mathbf{P} = \begin{bmatrix} 0 & 1 & 0 & 0 & 0 & 0 \\ 0 & 0 & 1 & 0 & 0 & 0 \\ 0 & p_{32} & 0 & p_{34} & 0 & 0 \\ 0 & 0 & 0 & 0 & 1 & 0 \\ 0 & p_{52} & 0 & p_{54} & 0 & p_{56} \\ 0 & 0 & 0 & 0 & 0 & 1 \end{bmatrix}. \tag{14.9}$$

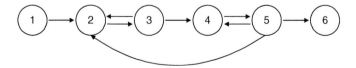

Figure 14.4 Flow graph of the possible state changes.

A number p_{32} is a transition probability from state 3, which means a control of a first stage of maintenance quality to state 2, denoting the first stage of operation. This means the probability that checking the service quality gives a negative result. A transition probability p_{34} denotes a probability of positive result of a maintenance quality control. Similarly, p_{52} denotes a probability that the final checking of the object's technical condition gives a negative result and the whole maintenance operations starts from the beginning, and p_{54} means the probability that the control gives a negative result but the object needs only a second stage of the maintenance. A transition probability p_{56} denotes a probability of a positive result of a final checking of the object's technical condition. Like in the previous model, we are interested in computing expected values and standard deviations of the random variables Θ_{i6}, $i = 1, \ldots, 6$, which denote a duration of the maintenance operation with an initial state i. In this case, the matrices from (14.6) and (14.7) are

$$I - P_{A'} = \begin{bmatrix} 1 & -1 & 0 & 0 & 0 \\ 0 & 1 & -1 & 0 & 0 \\ 0 & -p_{32} & 1 & -p_{34} & 0 \\ 0 & 0 & 0 & 1 & -1 \\ 0 & -p_{52} & 0 & -p_{54} & 1 \\ 0 & 0 & 0 & 0 & 1 \end{bmatrix}, \qquad (14.10)$$

$$\overline{\Theta}_{A'} = \begin{bmatrix} E(\Theta_{16}) \\ E(\Theta_{26}) \\ E(\Theta_{36}) \\ E(\Theta_{46}) \\ E(\Theta_{56}) \end{bmatrix}, \quad \overline{T}_{A'} = \begin{bmatrix} E(T_1) \\ E(T_2) \\ E(T_3) \\ E(T_4) \\ E(T_5) \end{bmatrix}, \quad \overline{\Theta}_{A'}^2 = \begin{bmatrix} E(\Theta_{16}^2) \\ E(\Theta_{26}^2) \\ E(\Theta_{36}^2) \\ E(\Theta_{46}^2) \\ E(\Theta_{56}^2) \end{bmatrix}, \quad B_A = \begin{bmatrix} b_{16} \\ b_{26} \\ b_{36} \\ b_{46} \\ b_{56} \end{bmatrix},$$

where

$$b_{i6} = E(T_i^2) + 2 \sum_{k \in A'} p_{ik} E(T_i) E(\Theta_{k6}), \quad i = 1, \ldots, 6.$$

For a SMP $\{U(x) : x \geqslant 0\}$, describing the profit from the maintenance operation, the matrices from (14.6) and (14.7) take the form

$$\overline{Z}_{A'} = \begin{bmatrix} E(Z_{16}) \\ E(Z_{26}) \\ E(Z_{36}) \\ E(Z_{46}) \\ E(Z_{56}) \end{bmatrix}, \quad \overline{C}_{A'} = \begin{bmatrix} E(C_1) \\ E(C_2) \\ E(C_3) \\ E(C_4) \\ E(C_5) \end{bmatrix}, \quad \overline{Z}_{A'}^2 = \begin{bmatrix} E(Z_{16}^2) \\ E(Z_{26}^2) \\ E(Z_{36}^2) \\ E(Z_{46}^2) \\ E(Z_{56}^2) \end{bmatrix}, \quad B_A = \begin{bmatrix} b_{16} \\ b_{26} \\ b_{36} \\ b_{46} \\ b_{56} \end{bmatrix},$$

where

$$b_{i6} = E(C_i^2) + 2 \sum_{k \in A'} p_{ik} E(C_i) E(Z_{k6}), \quad i = 1, \ldots, 6.$$

Table 14.4 **Parameters of random variables** Θ_{i6}, $i = 1, \ldots, 5$

State i	1	2	3	4	5
$E(\Theta_{i6})$ [h]	0.68	0.56	0.34	0.20	0.09
$D(\Theta_{i6})$ [h]	0.23	0.23	0.22	0.18	0.17

14.3.2 Numerical example

We apply the model presented above to operations consisting of the exchange of car wheels with winter tires for wheels with summer tires. For simplicity, we refer to winter wheels and summer wheels. We suppose that the states of the operation are:

1. Waiting to start exchanging wheels
2. Dismantling the winter wheels
3. Checking the pressure and balancing the summer wheels
4. Pumping, balancing, and mounting the summer wheels
5. Checking or fixing the summer wheels
6. Waiting for reuse

The parameters of the model are assumed to be

$$
P = \begin{bmatrix}
0 & 1 & 0 & 0 & 0 & 0 \\
0 & 0 & 1 & 0 & 0 & 0 \\
0 & 0.12 & 0 & 0.88 & 0 & 0 \\
0 & 0 & 0 & 0 & 1 & 0 \\
0 & 0.06 & 0 & 0.08 & 0 & 0.86 \\
0 & 0 & 0 & 0 & 0 & 1
\end{bmatrix},
$$

$$
\overline{T}_{A'} = \begin{bmatrix}
0.12 \\
0.22 \\
0.1 \\
0.11 \\
0.04
\end{bmatrix}, \quad
\overline{T}^2_{A'} = \begin{bmatrix}
0.016 \\
0.052 \\
0.012 \\
0.018 \\
0.002
\end{bmatrix}.
$$

The results that are achieved from (14.6) and (14.7) are shown in Table 14.4.

14.4 Conclusions

SMP theory provides the possibility to construct probability models of maintenance systems functioning. A maintenance system is called a *maintenance net*. A semi-Markov model of a maintenance net allows us to compute some important characteristics and parameters of a maintenance process. We apply linear equations for parameters of random variables denoting the first passage times from a given state of a SMP to a subset of states to calculate an expected value of duration of the maintenance operation and its standard deviation. In a similar way, we get an expected total profit from the maintenance operation.

Semi-Markov decision processes

15

Abstract

This chapter presents basic concepts and results of the theory of semi-Markov decision processes. The algorithm of optimization of a SM decision process with a finite number of state changes is discussed here. The algorithm is based on a dynamic programming method. To clarify it, the SM decision model for the maintenance operation is shown. The optimization problem for the infinite duration SM process and the Howard algorithm, which enables us to find the optimal stationary strategy, is also discussed here. To explain this algorithm, a decision problem for a renewable series system is presented.

Keywords: Semi-Markov decision processes, Optimization, Dynamic programming, Howard algorithm

15.1 Introduction

Semi-Markov decision processes (SMDPs) theory delivers methods that give us the opportunity to control the operation processes of systems. In such kind of problems, we choose the most rewarding process among the alternatives available for the operation. We discuss the problem of optimizing a SM decision process with a finite number of state changes. The problem is solved by an algorithm that is based on a dynamic programming method. We also investigate the infinite duration SM decision processes. The Howard algorithm modified by Main and Osaki is applied for finding an optimal stationary policy for the kind of processes. This algorithm is equivalent to the some problem of linear programming.

SMDPs theory was developed by Jewell [49], Howard [43–45], Main and Osaki [76], and Gercbakh [24]. Those processes are also discussed in Ref. [30].

15.2 Semi-Markov decision processes

SMDP is a SM process with a finite states space $S = \{1, \dots, N\}$ such that its trajectory depends on decisions that are made at an initial instant and at the moments of the state changes. We assume that a set of decisions in each state i, denoted by D_i, is finite. To take a decision $k \in D_i$, means to select the kth row among the alternating rows of the semi-Markov kernels:

$$\{Q_{ij}^{(k)}(t) : t \geqslant 0, k \in D_i, \ i, j \in S\}, \tag{15.1}$$

where

$$Q_{ij}^{(k)}(t) = p_{ij}^{(k)} F_{ij}^{(k)}(t). \tag{15.2}$$

Semi-Markov Processes: Applications in System Reliability and Maintenance. http://dx.doi.org/10.1016/B978-0-12-800518-7.00015-1

If an initial state is i and a decision (alternative) $k \in D_i$ is chosen at the initial moment, then a probabilistic mechanism of the first change of the state and the evolution of the system on the interval $[0, \tau_1^{(k)})$ is determined. The mechanism is defined by a transition probability (15.2). The decision $k \in D_i$ at some instant $\tau_n^{(k)}$ determines the evolution of the system on the interval $[\tau_n^{(k)}, \tau_{n+1}^{(k)})$. More precisely, the decision $\delta_i(n) = k \in D_i$ means that according to the distribution $(p_{ij}^{(k)} : j \in S)$, a state j is selected for which the process jumps at the moment $\tau_{n+1}^{(k)}$, and the length of the interval $[\tau_n^{(k)}, \tau_{n+1}^{(k)})$ is chosen according to distribution given by the cumulative distribution function (CDF) $F_{ij}^{(k)}(t)$. A sequence of decision at the instant $\tau_n^{(k)}$

$$\delta(n) = (\delta_1(n), \dots, \delta_N(n)) \tag{15.3}$$

is said to be a *policy* in the stage n. A sequence of policies

$$d = \{\delta(n) : n = 0, 1, 2, \dots\} \tag{15.4}$$

is called a *strategy*.

We assume that the strategy has the Markov property, which means that for every state $i \in S$ a decision $\delta_i(n) \in D_i$ does not depend on the process evolution until the moment $\tau_n^{(k)}$. If $\delta_i(n) = \delta_i$, then it is called a *stationary decision*. This means that the decision does not depend on n. The policy consisting of stationary decisions is called a *stationary policy*. Hence, a stationary policy is defined by the sequence $\delta = (\delta_1, \dots, \delta_N)$. Strategy that is a sequence of stationary policies is called a *stationary strategy*.

Optimization of the SMDP consists of choosing the strategy that maximizes the gain of the system.

15.3 Optimization for a finite states change

First, we consider a problem of semi-Markov process optimization for a finite state changes m, so we investigate the process at the time interval $[0, \tau_{m+1}^{(k)})$.

To formulate the optimization problem, we have to introduce the reward structure for the process. We assume that the system, which occupies the state i when a successor state is j, earns a gain (reward) at a rate

$$r_{ij}^{(k)}(x), \quad i, j \in S, \ k \in D_i$$

at a moment x of the entering state i for a decision $k \in D_i$. The function $r_{ij}^k(x)$ is called the "yield rate" of state i at an instant x when the successor state is j and k is a chosen decision [45]. A negative reward at a rate $r_{ij}^{(k)}(x)$ denotes a loss or a cost of that one.

A value of a function

$$R_{ij}^{(k)}(t) = \int_0^t r_{ij}^{(k)}(x)dx, \quad i,j \in S, \ k \in D_i \tag{15.5}$$

denotes the reward that the system earns by spending a time t in a state i before making a transition to state j, for the decision $k \in D_i$. When the transition from the state i to the state j for the decision k is actually made, the system earns a bonus as a fixed sum. The bonus is denoted by

$$b_{ij}^{(k)}, \quad i,j \in S, \ k \in D_i. \tag{15.6}$$

A number

$$u_i^{(k)} = \sum_{j \in S} \int_0^\infty (R_{ij}^{(k)}(t) + b_{ij}^{(k)})dQ_{ij}^{(k)}(t) \tag{15.7}$$

is an expected value of the gain that is generated by the process in the state i at one interval of its realization for the decision $k \in D_i$.

We denote by $V_i(d_m)$, $i \in S$ the expected value of the gain (reward) that is generated by the process during a time interval $[0, \tau_{m+1})$ under the condition that the initial state is $i \in S$ and a sequence of policies is

$$d_m = \{(\delta_1(\tau_n^{(k)}), \ldots, \delta_N(\tau_n^{(k)})\} : n = 0, 1, \ldots, m), \quad m = 0, 1, \ldots. \tag{15.8}$$

By $V_j(d_{m-1})$, $j \in S$ we denote the expected value of the gain that is generated by the process during a time interval $[\tau_1^{(k)}, \tau_{m+1}^{(k)})$ under the condition that the process has just entered the state $j \in S$ at the moment τ_1 and a sequence of policies

$$\{(\delta_1(\tau_n^{(k)}), \ldots, \delta_N(\tau_n^{(k)})) : n = 1, \ldots, m\}, \quad m = 1, 2, \ldots \tag{15.9}$$

is chosen.

The expected value of the gain during a time interval $[0, \tau_m^{(k)})$ under the condition that the initial state is $i \in S$, which is the sum of expectation of the gain that is generated by the process during an interval $[0, \tau_1^{(k)})$ and the gain that is generated by the process during the time $[\tau_1^{(k)}, \tau_m^{(k)})$. Because

$$V_i(d_m) = u_i^{(k)} + \sum_{j \in S} p_{ij}^{(k)} V_j(d_{m-1}), \quad i \in S. \tag{15.10}$$

Substituting (15.7) in this equality, we get

$$V_i(d_m) = \sum_{j \in S} \int_0^\infty (R_{ij}^{(k)}(t) + b_{ij}^{(k)})dQ_{ij}^{(k)}(t) + \sum_{j \in S} p_{ij}^{(k)} V_j(d_{m-1}), \quad i \in S. \tag{15.11}$$

The strategy (the sequence of policies) d_m^* is called optimal in gain maximum problem on interval $[\tau_0' \tau_{m+1})$ for the SMDP, which starts from state i, if

$$V_i(d_m^*) = \max_{d_m}[V_i(d_m)]. \tag{15.12}$$

This means that $V_i(d_m^*) \geqslant V_i(d_m)$ for all strategies d_m.

We can get the optimal strategy by using the dynamic programming technique, which uses the Bellman principle of optimality. In our case the principle can be formulated as follows:

B e l l m a n principle of optimality.
Let for any initial state and adopted in this state strategy process move to a new state. If the initial strategy is optimal, then its remaining part is also optimal for the process whose initial state is a new state that has been reached at the moment the first state changes.

This principle allows us to obtain an algorithm of computing the optimal strategy. The algorithm is defined by the following formulas:

$$V_i(d_n^*) = \max_{k \in D_i} \left[u_i^{(k)} + \sum_{j \in S} p_{ij}^{(k)} V_j(d_{n-1}^*) \right], \quad i \in S, \; n = 1, \dots, m, \tag{15.13}$$

$$V_i(d_0^*) = \max_{k \in D_i}[u_i^{(k)}], \quad i \in S. \tag{15.14}$$

To obtain the policy d_0^*, we start from (15.14). Based on formula

$$V_i(d_1^*) = \max_{k \in D_i} \left[u_i^{(k)} + \sum_{j \in S} p_{ij}^{(k)} V_j(d_0^*) \right], \quad i \in S, \tag{15.15}$$

we find strategy

$$d_1^* = ((\delta_1^*(\tau_n^{(k)}), \dots, \delta_N^*(\tau_n^{(k)})) : n = 0, 1)$$

in the next step. Continuing this procedure, we obtain the optimal strategy

$$d_m^* = ((\delta_1^*(\tau_n^{(k)}), \dots, \delta_N^*(\tau_n^{(k)})) : n = 0, 1, \dots, m).$$

15.4   SM decision model of maintenance operation

Now we develop the model from Chapter 14 dealing with the maintenance net without diagnostics. To construct the semi-Markov decision model corresponding to this model, we have to determine the sets of decisions D_i, $i = 1, \dots, 6$. Assume that

$$D_1 = \{1\}, \quad D_2 = \{1, 2\}, \quad D_3 = \{1, 2\}, \quad D_4 = \{1, 2\}, \quad D_5 = \{1, 2\}, \quad D_6 = \{1\}$$

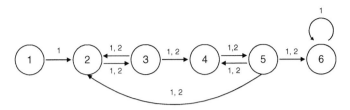

Figure 15.1 The possible state changes of the maintenance process.

where

D_2 : 1—normal maintenance, 2—expensive maintenance;
D_3 : 1—normal control,   2—expensive control;
D_4 : 1—normal maintenance, 2—expensive maintenance;
D_5 : 1—normal control,   2—expensive control.

The states of the decision maintenance process are as follows:

1. An object waiting for the beginning of the operation
2. The first stage of a normal or an expensive maintenance
3. Control of a maintenance quality
4. The second stage of a normal or an expensive maintenance
5. A normal or expensive control of the object's technical condition
6. The object waiting for reuse

The possible state changes of the decision maintenance process are shown in Figure 15.1.

A semi-Markov process with a set of states $S = \{1, 2, 3, 4, 5, 6\}$ is an appropriate stochastic model of the maintenance operation without diagnostics. The family of kernels

$$\boldsymbol{Q}^{(k)}(t) = \begin{bmatrix} 0 & Q_{12}^{(k)}(t) & 0 & 0 & 0 & 0 \\ 0 & 0 & Q_{23}^{(k)}(t) & 0 & 0 & 0 \\ 0 & Q_{32}^{(k)}(t) & 0 & Q_{34}^{(k)}(t) & 0 & 0 \\ 0 & 0 & 0 & 0 & Q_{45}^{(k)}(t) & 0 \\ 0 & Q_{52}^{(k)}(t) & 0 & Q_{54}^{(k)}(t) & 0 & Q_{56}^{(k)}(t) \\ 0 & 0 & 0 & 0 & 0 & Q_{66}^{(k)}(t) \end{bmatrix}, \quad k \in D_i \qquad (15.16)$$

is crucial in a SM decision problem. Assume that

$$Q_{ij}^{(k)}(t) = p_{ij}^{(k)} G_i^{(k)}(t), \quad i, j \in S = \{1, \ldots, 6\}, \ k \in D_i, \qquad (15.17)$$

where $G_i^{(k)}(t)$ is the CDF of a random variable $T_i^{(k)}$, which denotes a waiting time in a state i under decision $k \in D_i$. According to the well-known rule

$$p_{ij}^{(k)} = \lim_{t \to \infty} Q_{ij}^{(k)}(t),$$

the transition probability matrix $P = [p_{ij} : i, j \in S]$ of the SM maintenance process is

$$P = \begin{bmatrix} 0 & 1 & 0 & 0 & 0 & 0 \\ 0 & 0 & 1 & 0 & 0 & 0 \\ 0 & p_{32}^{(k)} & 0 & p_{34}^{(k)} & 0 & 0 \\ 0 & 0 & 0 & 0 & 1 & 0 \\ 0 & p_{52}^{(k)} & 0 & p_{54}^{(k)} & 0 & p_{56}^{(k)} \\ 0 & 0 & 0 & 0 & 0 & 1 \end{bmatrix}. \tag{15.18}$$

Suppose that

$$r_{ij}^{(k)}(x) = r_{ij}^{(k)} \in \mathbb{R}, \quad k \in D_i, \; i, j \in S = \{1, \ldots, 6\}. \tag{15.19}$$

From (15.5), we obtain

$$R_{ij}^{(k)}(t) = r_{ij}^{(k)} t, \quad i, j \in S, \; k \in D_i \tag{15.20}$$

and

$$u_i^{(k)} = \sum_{j \in S} p_{ij}^{(k)} (r_{ij}^{(k)} m_{ij}^{(k)} + b_{ij}^{(k)}), \tag{15.21}$$

where $m_{ij}^{(k)} = E(T_{ij}^{(k)})$ denotes the expectation of the holding time of the state i if the successor state is j. From assumption (15.17), we get

$$m_{ij}^{(k)} = E(T_{ij}^{(k)}) = E(T_i^{(k)}) = m_i^{(k)}.$$

Moreover, we suppose $b_{ij}^{(k)} = 0$, $i, j \in S$, $k \in D_i$. Now the equality (15.21) takes the form

$$u_i^{(k)} = m_i^{(k)} \sum_{j \in S} p_{ij}^{(k)} r_{ij}^{(k)}. \tag{15.22}$$

Thus, the algorithm of choosing an optimal strategy takes the following form:

The Algorithm

1. Compute

$$u_i^{(k)} = m_i^{(k)} \sum_{j \in S} p_{ij}^{(k)} r_{ij}^{(k)} \tag{15.23}$$

   for all $i \in S = \{1, \ldots, 6\}$ and $k \in D_i$.
2. Find d_0^* such that

$$V_i(d_0^*) = \max_{k \in D_i} [u_i^{(k)}] \tag{15.24}$$

   for all $i \in S = \{1, \ldots, 6\}$.
3. Find d_l^* such that

$$V_i(d_l^*) = \max_{k \in D_i} \left[u_i^{(k)} + \sum_{j \in S} p_{ij}^{(k)} V_j(d_{l-1}^*) \right] \tag{15.25}$$

   for $l = 1, 2, \ldots, m - 1$ and for all $i \in S$.

15.5 Optimal strategy for the maintenance operation

We apply the algorithm to obtain an optimal strategy of the maintenance operation discussed here. We determine the numerical data (Tables 15.1 and 15.2).

1. We compute a gain $u_i^{(k)}$ from the first stage ($n = 0$) of the operation using (15.23) for $i \in S = \{1, \ldots, 6\}$ and $k \in D_i$. For example,

$$u_5^{(1)} = 0.15 \cdot (0.12 \cdot (-40) + 0.16 \cdot (-10) + 0.72 \cdot 50) = 4.656,$$

$$u_5^{(2)} = 0.20 \cdot (0.08 \cdot (-46) + 0.10 \cdot (-30) + 0.82 \cdot 52) = 7.192.$$

In a similar manner, we obtain

$$u_1^{(1)} = 0.30, \quad u_2^{(1)} = 85.44, \quad u_2^{(2)} = 114.8, \quad u_3^{(1)} = 0.9936,$$

$$u_2^{(3)} = 0.96, \quad u_4^{(1)} = 100.28, \quad u_4^{(2)} = 124.2, \quad u_6^{(1)} = 0.15. \tag{15.26}$$

Table 15.1 **Transition probabilities and mean waiting times for the maintenance process**

State i	Decision k	$p_{i1}^{(k)}$	$p_{i2}^{(k)}$	$p_{i3}^{(k)}$	$p_{i4}^{(k)}$	$p_{i5}^{(k)}$	$p_{i6}^{(k)}$	$m_i^{(k)}$ [h]
1	1	0	1	0	0	0	0	0.15
2	1	0	0	1	0	0	0	1.78
	2	0	0	1	0	0	0	2.05
3	1	0	0.16	0	0.84	0	0	0.18
	2	0	0.02	0	0.98	0	0	0.20
4	1	0	0	0	0	1	0	2.18
	2	0	0	0	0	1	0	2.30
5	1	0	0.12	0	0.16	0	0.72	0.15
	2	0	0.08	0	0.10	0	0.82	0.20
6	1	0	0	0	0	0	1	0.15

Table 15.2 **Gain rate for the maintenance process**

State i	Decision k	$r_{i1}^{(k)}$ [$]	$r_{i2}^{(k)}$ [$]	$r_{i3}^{(k)}$ [$]	$r_{i4}^{(k)}$ [$]	$r_{i5}^{(k)}$ [$]	$r_{i6}^{(k)}$ [$]
1	1	0	2	0	0	0	0
2	1	0	0	48	0	0	0
	2	0	0	56	0	0	0
3	1	0	−18	0	10	0	0
	2	0	−48	0	6	0	0
4	1	0	0	0	0	46	0
	2	0	0	0	0	54	0
5	1	0	−40	0	−10	0	50
	2	0	−46	0	−30	0	52
6	1	0	0	0	0	0	1

2. We find the gain $V_i(d_0^*)$ for all $i \in S$ applying (15.24).

$$
\begin{aligned}
V_1(d_0^*) &= 0.30 && \text{for } k = 1 \in D_1, \\
V_2(d_0^*) &= 114.8 && \text{for } k = 2 \in D_2, \\
V_3(d_0^*) &= 0.9936 && \text{for } k = 1 \in D_3, \\
V_4(d_0^*) &= 124.2 && \text{for } k = 2 \in D_4, \\
V_5(d_0^*) &= 7.192 && \text{for } k = 2 \in D_5, \\
V_6(d_0^*) &= 0.1 && \text{for } k = 1 \in D_6.
\end{aligned}
\tag{15.27}
$$

Hence, the optimal policy at the moment 0 is represented by the vector of decisions $\delta^*(0) = (1, 2, 1, 2, 2, 1)$.

3. We have to find a policy $\delta^*(1) = (\delta_1^*(\tau_1^{(k)}), \ldots, \delta_6^*(\tau_1^{(k)}))$. According to (15.25), we get

$$
V_i(d_1^*) = \max_{k \in D_i} \left[u_i^{(k)} + \sum_{j \in S} p_{ij}^{(k)} V_j(d_0^*) \right]
\tag{15.28}
$$

for $i = 1, \ldots, 6$. For example,

$$
V_5(d_1^*) = \max \left[\begin{array}{l} 4.656 + 0.12 \cdot 114.8 + 0.16 \cdot 124.2 + 0.72 \cdot 0.1, \\ 7.192 + 0.08 \cdot 114.8 + 0.10 \cdot 124.2 + 0.82 \cdot 0.1 \end{array} \right]
\tag{15.29}
$$

$$
= \max[38.376,\ 28.878] = 38.386 \quad \text{for } k = 1 \in D_5.
$$

Using (15.28) for the data from (15.27) and from Tables 15.1 and 15.3, we obtain

(a)

$$
\begin{aligned}
V_1(d_1^*) &= 0.60 && \text{for } k = 1 \in D_1, \\
V_2(d_1^*) &= 115.7936 && \text{for } k = 2 \in D_2, \\
V_3(d_1^*) &= 124.972 && \text{for } k = 2 \in D_3, \\
V_4(d_1^*) &= 131.392 && \text{for } k = 2 \in D_4, \\
V_5(d_1^*) &= 38.376 && \text{for } k = 1 \in D_5, \\
V_6(d_1^*) &= 0.25 && \text{for } k = 1 \in D_6.
\end{aligned}
\tag{15.30}
$$

The optimal policy on the second stage ($n = 1$) is

$$
\delta^*(1) = (1, 2, 2, 2, 1, 1).
$$

(b) Now we are able to find the policy on the third stage ($n = 2$) of the maintenance operation. Using (15.25), we have

$$
V_i(d_2^*) = \max_{k \in D_i} \left[u_i^{(k)} + \sum_{j \in S} p_{ij}^{(k)} V_j(d_1^*) \right]
\tag{15.31}
$$

for $i = 1, \ldots, 6$. Using this formula, the result of (15.30), and the data from Tables 15.1 and 15.3, we get:

$$V_1(d_2^*) = 0.60 \quad \text{for } k = 1 \in D_1,$$
$$V_2(d_2^*) = 239.772 \text{ for } k = 2 \in D_2,$$
$$V_3(d_2^*) = 132.224 \text{ for } k = 2 \in D_3,$$
$$V_4(d_2^*) = 162.21 \quad \text{for } k = 2 \in D_4, \tag{15.32}$$
$$V_5(d_2^*) = 39.754 \quad \text{for } k = 1 \in D_5,$$
$$V_6(d_2^*) = 0.40 \quad \text{for } k = 1 \in D_6.$$

The optimal policy on the third stage ($n = 2$) is

$$\delta^*(2) = (1, 2, 2, 2, 1, 1).$$

The optimal strategy for three stages of the operation is

$$d_2^* = (\delta^*(0), \delta^*(1), \delta^*(2)).$$

If we assume that the initial state of the maintenance process is 1, then we should calculate optimal policies until the process achieves the state 6. It seems that in this case it is sufficient to calculate the optimum policies for the $n = 10$ stages. Of course it requires writing the appropriate computer program.

15.6   Optimization problem for infinite duration process

We formulate the optimization problem of a semi-Markov process on the infinite interval $[0, \infty)$. This problem was investigated by Howard [45] and by Mine and Osaki [76]. It is known as a decision problem without discounting.

We assume that the considered SMDP with a finite state space $S = \{1, \ldots, N\}$ satisfies the assumption of the limiting theorem (Theorem 3.6).

The criterion function

$$g(\delta) = \frac{\sum\limits_{i \in S} \pi_i(\delta) u_i^{(k)}}{\sum\limits_{i \in S} \pi_i(\delta) m_i^{(k)}} \tag{15.33}$$

means the gain per unit of time is a result of a long operating system. The numbers $\pi_i(\delta)$, $i \in S$ represent the stationary distribution of the embedded Markov chain of the semi-Markov process defined by the kernel

$$Q^{(\delta)}(t) = [Q_{ij}^{(k)}(t) : t \geqslant 0, i, j \in S, k \in D_i]. \tag{15.34}$$

This means that for every decision $k \in D_i$ those probabilities satisfy the following linear system of equations:

$$\sum_{i \in S} \pi_i(\delta) p_{ij}^{(k)} = \pi_j(\delta), \quad j \in S, \quad \sum_{i \in S} \pi_i(\delta) = 1, \tag{15.35}$$

where

$$p_{ij}^{(k)} = \lim_{t \to \infty} Q_{ij}^{(k)}(t), \quad i,j \in S. \tag{15.36}$$

The number

$$m_i^{(k)} = E(T_i^{(k)}) = \lim_{t \to \infty} \int_0^\infty t \, dG_i^{(k)}(t), \quad i \in S, \, k \in D_i \tag{15.37}$$

is an expected value of a waiting time in a state i for a decision (alternative) $k \in D$. Recall that a number

$$u_i^{(k)} = \sum_{j \in S} \int_0^\infty (R_{ij}^{(k)}(t) + b_{ij}^{(k)}) dQ_{ij}^{(k)}(t) \tag{15.38}$$

is an expected value of the profit (gain) that is generated by the process in the state i at one interval of its realization for the decision $k \in D_i$.

Definition 15.1. A stationary policy δ^* is said to be optimal if it maximizes the gain per unit of time:

$$g(\delta^*) = \max_d [g(\delta)]. \tag{15.39}$$

In Ref. [76], it is proved that the optimal stationary strategy exists there. In Refs. [45, 76], the Howard algorithm is presented, which enables us to find the optimal stationary strategy. Here we present the Howard algorithm using our own notation.

The Algorithm

1. *Data*
   - Sets of decisions (alternatives)

     $$D_i, \quad i \in S = \{1, 2, \ldots, N\},$$

   - Set of functions defining the SMDPs

     $$\{Q_{ij}^{(k)}(t) : t \geqslant 0, k \in D_i, i,j \in S\},$$

   - Sets of functions that define the unit gains

     $$\{r_{ij}^{(k)}(x) : x \geqslant 0, d_i \in D_i, i,j \in S\},$$

     $$\{b_{ij}^{(k)}, d_i \in D_i, i,j \in S\}.$$

2. *Initial calculation procedure*
   Compute according to (15.36)–(15.38)

   $$p_{ij}^{(k)}, \quad m_i^{(k)}, \quad u_i^{(k)}$$

   for each decision $k \in D_i, i,j \in S$.

3. *Policy evaluation*

For the present policy $\delta = (\delta_1, \ldots, \delta_N)$, $\delta_i = k \in D_i$, *calculate the gain* $g = g(\delta)$ and *solve the system of linear equations*

$$g m_i^{(k)} + w_i = u_i^{(k)} + \sum_{j \in S} p_{ij}^{(k)} w_j, \quad i \in S, \tag{15.40}$$

with $w_N = 0$ and the unknown weights $w_1, w_2, \ldots, w_{N-1}$.

4. *Policy improvement*

For each state $i \in S$, find the set of decisions (alternatives)

$$\Delta_i = \left\{ k \in D_i : \Gamma_i^{(k)} = \frac{u_i^{(k)} + \sum_{j \in S} p_{ij}^{(k)} w_j - w_i}{m_i^{(k)}} > g(\delta) \right\}. \tag{15.41}$$

If for each $i \in S$ the set $\Delta_i = \emptyset$, then the policy $\delta = (\delta_1, \ldots, \delta_N)$ is optimal and the strategy corresponding to it is also optimal. If there is at least one state $i \in S$ such that $\Delta_i \neq \emptyset$, then the policy is not optimal and it must be improved. Therefore, substitute the policy $\delta = (\delta_1, \ldots, \delta_N)$ with the policy $\delta' = (\delta'_1, \ldots, \delta'_N)$, where $\delta'_i = \delta_i$ if $\Delta_i = \emptyset$ and $\delta'_i \in D_i$ is any other decision if $\Delta_i \neq \emptyset$. Repeat procedures 3 and 4.

It is proved [30, 76] that $g(\delta') > g(\delta)$ and the optimal decision is achieved after a finite number of iterations.

Mine and Osaki [76] have formulated and proved the following theorem:

Theorem 15.1 (Mine and Osaki [76]). The problem: find δ^* such that

$$g(\delta^*) = \max_d [g(\delta)], \tag{15.42}$$

where the criterion function $g(\delta)$ is defined by (15.33), is equivalent to the following problem of linear programming: Let

$$y_j^{(k)} \geqslant 0, \quad j \in S, \ k \in D_i.$$

Find

$$\max_{y_j^{(k)}} \left[\sum_{j \in S} \sum_{k \in D_j} m_j^{(k)} u_i^{(k)} y_j^{(k)} \right] \tag{15.43}$$

under constraints

$$\sum_{k \in D_j} y_j^{(k)} - \sum_{i \in S} \sum_{k \in D_i} p_{ij}^{(k)} y_j^{(k)} = 0, \quad j = 1, \ldots, N - 1, \tag{15.44}$$

$$\sum_{j \in S} \sum_{k \in D_i} m_j^{(k)} y_j^{(k)} = 1. \tag{15.45}$$

We obtain the optimal policy using the rule

$$a_j^{(k)} = \frac{y_j^{(k)}}{\sum_{k \in D_i} y_j^{(k)}}, \tag{15.46}$$

where $a_j^{(k)}$ denotes a probability that in the state $j \in S$ a decision $k \in D_j$ has been taken. It is obvious that

$$\sum_{k \in D_i} a_j^{(k)} = 1, \quad 0 \leqslant a_j^{(k)} \leqslant 1, \, j \in S. \tag{15.47}$$

The best decision in the state $j \in S$ is such alternative $k^* \in D_j$ for which the probability $a_j^{(k^*)}$ is the greatest. The best decisions in states $1, \ldots, N$ form the optimal policy.

15.7 Decision problem for renewable series system

A system consists of two components that form a series reliability structure. We assume that a lifetime of component k, $k = 1, 2$ is represented by a random variable ζ_k with exponential probability density function

$$f_k(t) = \lambda e^{-\lambda_k t} I_{[0, \infty)}(t). \tag{15.48}$$

From the structure of the system it follows that the damage of the system takes place if a failure of any component occurs. A damaged component is renewed. We assume that the renewal time of kth component is a nonnegative random variable γ_k with a CDF

$$H_k(t) = P(\gamma_k \leqslant t), \quad k = 1, 2. \tag{15.49}$$

We know that the exponential probability distribution has memoryless property. Therefore, the renewal of a component means renewal of the whole system. We also assume that the random variables denoting successive times to failure of kth component and random variables denoting their consecutive renewal times are independent copies of the random variables ζ_k and γ_k, accordingly. We suppose that the random variables $\zeta_1, \zeta_2, \gamma_1, \gamma_2$ are mutually independent. Moreover, we assume that γ_1, γ_2 have the positive, finite expected values and variances.

We introduce the following states:

1. Renewal of a first component after its failure
2. Renewal of a second component after its failure
3. Work of the "up" system

The "down" states are represented by a set $A = \{1, 2\}$, while the "up" state is represented by one element set $A' = \{3\}$.

We assume that

$$D_1 = \{1, 2, \}, \quad D_2 = \{1, 2\}, \quad D_3 = \{1, 2, 3, 4\}$$

are the sets of decisions (alternatives) for the states $1, 2, 3$.

D_1 : 1—normal renewal of a first component,
        2—expensive renewal of a second component,

D_2 : 1—normal renewal of a second component,
      2—expensive renewal of a second component,
D_3 : 1—normal reliability of a first and a second components,
      2—normal reliability of a first component and higher of a second component,
      3—higher reliability of a first component and a normal of a second component,
      4—a higher reliability of the both components.

The semi-Markov decision model is determined by the family of kernels

$$
\boldsymbol{Q}^{(\delta)}(t) = \begin{bmatrix} 0 & 0 & H_1^{(k)}(t) \\ 0 & 0 & H_2^{(k)}(t) \\ \frac{\lambda_1^{(k)}}{\Lambda^{(k)}}\left(1 - e^{-\Lambda^{(k)}t}\right) & \frac{\lambda_2^{(k)}}{\Lambda^{(k)}}\left(1 - e^{-\Lambda^{(k)}t}\right) & 0 \end{bmatrix},
\tag{15.50}
$$

where

$$
\Lambda^{(k)} = \lambda_1^{(k_1)} + \lambda_2^{(k_2)}, \quad k \in D_3, \quad k_1 = 1, 2, \; k_2 = 1, 2.
$$

Assume that

$$
H_1^{(k)}(t) = 1 - \left(1 + \alpha_1^{(k)}t\right) e^{-\alpha_1^{(k)}t}, \quad t \geqslant 0, \, k \in D_1,
\tag{15.51}
$$

$$
H_2^{(k)}(t) = 1 - \left(1 + \alpha_2^{(k)}t\right) e^{-\alpha_2^{(k)}t}, \quad t \geqslant 0, \, k \in D_2,
\tag{15.52}
$$

$$
F_1(t) = 1 - e^{-\lambda_1^{(k_1)}t}, \quad t \geqslant 0, \, k_1 = 1, 2,
\tag{15.53}
$$

$$
F_2(t) = 1 - e^{-\lambda_2^{(k_2)}t}, \quad t \geqslant 0, \, k_2 = 1, 2,
\tag{15.54}
$$

where

$$
D_3 = \{1, 2, 3, 4\} = \{(1, 1), (1, 2), (2, 1), (2, 2)\}
$$

is the set of alternative for the state 3.

$$
\begin{aligned}
&\text{For } k = 1 \in D_3 \; \lambda_1^{(k)} = \lambda_1^{(1)}, \quad \lambda_2^{(k)} = \lambda_2^{(1)}, \\
&\text{for } k = 2 \in D_3 \; \lambda_1^{(k)} = \lambda_1^{(1)}, \quad \lambda_2^{(k)} = \lambda_2^{(2)}, \\
&\text{for } k = 3 \in D_3 \; \lambda_1^{(k)} = \lambda_1^{(2)}, \quad \lambda_2^{(k)} = \lambda_2^{(1)}, \\
&\text{for } k = 4 \in D_3 \; \lambda_1^{(k)} = \lambda_1^{(2)}, \quad \lambda_2^{(k)} = \lambda_2^{(2)}.
\end{aligned}
\tag{15.55}
$$

The matrix of transition probabilities of embedded Markov chain corresponding to the kernel (15.55) has the form

$$
\boldsymbol{P}^{(\delta)} = \begin{bmatrix} 0 & 0 & 1 \\ 0 & 0 & 1 \\ \frac{\lambda_1^{(k)}}{\Lambda^{(k)}} & \frac{\lambda_2^{(k_1)}}{\Lambda^{(k)}} & 0 \end{bmatrix}.
\tag{15.56}
$$

In this case, the solution of the system of equations (15.35) is

$$\pi_1(\delta) = \frac{\lambda_1^{(k_1)}}{2\Lambda^{(k)}}, \quad \pi_2(\delta) = \frac{\lambda_2^{(k_2)}}{2\Lambda^{(k)}}, \quad \pi_3(\delta) = \frac{1}{2}. \tag{15.57}$$

The expected values of waiting times in states $i \in S$ for decisions $k \in D_i$ are

$$\begin{aligned}
m_1^{(k)} &= \frac{2}{\alpha_1^{(k)}} \quad \text{for } k \in D_1 = \{1,2\}, \\
m_2^{(k)} &= \frac{2}{\alpha_2^{(k)}} \quad \text{for } k \in D_2 = \{1,2\}, \\
m_3^{(k)} &= \frac{1}{\Lambda^{(k)}} \quad \text{for } k \in D_3 = \{1,2,3,4\}.
\end{aligned} \tag{15.58}$$

In this case, the criterion function (15.33) takes the form

$$g(\delta) = \frac{(\lambda_1^{(k)} u_1^{(k)} + \lambda_2^{(k)} u_2^{(k)} + \Lambda^{(k)} u_3^{(k)})\alpha_1^{(k)}\alpha_2^{(k)}}{2\lambda_1^{(k)}\alpha_2^{(k)} + 2\lambda_2^{(k)}\alpha_1^{(k)} + \alpha_1^{(k)}\alpha_2^{(k)}}. \tag{15.59}$$

Like in Section 15.3, we suppose that

$$r_{ij}^{(k)}(x) = r_{ij}^{(k)} \in \mathbb{R}, \quad k \in D_i, \ i,j \in S = \{1,2,3\} \tag{15.60}$$

and

$$u_i^{(k)} = m_i^{(k)} \sum_{j \in S} p_{ij}^{(k)} r_{ij}^{(k)}, \quad i \in S. \tag{15.61}$$

We determine the numerical data.
Parameters of CDFs for alternatives $k \in D_1$:

$$\alpha_1^{(1)} = 0.2, \ \alpha_1^{(2)} = 0.125. \tag{15.62}$$

Parameters of CDFs for alternatives $k \in D_2$:

$$\alpha_2^{(1)} = 0.25, \ \alpha_2^{(2)} = 0.2. \tag{15.63}$$

Parameters of CDFs for alternatives $k \in D_3$:

$$\begin{aligned}
\lambda_1^{(1)} &= 0.008, \ \lambda_2^{(1)} = 0.009, \ \Lambda^{(1)} = 0.017, \\
\lambda_1^{(2)} &= 0.008, \ \lambda_2^{(2)} = 0.006, \ \Lambda^{(2)} = 0.014, \\
\lambda_1^{(3)} &= 0.004, \ \lambda_2^{(3)} = 0.009, \ \Lambda^{(3)} = 0.013, \\
\lambda_1^{(4)} &= 0.004, \ \lambda_2^{(4)} = 0.006, \ \Lambda^{(4)} = 0.010.
\end{aligned} \tag{15.64}$$

We have all data to start the iteration cycle of the Howard-Main and Osaki algorithm (Tables 15.3 and 15.4). Let $\delta = (1,1,1)$ be the initial policy. Now the rule has the form

$$g((1,1,1)) = \frac{(\lambda_1^{(1)} u_1^{(1)} + \lambda_2^{(1)} u_2^{(1)} + \Lambda^{(1)} u_3^{(1)})\alpha_1^{(k)}\alpha_2^{(1)}}{2\lambda_1^{(1)}\alpha_2^{(k)} + 2\lambda_2^{(1)}\alpha_1^{(1)} + \alpha_1^{(1)}\alpha_2^{(1)}}. \tag{15.65}$$

Table 15.3 The transition probabilities and the mean waiting times for the process

State i	Decision k	$p_{i1}^{(k)}$	$p_{i2}^{(k)}$	$p_{i3}^{(k)}$	$m_i^{(k)}$ [h]
1	1	0	0	1	5
	2	0	0	1	8
2	1	0	0	1	4
	2	0	0	1	5
3	1	0.47	0.53	0	58.82
	2	0.57	0.43	0	71.43
	3	0.31	0.69	0	76.92
	4	0.40	0.60	0	100.0

Table 15.4 The gain rate for the process

State i	Decision k	$r_{i1}^{(k)}$ [$\$$]	$r_{i2}^{(k)}$ [$\$$]	$r_{i3}^{(k)}$ [$\$$]	$u_i^{(k)}$
1	1	0	0	-54	-270
	2	0	0	-62	-496
2	1	0	0	-58	-232
	2	0	0	-64	-320
3	1	21	24	0	1328.74
	2	21	28	0	1715.03
	3	25	24	0	1869.93
	4	25	28	0	2680.00

Using the equality, we calculate the gain $g = g((1, 1, 1))$. For this gain we solve the system of equations (15.40). As the result we get the weights w_1, and w_2. The solution is determined by the rules

$$w_1 = \frac{g(p_{32}^{(k)}m_2^{(k)} + m_3^{(k)})}{p_{21}^{(k)}p_{32}^{(k)} + p_{31}^{(k)}}, \quad w_2 = \frac{g(p_{21}^{(k)}m_3^{(k)} - p_{31}^{(k)}m_2^{(k)})}{p_{21}^{(k)}p_{32}^{(k)} + p_{31}^{(k)}}. \tag{15.66}$$

Substituting the appropriate numerical values, we obtain

$$g = -5.70604, \quad w_1 = 688.367, \quad w_2 = 22.8242,$$

$$\Gamma_1^1 = -191.673, \quad \Gamma_1^2 = -148.046, \quad \Gamma_2^1 = -63.706, \quad \Gamma_2^2 = -68.5648,$$

$$\Gamma_3^1 = 28.296, \quad \Gamma_3^2 = 29.6404, \quad \Gamma_3^3 = 27.289, \quad \Gamma_3^4 = 29.6904.$$

Hence,

$$\Delta_1 = \emptyset, \quad \Delta_2 = \emptyset, \quad \Delta_3 = \{1, 2, 3, 4\}.$$

According to the algorithm, we substitute the policy $(1, 1, 1)$ with the policy $(2, 1, 4)$ and we repeat procedures. Now we get

$$g = 27.3125, \quad w_1 = -6664.25, \quad w_2 = -109.25,$$

$$\Gamma_1^1 = 1278.85, \quad \Gamma_1^2 = 771.0312, \quad \Gamma_2^1 = -30.6875, \quad \Gamma_2^2 = -42.15,$$

$$\Gamma_3^1 = -31.645, \quad \Gamma_3^2 = -29.8274, \quad \Gamma_3^3 = -3.52795, \quad \Gamma_3^4 = -0.5125.$$

From here,

$$\Delta_1 = \{1, 2\}, \quad \Delta_2 = \emptyset, \quad \Delta_3 = \emptyset.$$

Therefore, in a next step we substitute the policy $(2, 1, 4)$ with the policy $(2, 2, 4)$. In this case, we have

$$g = 27.7474, \quad w_1 = -175.661, \quad w_2 = -3.62188,$$

$$\Gamma_1^1 = -18.8678, \quad \Gamma_1^2 = -40.0424, \quad \Gamma_2^1 = -57.0945, \quad \Gamma_2^2 = -63.2756,$$

$$\Gamma_3^1 = 21.1537, \quad \Gamma_3^2 = 22.5864, \quad \Gamma_3^3 = 23.5696, \quad \Gamma_3^4 = 26.0756$$

and

$$\Delta_1 = \emptyset, \quad \Delta_2 = \emptyset, \quad \Delta_3 = \emptyset.$$

This means that the policy $\delta^* = (2, 2, 4)$ maximizes the criterion function $g(\delta)$, $\delta \in D_1 \times D_2 \times D_3$. Recall that

$2 \in D_1$—expensive renewal of a first component,

$2 \in D_2$—expensive renewal of a second component,

$4 \in D_3$—higher reliability of the both components.

15.8   Conclusions

SMDPs theory provides the possibility to formulate and solve the optimization problems that can be modeled by SM processes. In such kinds of problems, we choose the process that brings the most profit among other decisions available for the operation. The problem requires the use of terms such as decision (alternative), policy, strategy, gain, and criterion function. We can solve the problem of optimizing a SM decision process with a finite number of state changes by using the algorithm that is based on a dynamic programming method. The Bellman optimality principle plays a crucial role in the algorithm. If the semi-Markov process describing the evolution of the real system in a long time satisfies the assumptions of the limit theorem, we can use the results of the infinite duration SM decision processes theory. We can apply the Howard algorithm for finding an optimal stationary policy. This algorithm is equivalent to the same problem of linear programming.

Summary

We can observe the phenomena and systems that change randomly during the time in various practical situations. The theory of stochastic (random) processes allows the modeling of random evolution of systems through the time.

The main stochastic processes considered in this book are the continuous time semi-Markov processes (SMP) with a discrete set of states. SMP is characterized in that, a future states of the process and their sojourn times do not depend on the past states and their sojourn times if a present state is known. A holding time in a state i before passing to a state j and a waiting time in state i of SMP are nonnegative random variables with *arbitrary* cumulative distribution function, while for a Markov process these random variables are exponentially distributed. So, continuous time SMP with a discrete state space is a generalization of that type of Markov process. The Markov process can be treated as a special case of the SMP. This kind of SMP is constructed by the so-called Markov renewal processes that are the special case of the two-dimensional Markov sequences. The Markov renewal process is completely defined by the renewal kernel and an initial distribution or by another characteristics which are equivalent to it. To construct a semi-Markov model of a real random process, we have to specify the set of states to determine initial distribution and the kernel of the process. However, we must remember that the nature of the real random process that we intend to model by the SMP is consistent with its definition. In particular, we must be convinced that in moments of state changes is preserved the memoryless property of the process.

Concepts and equations of the SMPs theory which are discussed in Chapters 1-6 allow us to calculate the parameters and characteristics of the modeled processes, in particular in the area of reliability and maintenance. These calculated parameters and characteristics of the semi-Markov models enable mathematical analysis of considered processes.

The way of building various semi-Markov models is presented in Chapters 7-14.

Chapter 15 is devoted to application of semi-Markov decision processes. Semi-Markov decision processes theory gives the possibility to formulate and solve the optimization problems that can be modeled by SMP. In such kind of problems decision maker ought to chose the process which provides the greatest profit. Terms such as decision (alternative), policy, strategy, gain, criterion function are the basic concepts of the semi-Markov decision processes theory. The problem of optimization of a SM decision process with a finite number of state changes is solved by the algorithm which is based on a dynamic programming method. The Bellman optimality principle plays a crucial role in the algorithm. If the semi-Markov model satisfies the assumptions of the limit theorem we can apply the Howard algorithm for finding an optimal stationary policy. This algorithm is equivalent to the some problem of linear programming.

Bibliography

[1]   B.B. Anisimov, Limit theorems for semi-Markov processes, Prob. Theor. Math. Stat. 2 (1970) 3-15 (in Russian).
[2]   B.B. Anisimov, Multidimensional limit theorems for semi-Markov processes, Prob. Theor. Math. Stat. 3 (1970) 3–21 (in Russian).
[3]   S. Asmussen, Applied Probability and Queues, Wiley, Chichester, 1987.
[4]   T. Aven, Reliability evaluation of multistate systems with multistate components, IEEE Trans. Reliab. 34(2) (1985) 463–472.
[5]   M.C.J. Baker, How often should a machine be inspected, Int. J. Qual. Reliab. Manage. 7 (1990) 14–18.
[6]   R.E. Barlow, F. Proshan, Mathematical Theory of Reliability, Wiley, New York, London, Sydney, 1965.
[7]   R.E. Barlow, F. Proshan, Statistical Theory of Reliability and Life Testing, Holt, Rinchart and Winston, Inc., New York, 1975.
[8]   L.J. Bain, M. Engelhardt, Introduction to Probability and Mathematical Statistics, PWS-KENT Publishing Company, Boston, 1991.
[9]   P. Billingsley, Probability and Measure, John Wiley & Sons, New York, 1979.
[10] H.W. Block, T.H. Savits, A decomposition for multistate monotone systems, J. Appl. Prob. 19(2) (1982) 391–402.
[11] D. Bobrowski, Mathematical Models in Reliability Theory, WNT, Warsaw, 1985 (in Polish).
[12] S.M. Brodi, O.N. Vlasenko, Reliability of systems with many kind of work, Reliability Theory and Queue Theory, Nauka, Moscow, 1969, pp. 165–171 (in Russian).
[13] S.M. Brodi, J.A. Pogosian, Embedded Stochastic Processes in Theory of Queue, Naukova Dumka, Kiev, 1973 (in Russian).
[14] Z. Ciesielski, Asymptotic nonparametric spline density estimation, Prob. Math. Stat. 12 (1991) 1–24.
[15] E. Cinlar, Markov renewal theory, Adv. Appl. Prob. 1 (1969) 123–187.
[16] A. Csenki, Dependability for Systems with a Partitioned State Space Markov and Semi-Markov Theory and Computational Implementation, Springer-Verlag Inc., New York, 1994.
[17] J.L. Doob, Stochastic Processes, John Wiley & Sons/Chapman & Hall, New York/London, 1953.
[18] J. Domsta, F. Grabski, The first exit of almost strongly recurrent semi-Markov processes, Appl. Math. 23(3) (1995) 285–304.
[19] J. Domsta, F. Grabski, Semi-Markov Models and Algorithms of Reliability Renewable Standby Systems, Institute of Mathematics of Gdańsk University, 1996, Preprint (in Polish).
[20] V.A. Epanechnikov, Non-parametric estimation of multivariate probability density, Theor. Prob. Appl. 85(409) (1969) 66–72.

[21] W. Feller, On semi-Markov processes, Proc. Natl. Acad. Sci. U.S.A. 51(4) (1964) 653–659.

[22] W. Feller, An Introduction to Probability Theory and its Applications, vol. 1, third ed., John Wiley & Sons Inc., New York, London, Sydney, 1968.

[23] W. Feller, An Introduction to Probability Theory and its Applications, vol. 2, John Wiley & Sons Inc., New York, London, Sydney, 1966.

[24] I.B. Gertsbakh, Models of Preventive Service, Sovetskoe Radio, Moskva, 1969 (in Russian).

[25] I.B. Gertsbakh, Asymptotic methods in reliability theory: a review, Adv. Appl. Prob. 16 (1984) 147–175.

[26] I.I. Gichman, A.W. Skorochod, Introduction to the Theory of Stochastic Processes, PWN, Warsaw, 1968 (in Polish).

[27] J. Girtler, F. Grabski, Choice of optimal time to preventive service of the ship combustion engine, Zagadnienia Eksploatacji Maszyn 4 (1989) 481–489 (in Polish).

[28] B.W. Gniedenko, J.K. Bielajew, A.D. Sołowiew, Mathematical Method in Reliability Theory, WNT, Warsaw, 1968 (in Polish).

[29] F. Grabski, Analyze of random working rate based on semi-Markov processes, Zagadnienia Eksplostacji Maszyn 3–4(47–48) (1981) 294–305 (in Polish).

[30] F. Grabski, Theory of Semi-Markov Operation Processes, ZN AMW, No. 75A, Gdynia, 1982 (in Polish).

[31] F. Grabski, L. Piaseczny, J. Merkisz, Stochastic models of the research tests of the toxicity of the exhaust gases, J. KONES Int. Combust. Engine 6(1–2) (1999) 34–41 (in Polish).

[32] F. Grabski, Semi-Markov Models of Reliability and Operation, IBS PAN, Warsaw, 2002 (in Polish).

[33] F. Grabski, Some Problems of Modeling Transport Systems, ITE, Warsaw, Radom, 2003 (in Polish).

[34] F. Grabski, Application of semi-Markov processes in reliability, Electron. J. Reliab. Theor. Appl. 2(3-4) (2007) 60–75.

[35] F. Grabski, K. Kołowrocki, Asymptotic reliability of multistate system with semi-Markov states of components, in: Proceeding of the European Conference on Safety and Reliability–ESREL'99, Safety and Reliability, A.A. Balakema, Roterdam, Brookfield, 1999, pp. 317–322.

[36] F. Grabski, The reliability of an object with semi-Markov failure rate, Appl. Math. Comput. 135 (2003) 1–16.

[37] F. Grabski, Semi-Markov failure rate process, Appl. Math. Comput. 217 (2011) 9956–9965.

[38] F. Grabski, J. Jaźwiński, Functions of Random Variables in Problems of Reliability, Safety and Logistics, WKŁ, Warsaw, 2009 (in Polish).

[39] M. Gyllenberg, D.S. Silvestrov, Quasi-stationary phenomena for semi-Markov processes, in: J. Janssen, N. Limnios (Eds.), Semi-Markov Models and Applications, Kluver, Dordrecht, 1999, pp. 33–60.

[40] M. Gyllenberg, D.S. Silvestrov, Quasi-Stationary Phenomena in Nonlinearly Perturbed Stochastic System, De Gruyter Expositions in Mathematics, vol. 44, Walter de Gruyter, Berlin, 2008, XII+579 pp.

[41] M.A. Harriga, A maintenance inspection model for a single machine with general failure distribution functions, Pacific J. Math. 31(2) (1996) 403–415.

[42] J.M. Hoem, Inhomogeneous semi-Markov processes, select actuarial tables and duration-dependence in demography, in: T.N.E. Greville (Ed.), Population, Dynamics, Academic Press, New York, 1972, pp. 251–296.

[43] R.A. Howard, Dynamic Programing and Markov Processes, MIT Press, Cambridge, MA, 1960.
[44] R.A. Howard, Research of semi-Markovian decision structures, J. Oper. Res. Soc. Jpn. 6 (1964) 163–199.
[45] R.A. Howard, Dynamic probabilistic system, Vol. II: Semi-Markow and Decision Processes, Wiley, New York, London, Sydney, Toronto, 1971.
[46] A. Iosifescu-Manu, Non homogeneous semi-Markov processes, Stud. Lere. Mat. 24 (1972) 529–533.
[47] M. Iosifescu, Finite Markov Processes and Their Applications, John Wiley & Sons, Ltd., Chichester, 1980.
[48] J. Jensen, R. De Dominicisis, Finite non-homogeneous semi-Markov processes, Insurance Math. Econom. 3 (1984) 157–165.
[49] W.S. Jewell, Markov-renewal programming, Oper. Res. 11 (1963) 938–971.
[50] T. Kato, Perturbation Theory for Linear Operators, Springer, Berlin, 1966, 1976, 1995.
[51] K. Kołowrocki, On limit reliability functions of large multi-state systems with ageing components, Appl. Math. Comput. 121 (2001) 313–361.
[52] K. Kołowrocki, Reliability of Large Systems, Elsevier, Amsterdam, Boston, Heidelberg, London, New York, Oxford, Paris, San Diego, San Francisco, Singapore, Sydney, Tokyo, 2004.
[53] K. Kołowrocki, J. Soszyñska-Budny, Asymptotic approach to reliability of large complex systems, Electron. J. Reliab. Theor. Appl. 2(4) (2011) 103–128.
[54] K. Kołowrocki, E. Kuligowska, Monte Carlo simulation application to reliability evaluation of port grain transportation system operating at variable conditions, J. Pol. Saf. Reliab. Assoc. Summer Saf. Reliab. Semin. 4(1) (2013) 73–82.
[55] B. Kopociński, Outline of the Renewal Theory and Reliability, PWN, Warsaw, 1973 (in Polish).
[56] I. Kopocińska, B. Kopociński, On system reliability under random load of elements, Appl. Math. XVI(1) (1980) 5–14.
[57] I. Kopocińska, The reliability of an element with alternating failure rate, Appl. Math. XVIII(2) (1984) 187–194.
[58] B. Kopociński, A private letter to F. Grabski, 1987 (in Polish).
[59] E. Korczak, Reliability analysis of non-repaired multistate systems, in: Advances in Safety and Reliability: ESREL'97, 1997, pp. 2213–2220.
[60] V.S. Korolyuk, A.F. Turbin, Semi-Markov Processes and Their Applications, Naukova Dumka, Kiev, 1976 (in Russian).
[61] V.S. Korolyuk, A.F. Turbin, Markov Renewal Processes in Problems of Systems Reliability, Naukova Dumka, Kiev, 1982 (in Russian).
[62] V.S. Korolyuk, A.F. Turbin, Decompositions of Large Scale Systems, Kluwer Academic, Singapore, 1993.
[63] V.S. Korolyuk, A. Swishchuk, Semi-Markov Random Evolutions, Kluwer Academic Publisher, Dordrecht, Boston, London, 1995.
[64] V.S. Korolyuk, L.I. Polishchuk, A.A. Tomusyak, A limit theorem for semi-Markov processes, Cybernetics 5(4) (1969) 524–526.
[65] V.S. Korolyuk, N. Limnios, Stochastic System in Merging Phase Space, World Scientific, Singapore, 2005.
[66] T. Kniaziewicz, L. Piaseczny, Simplified research tests of toxic compounds emission's for marine diesel engines, J. Pol. CIMAC 5(2) (2010) 99–106.
[67] L. Knopik, Some remarks on mean time between failures of repairable systems, Summer Safety and Reliability Seminars, vol. 2, Maritime University, Gdynia, Poland, 2007, pp. 207–211.

[68] G. Krzykowski, Equivalent conditions for the nonparametric spline density estimators, Prob. Math. Stat. 13 (1992) 269–276.
[69] T.C. Lee, G.G. Judge, A. Zellner, Estimating the Parameters of the Markov Probability Model from Aggregate Time Series Data, NHPC, Amsterdam, London, 1970.
[70] P. Levy, Proceesus semi-Markoviens, in: Proc. Int. Cong. Math., Amsterdam, 1954, pp. 416–426.
[71] J. Letkowski, Application of the Poisson Process Distribution, PDF file, 2013, www.aabri.com/SA12Manuscripts/SA12083.
[72] N. Limnios, G. Oprisan, Semi-Markov Processes and Reliability, Birkhauser, Boston, 2001.
[73] A. Lisnianski, I. Frankel, Non-homogeneous Markov reward model for an ageing multi-state system under minimal repair, Int. J. Perform. Eng. 5(4) (2009) 303–312.
[74] A. Lisnianski, G. Levitin, Multi-State System Reliability: Assessment, Optimization and Applications, World Scientific; NJ, London, Singapore, 2003
[75] E.H. Moore, R. Pake, Estimation of the transition distributions of a Markov renewal process, Ann. Inst. Stat. Math. 20 (1968) 411–424.
[76] H. Mine, S. Osaki, Markovian Decision Processes, AEPCI, New York, 1970.
[77] E. Olearczuk, Outline of Theory of Technical Device Using, WNT, Warsaw, 1972 (in Polish).
[78] B. Ouhbi, N. Limnios, Nonparametric estimation for semi-Markov kernels with application to reliability analysis, Appl. Stoch. Model Data Anal. 12 (1996) 209–220.
[79] Y. Taga, On the limiting distributions in Markov renewal processes with finitely many states, Ann. Inst. Stat. Math. 15 (1963) 1–10.
[80] A.A. Papadopoulou, P.-C.G. Vassiliou, Asymptotic behaviour of non homogenous semi-Markov systems, Linear Algebra Appl. 210 (1994) 153–198.
[81] A. Papadopoulou, P.-C.G. Vassiliou, Continuous time non-homogeneous semi-Markov systems, Semi-Markov Models and Applications, Kluwer Acad. Publ., Dordrecht, 1999, pp. 241–251.
[82] A. Papoulis, Probability, Random Variables, and Stochastic Processes, third ed., McGraw-Hill Inc., New York, 1991.
[83] I.V. Pavlov, I.A. Ushakov, The asymptotic distribution of the time until a semi-Markov process gets out of the kernel, Eng. Cybern. 2(3) (1978) 68–72.
[84] S. Piasecki, Optimization of Maintenance Systems, WNT, Warsaw, 1972 (in Polish).
[85] R. Pyke, Markov renewal processes: definitions and preliminary properties, Ann. Math. Stat. 32 (1961) 1231–1242.
[86] R. Pyke, Markov renewal processes with finitely many states, Ann. Math. Stat. 32 (1961) 1243–1259.
[87] R. Pyke, R. Schaufele, Limit theorems for Markov renewal processes, Ann. Math. Stat. 32 (1964) 1746–1764.
[88] A.N. Shiryayev, Probability, Springer-Verlag, New York, Berlin, Heidelberg, Tokyo, 1984.
[89] W.D. Shpak, On some approximation for calculation of a complex system reliability, Cybernetics 10 (1971) 68–73 (in Russian).
[90] D.C. Silvestrov, Semi-Markov Processes with a Discrete State Space, Sovetskoe Radio, Moskow, 1980 (in Russian).
[91] D.C. Silvestrov, Nonlinearly perturbed stochastic processes, Theor. Stoch. Process. 14(30)(3–4) (2008) 129–164.
[92] W.L. Smith, Regenerative stochastic processes, Proc. R. Soc. Lond. A 232 (1955) 6–31.

[93] A.D. Solovyev, Analytical Methods of Reliability Theory, WNT, Warsaw, 1979 (in Polish).

[94] L. Takács, Some investigations concerning recurrent stochastic processes of a certain type, Magyar Tud. Akad. Mat. Kutato Int. Kzl. 3 (1954) 115–128.

[95] L. Takács, On a sojourn time problem in the theory of stochastic processes, Trans. Am. Math. Soc. 93 (1959) 531–540.

[96] P.-C.G. Vassiliou, A.A. Papadopoulou, Non homogenous semi-Markov systems and maintainability of the state sizes, J. Appl. Prob. 29 (1992) 519–534.

[97] J. Xue, On multi-state system analysis, IEEE Trans. Reliab. 34 (1985) 329–337.

[98] J. Xue, K. Yang, Dynamic reliability analysis of coherent multi-state systems, IEEE Trans. Reliab. 4(44) (1995) 683–688.

[99] J. Xue, K. Yang, Symmetric relations in multi-state systems, IEEE Trans. Reliab. 4(44) (1995) 689–693.

[100] M. Zając, Reliability model of the inter-model system, Ph.D. Thesis, Wroclaw University of Technology, Wroclaw, 2007.

Notation

$\mathbb{N}_0$	set of natural numbers $\{0, 1, 2, \ldots\}$
$\mathbb{N}$	set of positive natural numbers $\{1, 2, \ldots\}$
$\mathbb{R}$	set of real numbers
$\mathbb{R}_+$	set of nonnegative real numbers $[0, \infty)$
$(\Omega, \mathcal{F}, P)$	probability space
S	state space or phase space
$\{X(t) : t \in \mathbb{T}\}$	stochastic process or random process
$\mathbb{T}$	set of parameters
$\{x(t) : t \in \mathbb{T}\}$	trajectory or realization of the random process
$p_{ij}(u, s)$	transition probabilities from the state i at the moment u to the state j at the moment s of the Markov process
$p_{ij}(t)$	transition probability from the state i to the state j during the time t of the homogeneous Markov process
MC	Markov chain
HMC	homogeneous Markov chain
EMC	embedded Markov chain
p_{ij}	transition probabilities of HMC
P	matrix of transition probabilities of HMC
π_i	stationary probability of the state i of HMC
π	stationary probability distribution of HMC
MRP	Markov renewal process
$\{(\xi_n, \vartheta_n) : n \in \mathbb{N}_0\}$	MRP
$Q(t)$	renewal matrix
$Q_{ij}(t)$	element of the renewal matrix
$q_{ij}(t)$	density of the renewal matrix element
p	initial distribution of MRP
p_i	initial probability of the state i of MRP
SM	semi-Markov
SMP	semi-Markov process
$\{X(t) : t \geqslant 0\}$	SMP
$\{(\xi_n, \tau_n) : n \in \mathbb{N}_0\}$	two-dimensional Markov chain
$\{X(\tau_n) = \xi_n : n \in \mathbb{N}_0\}$	embedded Markov chain (EMC) of SMP
$\{\vartheta_n : n \in \mathbb{N}_0\}$	sequence of sojourn times in states of SMP
$\{\tau_n : n \in \mathbb{N}_0\}$	sequence of jump moments of SMP
CDF	cumulative distribution function

$F_{ij}(t)$	CDF of a holding time in the state i if the next state is j
$G_i(t)$	CDF of a waiting time in the state i
$sms(t)$	sample path of SMP in the interval $[0, t]$
$\hat{P}_{ij}(t)$	likelihood estimator of the transition probability p_{ij}
$N_{ij}(t)$	direct number of transitions from the state i to j in the interval $[0, t]$ of SMP
$N_i(t)$	number of jumps to the state i of SMP in the interval $[0, t]$
NHSMP	nonhomogeneous semi-Markov process
$Q(t, x)$	the kernel of the NHSMP
$Q_{ij}(t, x)$	element of the NHSMP kernel
$T_{ij}(t)$	holding time of NHSMP
$F_{ij}(t, x)$	CDF of the holding time in the state i if the next state is j for NHSMP
$G_i(t, x)$	CDF of the waiting time in the state i of NHSMP
$P_{ij}(t, s)$	interval transition probabilities of NHSMP
Δ_A	number of jumps to the first arrival at the set of states $A \subset S$ of the SMP
$f_{iA}(m)$	conditional probability of m SMP jumps to the first arrival at the subset of states A of SMP, if the initial state is i
Θ_{iA}	first passage time from the state $i \in A'$ to the subset A
$\Phi_{iA}(t)$	CDF of the random variable Θ_{iA}
$\tilde{q}_{ij}(s)$	Laplace-Stieltjes (L-S) transform of the function $Q_{ij}(t)$
$\tilde{q}(s)$	L-S transform of the renewal kernel $Q(t)$
$\tilde{q}_{A'}(s)$	L-S transform of the renewal kernel submatrix $[\tilde{q}_{ij}(s) : i, j \in A']$
$P_{A'}$	submatrix $[p_{ij} : i, j \in A']$ of transition probabilities of EMC
$E(\Theta_{iA})$	expected value of the random variable Θ_{iA}
$E(\Theta_{iA})$	second moment of the random variable Θ_{iA}
Θ_{jj}	first return time to the state j of the SMP
$\Phi_{jj}(t)$	CDF of the random variable Θ_{jj}
$P_{ij}(t)$	interval transition probability from the state i to j
$P_j(t)$	probability of the state j at the moment t of PSM
P_j	limiting probability of the state j of SMP
$Q^\varepsilon(t)$	renewal kernel of perturbed SMP
$V_{ij}(t)$	number of visits of SMP in the state $j \in S$ on the interval $[0, t]$ if the initial state is $i \in S$
$W_{ij}(t, n)$	probability distribution of the process $\{V_{ij}(t) : t \in \mathbb{R}_+\}$
$H_{ij}(t)$	expected value of the random variable $V_{ij}(t)$
$\{K(t) : t \geqslant 0\}$	additive functional of the alternating process
$T(x)$	moment of achieving the level x by the random variable $K(t)$
$\{K_{ij}(t) : t \geqslant 0\}$	global (cumulative) sojourn time of the state j in the interval $[0, t]$ if an initial state is i
$T_{ij}(x)$	moment of exceeding a level x by a summary sojourn time of the state j of SMP until an instant t if the initial state is i

$\{L(t) : t \geqslant 0\}$ integral functional or cumulative process of SMP

$I_A(x)$ indicator or characteristic function of a subset A

$\mathcal{N}(m, \sigma)$ normal distribution with expectation m and standard deviation σ

ψ system structure function

$T_{[l]}^{(k)}$ l-level lifetime or lifetime to level l of the kth component

$R_{z_k[l]}^{(k)}(t)$ reliability function to level l of the kth component

$\boldsymbol{R}^{(k)}(t)$ multistate reliability function of the kth component of the system

$\boldsymbol{R}^{(k)}$ multistate reliability function of the kth component of the system

$\boldsymbol{R}(t)$ multistate reliability function of the system

$X_{kr}(t)$ the binary random variables

ψ_j the system level indicators

$SMDP$ semi-Markov decision process

$\delta_i(n)$ decision in the stage n of SMDP

D_i set of decisions $\delta_i(n) = k$ in each state $i \in S$ of SMDP

$\delta(n)$ policy in the stage n of SMDP

d strategy of SMDP

$Q_{ij}^{(k)}(t)$ element of the SMDP kernel for decision $k \in D_i$

$r_{ij}^k(x)$ "yield rate" of state i at an instant x when the successor state is j and k is a chosen decision

$R_{ij}^{(k)}(t)$ reward that the system earns by spending a time t in a state i before making a transition to state j, for the decision $k \in D_i$

$b_{ij}^{(k)}$ bonus in a state i before making a transition to state j, for the decision $k \in D_i$

$u_i^{(k)}$ expected value of the gain that is generated by the process i in the state at one interval of its realization for the decision $k \in D_i$

Printed and bound by CPI Group (UK) Ltd, Croydon, CR0 4YY

08/05/2025

01864799-0001